Privatization and Deregulation of Transport

Studies in Regulation

General Editor: **George Yarrow**, Director, Regulatory Policy Institute and Regulatory Policy Research Centre, Hertford College, Oxford

Government regulation of business activity is a pervasive characteristic of modern economies, including those most committed to free markets. For good or ill regulation has far reaching implications for economic performance, and understanding the processes at work is an important task for anyone seeking to analyse the determinants of performance. Quite frequently, however, analysis is restricted to a very specific aspect of business activity or to a particular sector of the economy, an approach that serves to limit the insights into regulatory issues that may be gained.

A guiding principle behind this series of books is that regulatory processes exhibit a number of common features that are likely to manifest themselves in a range of different circumstances. A full understanding of the motives for and effects of regulation therefore requires study of these common features, as well as the specifics of particular cases of government interventions. Thus, it is possible to learn something relevant about, say, the regulation of utilities from the study of financial services regulation, or about industrial policy from the study of environmental regulation.

This focus on regulatory processes in general, as well as on specific aspects of particular interventions, also points to the value of interdisciplinary analysis. Policy formulation, development and implementation each have political, legal and economic aspects, and the boundaries between traditional academic disciplines can be obstacles to progress in regulatory studies. In this series, therefore, a wide range of different perspectives on regulation and on regulatory processes will be presented, with the aim of contributing to the development of new insights into important policy issues of the day.

Titles include:

Bill Bradshaw and Helen Lawton Smith (*editors*)
PRIVATIZATION AND DEREGULATION OF TRANSPORT

Helen Lawton Smith and Nick Woodward (*editors*)
ENERGY AND ENVIRONMENT REGULATION

Privatization and Deregulation of Transport

Edited by

Bill Bradshaw
Centre for Socio-Legal Studies
University of Oxford

and

Helen Lawton Smith
Reader
Local Economic Development
Coventry Business School
and
Senior Research Associate
School of Geography
University of Oxford

First published in Great Britain 2000 by
MACMILLAN PRESS LTD
Houndmills, Basingstoke, Hampshire RG21 6XS and London
Companies and representatives throughout the world

A catalogue record for this book is available from the British Library.

ISBN 0–333–79039–1

First published in the United States of America 2000 by
ST. MARTIN'S PRESS, INC.,
Scholarly and Reference Division,
175 Fifth Avenue, New York, N.Y. 10010

ISBN 0–312–23273–X

Library of Congress Cataloging-in-Publication Data
Privatization and deregulation of transport / edited by Bill Bradshaw and Helen Lawton Smith.
p. cm. — (Studies in regulation)
Papers presented at a conference.
Includes bibliographical references and index.
ISBN 0–312–23273–X (cloth)
1. Transportation—Great Britain—Congresses. 2. Privatization—Great Britain—Congresses. I. Bradshaw, Bill, 1936– II. Lawton Smith, Helen. III. Series.

HE243.A2 P75 2000
388'.049—dc21

99–088250

This book is printed on paper suitable for recycling and made from fully managed and sustained forest sources.

10 9 8 7 6 5 4 3 2 1
09 08 07 06 05 04 03 02 01 00

Printed and bound in Great Britain by
Antony Rowe Ltd, Chippenham, Wiltshire

Contents

Contributors

Julian Allen	Transport Studies Group, University of Westminster
Alfred J. Baird	Department of Business Studies, Napier University
Chris Bolt	Chief Economic Adviser and Director, Economic Regulation Group, Office of the Rail Regulator
Bill Bradshaw	Centre for Socio-Legal Studies, Wolfson College, University of Oxford
Michael Browne	Professor, Transport Studies Group, University of Westminster
Andrew Burchell	Director, Strategy and Analysis, Department of Transport
Kenneth Button	Distinguished Research Professor, George Mason University, USA. Formerly of Loughborough University
Chris Castles	Partner, Coopers & Lybrand
Stuart Condie	BAA
Richard Davies	Senior Consultant, KPMG.Until the end of 1996, Financial Planning Manager, Privatisation, at British Railways Board Headquarters
Russ Haywood	School of Urban and Regional Studies, Sheffield Hallam University
Helen Lawton Smith	Research Fellow in the Centre for Local Economic Development (CLED), Coventry University; Director of Science Policy Studies of the Regulatory Policy Research Centre (RPRC) at Hertford College, Oxford University
Fumitoshi Mizutani	Associate Professor, School of Business Administration, Kobe University, Japan
Kiyoshi Nakamura	Professor of Industrial Economics, School of Commerce, Waseda University, Japan
Chris Nash	Institute for Transport Studies, University of Leeds
Philip O'Donnell	Head of Competition and Regulatory

	Relations, Railtrack
Mike Parr	Associate, London Economics
Amanda Root	Environmental Change Unit, University of Oxford
Michael Schabas	Founder Director (non-executive), GB Railways Group, Plc
Bob Stannard	Chief Economist, Office of Passenger Rail Franchising
David Starkie	Director, Economics Plus Ltd
Stefan Szymanski	Imperial College Management School
Peter White	Transport Studies Group, University of Westminster

PART 1

Introduction

CHAPTER 1

Introduction

Bill Bradshaw and Helen Lawton Smith

The Regulatory Policy Research Centre Seminar in September 1997 where the papers which make up the chapters of this book were given, took place following the election of the first Labour government for 17 years and following an announcement of the consultation process leading up to the publication of a Transport White Paper.[1] The seminar sought to address the somewhat contradictory situation of a government of the left, inheriting a transport industry largely privatised and deregulated, but unwilling to commit itself either to the expense of re-nationalisation or to finding the large sums of money needed for investment in public infrastructure and facilities.

The book follows the structure of the seminar by taking each of the bus, train, freight, and airline industries separately, with a chapter each on ports and cross-channel competition completing the picture. The chapters reflect a mix of academic, regulatory body and industry viewpoints. Most, but not all, of the papers start from an economic regulation perspective.

The book begins with an overview of the economic regulatory arrangements within the transport sector by Andrew Burchell from the Department of Environment, Transport and the Regions. He summarises the key features across different modes of transport and draws out the differences and similarities between them.

Buses

Three chapters comprise the section on buses. The problems of the long-term decline in local bus appears as the common theme. Peter White's chapter draws some conclusions about the effects of deregulation and the transfer of ownership. It is evident, even in a relatively uncomplicated industry like the bus business with a low proportion of fixed costs, that the upheavals set in train by the Transport Act 1985 have yet to work themselves out. Consolidation continues among the larger operators and the development of the concept of Quality Partnerships between operators and local authorities perhaps shows the industry reaching for a plateau of stability. Although bus deregulation has brought gains, notably a reduction in public subsidy and rises in efficiency, there have been significant losers with greatly increased fares in some metropolitan areas and a sharp fall in the earnings of bus drivers relative to other groups. The long term decline in bus use outside London over the last forty years, apart from some areas such as Oxford, Brighton, Fife and Bristol, has not been halted in the face of the rise in the ownership and use of cars. Mike Parr examines the use of competitive markets to provide local bus services. His focus is the experience since the 1985 Transport Act and other related legislation became effective. He assesses the current economic organisation of, and considers the degree of competition within, local bus services in the context of the decline in popularity of local bus transport. The problems of maintaining both demand and supply is particularly acute in rural areas. Amanda Root paints a particularly bleak picture of the use of public transport in a study of two large rural villages in Oxfordshire.

Railways

If the effects of the re-organisation of the bus industry have not matured after ten years, the radical shake-up accorded to the railways under the 1993 Railways Act had only just been completed at the time of the seminar so it is too early to draw any firm conclusions. The section on railways is the most comprehensive in the book consisting of eight chapters exploring a range of issues. Richard Davies describes the process of rail privatisation as seen from the inside of British Railways Board and offers some views on future developments. The prospect of the Review by the Rail Regulator of Railtrack's track access charges and of the case for introducing more on rail competition promise a continuing revolution in the industry perhaps more enduring than those promised by Trotsky or practised by Mao. Many important issues are discussed by Chris Bolt, of the Office of the Rail Regulator, who draws attention to the paradox of an industry which has been restructured to provide vertical separation and competition between operators and suppliers but where cooperation between these same players is needed to deliver integration, network benefits and other elements of the seamless journey. Bolt also draws attention to the pressures in the industry to reintegrate the business. First there has been the acquisition by Stagecoach, the franchisee of one of the Train Operating Companies, South West Trains, of the Rolling Stock Company, Porterbrook. Since the seminar we have seen the approach to the Rail Regulator by Virgin Trains, the holders of the franchise for the West Coast Main Line, and Railtrack, to enter a profit sharing agreement to cooperate on a proposal for modernising the signalling on the route which would lead to a radical improvement of journey times. The Periodic Review by the Rail Regulator of Railtrack's access charges is contrasted by Bolt with similar reviews carried out by other utility regulators but he draws attention to some unique features of railways which include the continuing need for subsidy, the mismatch between the short term nature of franchises and the long lives of rolling stock and infrastructure assets, and the uncertainty about the conduct of the next franchising round which lies beyond the next general election.

Philip O'Donnell, from Railtrack, examines the challenges and opportunities to the railway industry arsing from the development of liberalised access to the British railway network. His chapter puts the current regulatory framework into its political context, discussing the different stages in its development. He then explores what the system means in practice and considers the issues related to fully liberalised access in 2002. He concludes that liberalisation appears to bring more complex regulation rather than less.

Chris Nash looks at models of railway privatisation adopted elsewhere and considers in more detail some issues arising in Britain. He deals with the question of the incentives which Railtrack has to invest in the network. Since the seminar and because of these misgivings about the commitment of Railtrack to investment, the Rail Regulator has modified Railtrack's Licence to require the company to formulate plans to develop the network (in addition to the existing obligation to maintain and renew the assets in modern form). Railtrack's plans must be published in the Network Management Statement published by the company each year. The Regulator may be said to be pushing Railtrack in the direction of acting as a strategic rail authority by casting the company in the role of steward of the network.

The number of different parties involved in a decision to upgrade an important railway route is highlighted as is the fact that the benefits will not be shared equally. Nash also draws attention to the fact that Railtrack has the advantages of a monopolist in determining the cost of investments and how benefits should be shared between themselves and users. Unravelling the costs and benefits of investment plans represents a significant challenge to the Rail Regulator. The railways in Britain have been fragmented and Nash discusses the privatisation process and its immediate financial outcome. He concludes, contrary to the foreboding of many commentators, that the system still functions reasonably well. Although it is too early to draw final conclusions, he foresees that tight regulation in the public interest remains necessary.

Control over fares and some other issues relating to franchises are discussed by Bob Stannard (job) and in the following chapter by Micahel Schabas. Since the seminar there has been some dissatisfaction expressed about fare rises well above inflation on services which have met contractual performance standards but where customers' complaints are reported to be at a high level. This raises the question of the adequacy of the performance standards and the means by which these are monitored. The length of franchise terms is a matter of current debate with some operating companies arguing that re-negotiation of franchises will be necessary if investment to deal with growing patronage is not to be held up. With the first franchises not due for renewal before 2003, there will be no opportunity for the present government to intervene to alter the terms of existing franchises unless primary legislation is passed to allow this or the government decides to open negotiations to extend the life of existing franchises.

Michael Schabas, from GB Railways, writes from the point of view of private sector interests and is enthusiastic about the level of competition within the new rail industry. He examines the changing structure of the industry including the passenger train operating companies, the companies which maintain tracks and trains and those which operate freight services. He then discusses regulatory responses to such issues as anti-competitive behaviour and barriers to entry.

Fumitoshi Mizutani and Kiyoshi Nakamura's chapter on Japan railways since privatisation in 1987 provides a contrast with the UK experience. They provide a summary of the effects of privatisation and some of the implications for policy. They conclude that privatisation of the Japan National Railway (JNR) has brought about improved efficiency. However, there remains a legacy of an increasing burden of debt from old JNR, some of which will have to be paid for by the taxpayer.

Looking at a different aspect of regulation from an academic point of view, Russ Hayward, considers the changes to the planning system which occurred in the name of deregulation during the early 1980s. These allowed developments such as offices and shops to be moved away from traditional town centre locations, which are the focus of most public transport networks, often to sites on motorways and major trunk roads. These out-of-centre roadside developments feature low density development with a great deal of car parking space, usually free to users. Hayward plotted developments around Manchester to show the dispersion of employment which has taken place. He subsequently examined planning applications along the route of the Sheffield Supertram. He then compared these applications with those for sites with convenient car access. The results show that the sites in the road corridors, remain the most attractive to developers. Hayward argues that the planning system will need to be used to steer development towards corridors, which can be served by public transport, particularly by railways.

Freight

The one chapter in the book on freight is by an academic, Michael Browne. He compares the deregulation of the road haulage markets in Britain, Europe and the United States. This has been less controversial than deregulation in other sections of the transport industry and the industry has matured and expanded into providing a whole range of logistical services. In Britain deregulation of road haulage has been the rule, for nearly 30 years. The challenges faced by the industry arise principally from congestion, particularly in cities where various bans and controls are expected. Browne also expects more emphasis on internalising the costs of the environmental damage caused by transport and more strict enforcement to limit illegal operations.

Airline industry

The four papers on the airline industry consider different aspects of its operation. Kenneth Button, examines the ownership and regulation of international operations; Chris Castles considers the alternatives to present methods of slot allocation at airports and the possible implications of introducing monetised systems of slot trading; David Starkie takes a different approach to the same issue by discussing a possible role for the market in allocating airport slots; while Stuart Condie provides a critique of today's economic regulation of UK airports.

Button considers the economic implications of airline mergers from the point of view of the carrier and the user. One is impressed in the papers by both Button and Castles of the difficulties of promoting competition without inflicting damage on other interests such as smaller airlines, the airlines of smaller countries or blocks of countries and services from less popular centres. One of the problems for the airlines is inadequate runway capacity defined in terms of the current competition for take-off and landing slots. David Starkie, an economist, argues that his preferred solution to this problem is the building of more runways. The alternative would be to increase the price charged for landing aircraft. At the present time runway capacity is charged sub-optimally which means that incumbent airlines with long-standing access gain economic rent and a yield premium on fares charged to passengers arising from the economic rent associated with the slot. To raise the price of landing charges would put incumbent and entrant on a more equal footing.

In a complementary paper, Stuart Condie provides an alarming view of the present day difficulties of the regulation of airports. He makes the point that the UK model of regulation of airport charges is approaching crisis point by the bizarre conjuntion of all time low charges at congested airports and the high capital cost of significant capacity expansion. He argues for the need for regulatory reform encompassing a package of revisions to existing regulations including those to pricing policies and to anti-competitive behaviour.

Ports and channel tunnel

The final section of the book examines regulation of ports and the Channel Tunnel. Privatisation of ports in Britain is unique in that there are examples where all aspects of port operation - regulation, landlords and responsibilities for port operations - have been transferred to the private sector. Alfred Baird's chapter analyses different models of port privatisation. He draws attention to the large financial gains of new owners but doubts their willingness to invest in what is very expensive infrastructure.

The final chapter is on the traffic forecasts made about cross channel traffic in Eurotunnel prospectuses. Stefan Szymanski's contribution is of particular interest in light of the recent collapse of the London and Continental Railways bid to build the Channel Tunnel Rail Link. The analysis will also be useful in considering the traffic forecasts some Train Operating Companies have made in support of their franchise bids. Deregulation and competition are powerful weapons in stimulating innovation, driving down costs and improving the choice and prices facing users. Like all weapons, as well as hitting targets, some collateral damage appears to be unavoidable. On the debit side assets are sold too cheaply, sometimes because the market is immature and sometimes because of the haste in which the sale is conducted. People lose their jobs sometimes undeservedly. Prices fall where there is competition and good regulation but rise wherever monopoly power can be exerted. Efficiency sometimes rises dramatically, particularly when union power is weakened (such as the abolition of the National Dock Labour Scheme) or in conditions of real competition (railway track maintenance). Services also tend to improve where there is real and sustained competition (air services between London and Scotland and Cross Channel vehicle trips) and buses in Oxford.

The academic approach is most useful when it can isolate the circumstances where the outcomes of deregulation and competition appear to be beneficial and suggest ways to avoid damaging outcomes. In this vein Szymanski suggests the MMC may have been wrong to allow the mergers of the ferry companies on the short sea routes and Castles warns against rushing into a wholesale deregulation of the market in airport slots.

Conclusions

In approaching the legislation which will follow the White Paper the government has to balance a range of difficult issues including a review of land use planning and housing allocation and the commitments entered into at Kyoto. Both of these huge challenges relate very closely to transport. At the same time government hopes to persuade the private sector to invest in major transport projects such as the upgrading of the London Underground. The regulatory packaging which is used to surround these issues must create incentives to long term investment but requires the containment of risk. Land use, environmental and transport objectives can be integrated but not within a regulatory or political framework which is capricious. The chapters in this book raise issues which policy makers must address but no author suggests the situation is hopeless and all have prescriptions for improvement.

Notes

1. New Deal for Transport White Paper, published 20 July 1998

CHAPTER 2

Regulation of transport: An overview

Andrew Burchell

Introduction

For the purposes of this paper I shall define the transport industry as comprising transport operations and the provision of non-roads infrastructure. Apart from the regulation of some tolled roads and crossings, and in the future some privately financed road schemes, the road network remains publicly owned and is not subject to economic regulation. That said, moves towards greater commercialisation of roads as advocated by some commentators carry with them a requirement to examine regulatory arrangements for roads given the monopoly characteristics of at least parts of the network.

Over the last twenty years the trend worldwide has been to move away from public ownership of industry in favour of the private sector. This has been particularly true in the UK, which under the Conservative government was among the standard bearers for privatisation. With the notable exception of London Underground, very little of the UK transport industry is now publicly owned, but the state continues to play a major role in the sector through regulation and subsidy with a view to correcting perceived market failures. In general, where the potential for market failure is believed to be small such as on bus and coach the mode is lightly regulated and subsidy is strictly limited. Where market failure is more prevalent, regulation is more intrusive, as is the case with Railtrack and BAA. And the passenger railway, which is perceived to bring large external, non-user benefits, receives large amounts of subsidy and is regulated accordingly. There is, in addition, a significant effort dedicated to the oversight and regulation of safety within the transport sector reflecting concerns over market failure associated with asymmetry of information between consumers and producers.

The incoming Labour government made several commitments in its manifesto which impact on the regulatory arrangements in the transport sector. Firstly, a commitment to bring forward changes to competition law to prohibit, in a more pro-active manner, anti-competitive practices. This will involve measures which will impact on all companies generally. Secondly, a commitment to review the regulation of the privatised utilities in general, and the regulation of railways and buses in particular. Roles and responsibilities will also be affected by the government plans for a Scottish Parliament, Welsh Assembly and the creation of Regional Development Agencies.

Reviews of the regulatory arrangements in transport form part of the more general, fundamental review of transport policy announced in June 1997 to underpin the development of an integrated transport policy. A White Paper on transport policy is promised for Spring 1998. This paper concentrates on providing a background overview of the economic regulatory arrangements within the transport sector; summarising the key features across the modes; and attempting to draw out differences and similarities between them. Safety regulation is not covered in this paper, meriting a conference and book of proceedings in its own right.

Bus and coach

Outside of London

The British bus industry has changed dramatically since the early 1980s, with less than 5 per cent of local services now provided directly by the public sector. Regulation, at least outside of London, is very light, with no single overall regulator: the operator licensing system overseen by Traffic Commissioners fulfils largely a safety function. This approach was adopted because buses were thought to be in a good position to benefit from the competition that privatisation and deregulation could bring. Barriers to entry and withdrawal are low and economies of scale relatively small. There has been significant reagglomeration in the industry, with some commentators questioning whether the market is sufficiently contestable, acting as a discipline on incumbent operators to behave efficiently and innovatively to prevent new entrants from making significant inroads into their business.

Privatisation and deregulation has brought some significant benefits. Vehicle mileage has increased, reversing the previous long term decline, and new types of service with more appropriate buses have been introduced. Operating costs have fallen sharply and subsidy levels are also down, with over 80 per cent of local mileage outside London now run commercially. Against this, passenger patronage has continued to decline and average fares have risen (see Figures in Appendix 1). There is evidence of a lack of coordination between separate bus companies and attempted 'cherry-picking', resulting in an unnecessary proliferation of buses and bunching of services. Concern has also been expressed about the poor provision of information on bus services.

The Government is keen to promote the use of public transport, and for bus in particular to play its full part as an alternative to the car in an integrated transport policy and is examining possible options to promote the profile of the bus and to increase bus patronage. Options include examination of the regulatory arrangements, the role of 'quality partnerships' between operators and local authorities. Expansion of the 'quality partnership' approach is already being introduced in some areas - including Birmingham, Manchester and Northampton. There are a variety of different arrangements described as quality partnerships, but typically they involve local agreement with operators on quality of service (perhaps with low floor, low emissions vehicles) in exchange for local authority initiatives to improve the operating environment for buses - for example better bus stops and shelters and measures to give priority to bus traffic. The question is whether, and how, this approach should be strengthened by new powers for local authorities or some other body (e.g. Traffic Commissioners).

London

While the rest of the country's bus market was privatised and deregulated in the 1980s, London's buses were not privatised until 1994, with the market continuing to be regulated through route franchising. Bus routes are tendered, with companies competing for the market rather than in the market. London Transport no longer operates the buses, but continues to set fares and plan routes. This difference between London and the Provinces provides useful comparative information which is available to the review of bus regulation to draw upon.

As in the rest of the country, operating costs per vehicle-km on London's buses have almost halved, while Government support is also sharply down if anything by more in London than elsewhere, helped by a significantly greater increase in fares. Despite this, the number of passenger journeys in London has actually increased (by 5 per cent since 1986), reflecting in part at least, the comparative advantage of public transport for some journeys in the capital.

Coaches

Coaches are one of the least regulated modes of transport. Both National Express and the Scottish Bus Group were privatised by 1992, and entry is now free of all economic regulation. However, although competition and contestability have been encouraged in the coach industry, there is evidence to suggest that a few firms now have a secure grip on the bulk of the market. This has attracted the attention of the competition authorities and the proposed amendments to general competition law are of most direct relevance to the bus and coach sector.

Rail

The privatisation of British Rail was one of the more complicated sell-offs attempted in the UK. In part, this is because of the disaggregated structure chosen for the sale, which was designed to allow the introduction of competition in the industry where possible. However, rail privatisation was also complicated by the high level of government subsidy required to support the passenger side of the industry (£1.7 billion to the 25 passenger train operating companies in 1997/98).

Regulation of the railway industry reflects the complex nature of the sale. The rolling stock leasing and the infrastructure maintenance companies are primarily regulated by the normal competition authorities. This is also true of rail freight, which is dominated by one company, EW&S, but operates against fierce competition from road haulage companies. None of these areas receive significant support directly from government

The passenger train companies are much more heavily regulated. In a similar approach to that adopted for London's buses, companies bid for franchises, agreeing to operate services above a minimum level, to ensure that passenger loads do not breach set limits and, in some cases, to invest in agreed schemes. Recognising the lack of competition, the Franchise Director also controls the price level of certain fares.

The Regulator has imposed tight restrictions with respect to direct, on-track competition, which he will review in 2001. At present, significant cross-subsidisation occurs, with companies subsidising loss-making routes with profits made elsewhere. The introduction of competition would threaten this, with competing train operators forcing margins down on the currently profitable lines. While this would produce lower fares and improved frequency on some routes, it would also inevitably lead to a higher subsidy bill and/or closures of some loss-making services.

Railtrack, the infrastructure provider, is regulated in a similar fashion to the other privatised utilities, with a price cap based on RPI-X set by the Rail Regulator at regular intervals, the next being in 2001. The price controls operate differently from the other regulated utilities insofar as the Regulator exercises control through the approval of access charges in individual access agreements between Railtrack and the operators. Reflecting the uncertainty over the value of the property portfolio being transferred to the private sector, a profit-sharing arrangement was introduced, with access charges to operators being reduced prior to the next price review if Railtrack's property income exceeds the level assumed when access charges were set in 1995. Another is the performance regime which sees Railtrack rewarded for improving punctuality and reliability of services through the way in which it manages the infrastructure.

In general, however, Railtrack's regulatory system raises similar issues surrounding the form and frequency of price controls, incentives or requirements to invest and the balance between producer and consumer interests, to those experienced in the regulation of public utilities in general. These concerns were highlighted by the Government in its manifesto in which it made several commitments, in particular its intention to establish a new rail authority, combining the functions carried out by the Franchising Director and the Department of the Environment, Transport and the Regions, to provide a clear, coherent and strategic programme for the development of the railways, so that passengers' legitimate expectations are met. The Government wishes to establish more effective and accountable regulation of the railway, enhancing the services for passengers and promoting growth in the use of the network, serving the community, benefiting the environment and getting good value for taxpayer's money.

Integrated public transport

More generally the Government is keen to see an increase in the use of public transport as an alternative to the car and recognises the need to make multi-modal public transport journeys more attractive to those who currently use cars.

There are several areas where multi-stage journeys could be made easier; for example ticketing/fares, with more through ticketing and greater use of travelcards and smartcards; interchange facilities - better facilities when interchanging between or within modes; timetabling promotes greater interconnection of services and service coordination; and better information - covering the availability and provision of timetable, route planning, and fares information. These areas are being examined as part of the fundamental review of policy and include the role regulation might play in promoting a more integrated public transport system.

Aviation

As with Railtrack, the major UK airports those owned by BAA and Manchester, which remains in local authority ownership are regulated using a RPI-X price cap. A single till ensures that receipts from airport retailing are taken into account when setting airport charges.

Regulation is complicated by the existence of significant excess demand for airport slots, particularly at Heathrow where the single till contributes to keeping airport charges low and price cap regulation prevents prices rising to market clearing levels. Scarcity rents accrue to operators rather than the infrastructure provider. Action in this area is difficult: increasing capacity has to have regard to the environmental impacts this would have; and changes to the traffic distribution rules and moves towards slot auctions are heavily constrained by international agreements.

Some commentators have remarked that the efficacy of regulation would be improved if BAA's effective monopoly of the London airports was reduced through the forced divestment of Gatwick or Stansted. To date the regulatory authorities have considered that any benefits competition might bring are outweighed by the benefits gained by the London airports operating as a system.

As with airports, airline regulation is heavily constrained by international agreements. The trend here is towards increased liberalization to promote competition, particularly within Europe, although this is complicated by the difficulty of opening up sufficient slots to provide effective competition. Consequently, the European Commission continues to lay down guidelines for fare regulation on intra-EEA routes, and the Civil Aviation Authority follows a broadly similar line on routes outside of the EEA. Fare regulation is relatively light and is based on cost-plus, rather than the price cap system commonly employed in UK utility regulation.

Shipping

Cartelisation has long been tolerated for scheduled deep sea cargo services in the interests of tariff and scheduling certainty for shippers and freight forwarders. Competition from Eurotunnel has now led to the DTI removing restrictions on similar behaviour between cross-Channel ferry companies.

Ports, on the other hand, enjoy a high level of competition and there has also been a historic tendency to excess capacity in the industry. Both of these together limit the potential for abuse of monopoly power and market forces are relied upon to discipline incumbent operators, although there are backstop procedures for appeals to the Secretary of State for Transport over the level of charges. So far as the future is concerned, port ownership is beginning to concentrate in fewer hands and ship operators are also now looking to vertically integrate upstream. This may have implications for the future contestability of the market although one would anticipate at this stage reliance on general competition law to protect consumer interests.

Concluding remarks

Few areas in transport escape economic regulation. This generates a variety of models which reflect in good part the different characteristics of the market within which the different services providers operate. That said, the Government believes that the regulatory regimes in some areas, particularly rail and bus may work against the development of an integrated transport policy and is examining the arrangements as part of its fundamental policy review.

Appendix 1

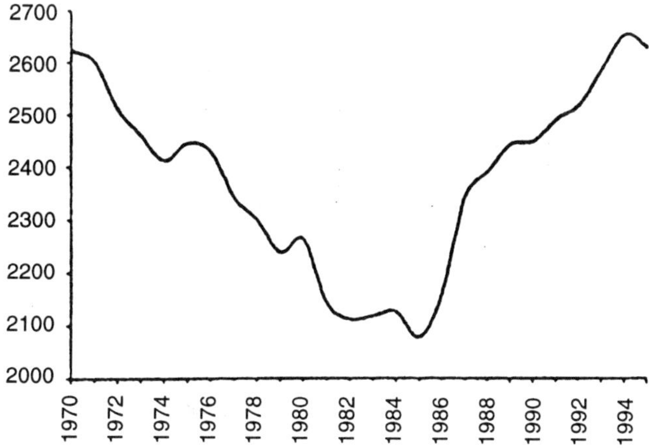

Figure A1.1 Local bus vehicle mileage 1970-1995 (million kms p.a.)

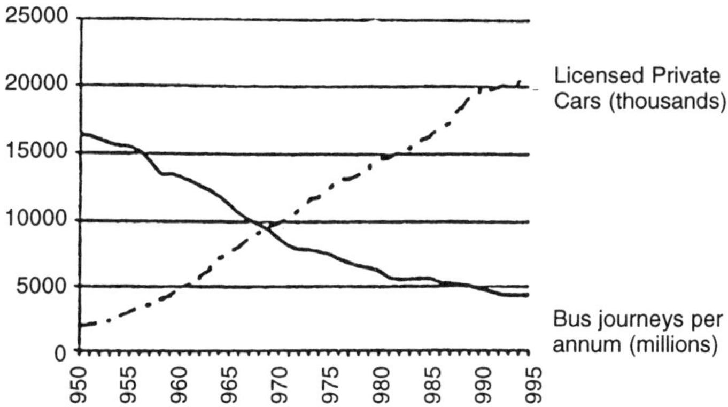

Figure A1.2 Bus patronage and car ownership 1950-1995

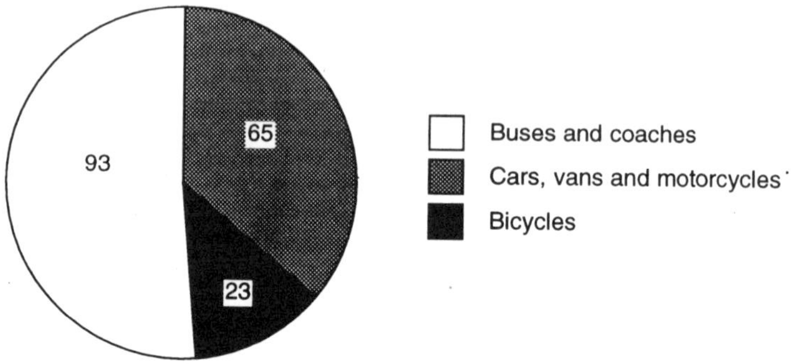

Figure A1.3 Modal share 1952 (passenger kilometres, billions)

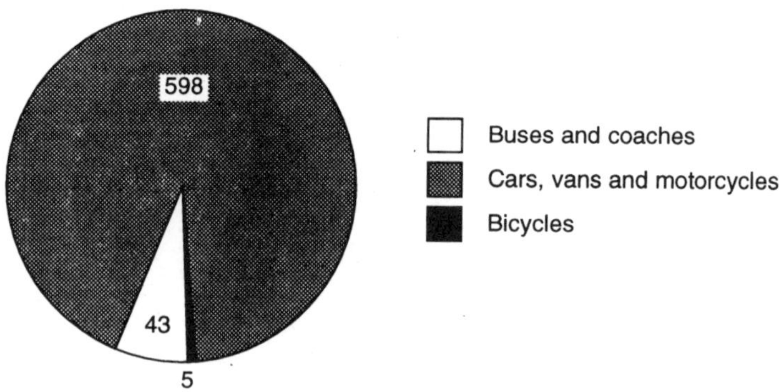

Figure A1.4 Modal share 1995 (passenger kilometres, billions)

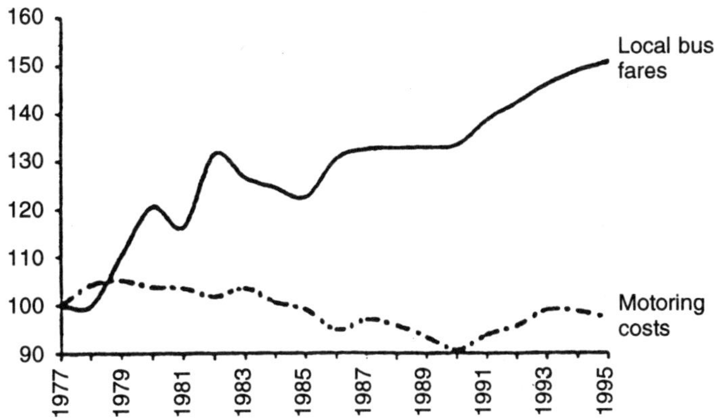

Figure A1.5 Passenger transport price indices (1977-1995, constant prices (1997=100)

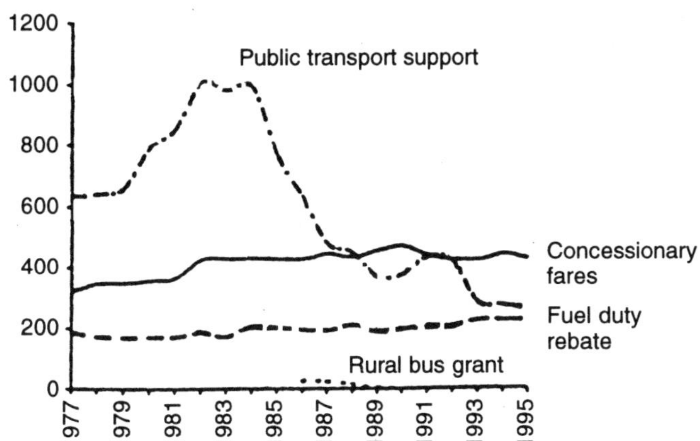

Figure A1.6 Support for bus services 1977-1995 (3m, 1995/96 prices)

PART 2

Bus Industry

CHAPTER 3

Experience in the UK bus and coach industry

Peter White

Introduction

The paper begins by reviewing the major changes which have occurred in the industry since 1980. As these are now well-documented only a brief outline is given, with references to other published work. Certain aspects are then examined in more detail, including variations in trends around the average for the industry as a whole, financial performance, the extent to which net savings in public spending have been made, and contrasts with the approach taken in rail privatisation. Some suggestions for future policy are then offered.

The overall picture

Prior to 1980, the bus and coach industry in Britain was strongly regulated, and - so far as scheduled services were concerned - largely in public ownership. Each service required a road service licence, specifying the exact route, timetable and fares to be charged. This procedure generally protected the incumbent operator, whether on local ('stage carriage') services or express and tour work.

About two-thirds of the entire bus and coach fleet was in public ownership, either in urban undertakings controlled through elected local authorities (London Transport, the seven Passenger Transport Executives [PTEs], and under district and regional councils elsewhere), or regional companies owned by the two nationalised holding companies (the National Bus Company in England and Wales, the Scottish Bus Group in Scotland). The remaining third was owned by the private sector, generally in the form of small local operators, often based in rural areas, or specialising in 'coach' operation. These dominated the private hire and contract market, but provided less than 10 per cent of scheduled public services.

In addition to their role as owners of many urban fleets, local authorities had become more heavily involved in local bus operation through providing financial support for concessionary fares (mainly to the elderly), and toward fares and service levels beyond those which would apply on a purely commercial basis. Although the latter had begun mainly in rural areas (under the Transport Act of 1968), their greatest extent was in larger urban areas, notably the PTEs and London. A fuller description of earlier trends in the industry is given in the paper by Mike Parr elsewhere in this volume.

One of the first moves toward deregulation under the Conservative government elected in 1979 was to pass the Transport Act of 1980, which removed quantity and price control from the coach industry, both in respect of scheduled express services (defined as those where all passengers travelled at least 30 miles), and excursion and tour operation (in contrast to some other countries, private hire and contract was never subject to quantity or price control). At the same time, 'quality' (as distinct from 'quantity') regulation was tightened up by the Operator Licence (commonly, 'O-licence') procedure, in which the person or firm operating buses or coaches is required to hold a licence. Now of unlimited duration, this specifies the maximum number of vehicles that can be operated, based on several criteria, notably adequacy of maintenance facilities. This appears to have been succcessful in ensuring that the safety record of the industry did not worsen following deregulation, as many had feared (White, Dennis and Tyler 1995).

Express coach deregulation

Scheduled long-distance coach passenger journeys grew by about 50 per cent between 1980 and 1985. However, since 1984 little data has been collected by the government from operators. Nonetheless, it is clear that the great majority of trips are handled by National Express, whose ridership may thus be used as a proxy for the industry as a whole (White 1997a). These indicate that passenger numbers fell back by 1993 to about the same level as in 1980, but since then have recovered slightly, probably to around 11-12 million per year (excluding Scottish Citylink, which is now also owned by National Express). Approximate annual totals are given in Table 3.1.

A notable outcome of coach deregulation was the limited degree of inter-coach operator competition, with only a few independents managing to establish all-year-round daily services. Their number has diminished recently, leaving only about ten such services at present. However, much more extensive competition between coach and rail continues, covering the great majority of towns served by rail. This has focused mainly on price, given the lower speeds which coaches offer in most cases. The coach industry has been more successful in establishing direct links to major airports (notably Heathrow), which rail does not offer. Here, a higher-fare market can be attracted, and substantial growth continues.

Figure 3.1 shows the National Express network (in diagrammatic form), taken from its National timetable, almost all points on which are served at least once daily, all year round. Figure 3.2 shows the independent services on the same definition, at July 1996. Since it was compiled, that from Grimsby has ceased, while that from Ripon and Leeds now runs on a less-than-daily frequency. The principal route from Scotland now runs from Glasgow only, and its operator spent a period of several months in administrative receivership. Most services are concentrated in the south west region, from which two formerly separate independent services from the Bristol area merged in July 1997.

There appear to be substantial benefits obtained through marketing of express coach services as a single network, in sales outlets, advertising, and provision of inter-connecting services, although Thompson and Whitfield (1995), estimate the effective price premium enjoyed by National Express at only 6 to 8 per cent. In practice, almost all NE's services are run by contractors, many from the independent coach sector, who presumably find this mode of operation more attractive than running directly under their own names with the associated marketing and setting-up costs.

Table 3.1 National Express passenger trips 1978 - 1996 (millions, approximate)

Year	Passengers
1978	10.7
1979	10.4
1980	9.2
1981	12.5
1982	14.0
1983	13.5
1984	15.0
1985	15.4
1986	14.7
1987	13.1
1988	13.1
1989	13.9
1990	13.5
1991	12.0
1992	10.5
1993	9.8
1994	10.9
1995	11.0
1996	11.7

Sources (to 1994): National Express, and technical press reports.

a. Includes Stagecoach Scottish services (trunk and interurban) from takeover in 1989, but not Citylink services in Scotland from 1993.
b. 1982 and 1994 traffic was boosted by rail strikes.
c. No specific figures published after 1994. 1995 and 1996 figures based on statement by Chief Executive that NEG coach market has grown by 6% per annum for last three years (*Transit* 11.6.97, p 9) (assuming that this applies to trips, not revenue), and applying it to 1993 base.

Figure 3.1 The National Express Network

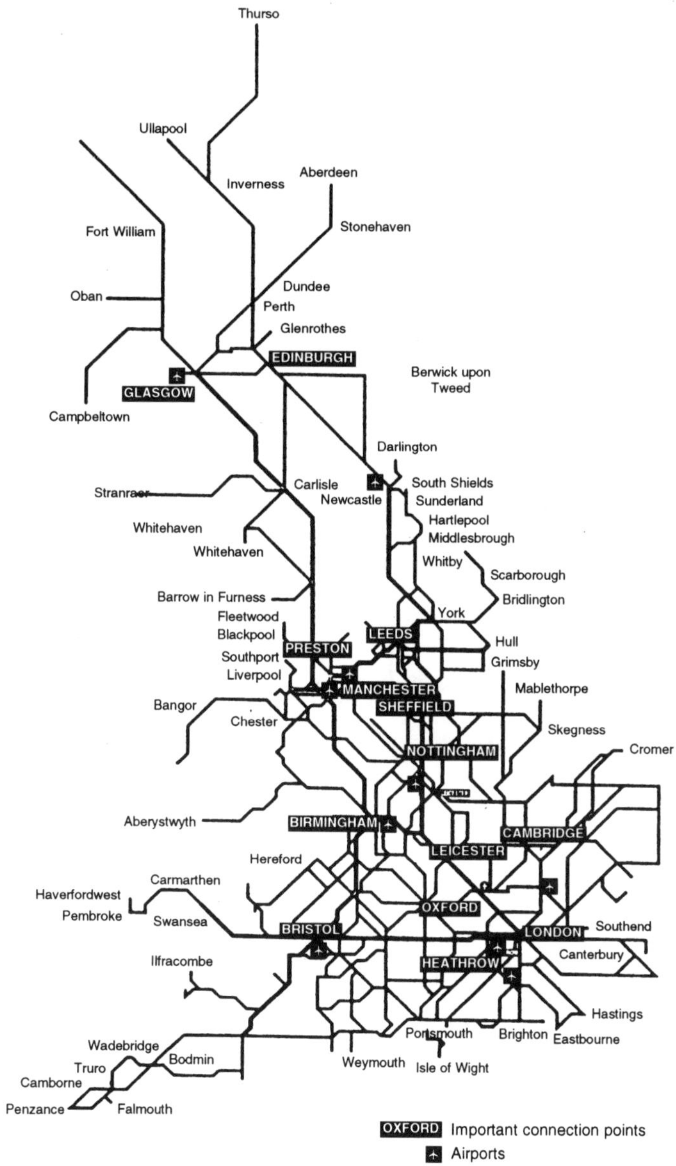

Figure 3.2 Independent services at July 1996

Changes in ownership

Apart from some expansion of independent coach operation, little change in ownership patterns occurred before the Transport Act of 1985, which both deregulated local bus services (except in London and Northern Ireland), and privatised the National Bus Company (including National Express). The latter process was completed in 1988, and by 1992 was followed by the Scottish Bus Group (SBG). Privatisation of local authority urban operations has also been encouraged. All the PTE-owned fleets have been privatised, and only about 18 operators now remain under district or unitary council control.

In the London case, control of London Transport was transferred from the then Greater London Council to the Department of Transport under the London Regional Transport Act of 1984. Subsequently, all of its subsidiary bus operators were privatised in the period 1993-1995. The London network thus offers an example of a centrally-planned, regulated system run entirely by privately-owned operators on a contract basis.

Whereas deregulation effects can be examined for a specific 'before and after' period (i.e. with respect to 26 October 1986, when the main provisions came into effect), this is not the case for bus and coach industry privatisation, which although largely complete (the remaining publicly-owned operators represent less than 10 per cent of output), was phased over a period of about ten years.

Trends in local bus services since deregulation

Very large unit cost reductions - of over 40 per cent - have occurred (in terms of the real operating cost per bus-kilometre), both in the deregulated regions and London. As a result of general support to fares and service levels being removed, public spending on support to services (now through competitive tendering of specific routes) has fallen substantially. Outside London, this drop has occurred mainly in the former Metropolitan counties (the English PTEs), and there has been a fall of 59 per cent nationally between 1985/86 and 1995/96, and by 69 per cent in the Mets. as such.[1]

However, deregulation has seen both a growth in bus-kilometres run (by 25-30 per cent overall, varying by area), and a drop in ridership. The latter, outside London, followed the previous long-run trend of about 3 per cent per annum. Hence, between 1985/86 and 1995/96, passenger trips (i.e. users boarding a bus to buy a ticket or showing a pass) fell by 29 per cent outside London. Combined with the effects of increased bus-km run, average passengers boarding per bus-km fell by 43 per cent. The drop in real operating cost per bus-km was almost totally offset by this factor, such that average cost per passenger trip remained almost unchanged, falling by only 2 per cent.[2] The drop was less marked in terms of passenger-km (the total distance travelled), as average trip length outside London rose by about 13 per cent between 1985/86 and 1993-95,[3] but even taking this into account, average load fell by about 34 per cent.

Conversely, in London, ridership has remained broadly stable, growing slightly in terms of passenger trips, while stable marginally in terms of passenger-km (i.e. average length of ride fell slightly). Hence, average operating cost per passenger trip fell sharply, by 34 per cent [4] and about 30 per cent per passenger-km. The better ridership trend in London was not due to falling real fares (which rose significantly). Hence, real total passenger revenue increased. Combined with a similar real total cost reduction to that elsewhere (since unit cost per bus-km fell, and bus-km run rose, by very similar percentages to the averages outside London), net public expenditure fell more rapidly than in the metropolitan areas - by 81 per cent between 1985/86 and 1995/96.[5]

Table 3.2 shows trends for each area category for which statistics are published, between 1985/86 and 1995/6. At the time of writing, these are the latest available, except for London, the 1996/7 annual report for which (London Transport 1997) indicates a further 3 per cent growth in passenger numbers over the previous year. Its 1996/7 passenger volume thus lies at 8 per cent above its 1985/86 figure in terms of passenger trips, and 0.4 per cent above in terms of passenger-km. Elsewhere, between 1985/86 and 1995/96, trips fell in aggregate by 29 per cent, and passenger-km by about 20 per cent., implying an annual average decline of about 3.4 per cent in trips and 2.1 per cent in passenger-km.

These statistics (except for trip length, derived from the National Travel Survey, NTS) are dependent upon operator returns made to the Department of Transport, which obtains such data from a large sample, covering 100 per cent of operators with 15 or more vehicles (DoT 1996), who in turn operate the vast majority of scheduled local services. However the quality of data recorded may be variable. There is little reason to believe that any incentives exist systematically to over- or under-estimate (except possibly for kilometres run, given the fuel duty rebate payable). Even allowing for some margin of error, however, the differences between London, the Metropolitan areas, and the rest of the country, are clearly very large.

Table 3.2 Trends in local bus passenger trips: bus-Kms operated, real fares and cost per bus-Km since deregulation

Percentage changes, 1985/86 to 1995/96 inclusive.
(to nearest whole number)

Area	Total passenger trips[a]	DoT real fares index[b]	Total bus-km[c]	Real op. cost/bus -km[d]
Metropolitans	-38	+53	+21	-49
English shires	-20	+11	+30	-42
Wales	-22	n/a	+29	-42
Scotland	-26	+6	+23	-41
Average for deregulated areas	-29	+21	+26	-44
London	+5	+35	+29	-44
Average for mainland Britain	-22	+23	+26	-44
Northern Ireland	-5	n/a	+29	n/a

Sources: Bus & Coach Statistics 1995/6,(DoT/HMSO, October 1996)

a. Table 2.1
b. Table 3.1.
c. Table 1.1
d. Table 6.1(b) : excluding depreciation
e. Northern Ireland data is the combined total for for Ulsterbus and Citybus, from the operator (for 1994/95). Note that the bus-kilometres figure includes work on all types of service, but the passenger trips figure is for scheduled services only. There is also evidence that car ownership is not a wholly exogenous variable - for example, it rose faster than before in the Mets. in the late 1980s, but much less than the national average in London since around 1990, possibly reflecting public transport provision and quality.

A good deal of the decline may be attributed to rising car ownership. For example, using data from the NTS, it may be shown that the annual bus trip rate by those in non-car-owning households in London, and the rest of Britain (outside the metropolitan areas), was almost unchanged between 1985/6 and 1992-94, although that in car-owning households did fall slightly, partly due to the growing proportion of two-car households (White 1997b). In the Metropolitan areas, trip rates in both categories fell, especially among car-owning household members, associated with the rapid growth in fares, and highly unstable patterns of service, especially in the period just after deregulation. The industry has thus become relatively more dependent upon its 'captive' (non-car-owning) market, an ever-diminishing percentage of the total population.

It may thus be argued that the drop in ridership outside London is largely what might have been expected, allowing for the effects of rising car ownership and real fares (i.e. the net effect of the growth in bus-km run seems to have been very small, despite some specific examples of successful conversion to high-frequency minibus operation). However, taken in combination with the growth in bus-km run, the average load outside London is now around 8.5 people (all day, both directions), a figure which lessens the usual advantage that buses have over private cars in use of energy and road space.

It should be noted that these trends do not necessarily relate to the same individuals. Each year, there is a 'turnover' of around 5 per cent to 10 per cent in the public transport market in an urban area, as individuals join or leave the system (through birth, death, changes in household structure, workplace, etc.). On a specific route this can be as high as 15-20 per cent. Hence, even to retain the 'captive' (non-car-owing) market, a continuous marketing and information effort may be needed for operators to remind potential users of the services available.

The London case

The London case can also be seen as one in which cost reductions came about initially through competitive tendering, rather than 'on the road' competition. A fuller description is given by Kennedy (1996). Until 1993, the system was based on 'gross cost' tendering, in which the operator was paid a sum for a specific service to be operated, with revenue being retained by London Transport (LT). Cost per bus-km on routes subject to competitive tendering fell by about 18 per cent in real terms (or 14 per cent after additional administrative costs), while service quality also improved through a greatly reduced proportion of lost mileage, and the ability of LT to 'buy' more service-km at lower unit cost, thus raising frequencies. Hence, substantial public expenditure savings were obtained, while users also benefitted.

The level of reliability in London was very poor prior to the mid-1980s, and the tendered service quality improvement may help to explain the greater impact of increased bus-km on ridership on London than elsewhere, since there is evidence of a higher user sensitivity to variations in bus-km from the schedule than in scheduled km.

Costs were reduced due to greater labour productivity, and using cheaper sites than the traditional purpose-built operating depots of LT. Staff numbers fell substantially, despite high bus-km being run, and lower wages were paid. Hence, an evaluation of the change in terms of net welfare might reveal much lower savings than the financial cost to LT as such, dependent upon assumptions concerning the extent to which labour made redundant was re-employed elsewhere, and the transfer effect of lower wages.

Since 1992, the distinction between tendered and directly-operated services has become less meaningful, as major cost reductions were made by existing LT-owned companies prior to privatisation, and the impact of the early 1990s recession enabled all operators to recruit staff at lower wages than before (as well as reducing labour costs through higher productivity). The tendering system also switched to a 'net cost' basis from 1993, in which revenue risk is taken by the operator, rather than LT.

As indicated in Table 3.2, the overall effect is to give a unit cost reduction of over 40 per cent over the whole period, a mix of tendering and direct cost reductions. A further cost reduction is reported between 1995/96 and 1996/7, resulting in the net subsidy per bus-km falling from 9.1p to 3.3p in this period (London Transport 1997, page 8). By 2001 all services will have been tendered out individually. However, staff costs may begin to rise again as recruitment becomes more difficult.

Variations in trends

The data presented so far relates to the industry as a whole, or highly aggregated levels. Critical indicators such as passenger trips carried, and bus-km run, are obtained from a large sample of operators by the Department of Transport, but details for individual operators are treated as confidential, the published data being aggregated to major categories (i.e.

London; the former metropolitan counties; rest of England; Wales; and Scotland - as in Table 3.2).

Conversely, detailed financial data is available for individual companies, specifying turnover, profitability and range of other indicators, as reported in the fortnightly journal 'Transit' and the annual 'Bus Industry Monitor' of TAS Ltd of Preston. However, this data is also aggregated in the sense that all forms of operator income are simply shown as 'turnover' or 'sales' (whether from passenger ticket sales, or concessionary fares compensation, for example), and physical indicators (passenger trips, bus-kilometres) are not published at company level. Given the fairly low average price elasticity which has traditionally applied (around -0.4), an increase in turnover may simply reflect fewer passengers at higher fares, not an increase in physical volume.

It is clear that averages give only part of the picture. Not all operators show the decline in ridership indicated at national level. However, comparing specific cases is difficult. In the absence of any requirement to publish ridership figures at operator level, it is only those operators who have disclosed figures (often in percentage change terms, rather than absolute figures) whose performance is known. It may also be difficult to compare trends, in that such data as is quoted is often for periods of varying length, and in some cases for specific routes rather than entire networks. Having said this, some cases seem fairly well-established:

a) Growth in total network ridership, from the mid-1980s (i.e. around the time of deregulation) to the present, or a recent year, e.g. Exeter, Bristol, Inverness, Oxford.
b) Growth in network ridership, but over a shorter period, which may reflect some recovery from ridership losses since 1985/86 rather than a net increase vis a vis that year, e.g. Brighton, also Carlisle, and some other parts of the Stagecoach network.
c) Growth on some routes, or substantial parts of networks, associated with minibus conversion, and/or quality improvement. Swansea (minibus conversion in the 1980s), Trent Buses (who run an interurban network around Derby and Nottingham), Interurban express routes of Stagecoach (in several parts of Britain), Guided busways in Leeds and Ipswich.

It also follows, since the overall average decline in ridership for the industry is known, that there must be cases worse than the average, but such operators seem unwilling to volunteer evidence.

It is clearly worth considering whether these exceptions are simply part of the scatter of results that one might expect, or display some systematic features that might be adopted elsewhere.

On closer examination, one phenomenon that may be discerned is that of 'convergence', i.e. some of the ridership growth represents cases that had untypically low ridership before change occurred. Likewise, some of the greatest decline may be from exceptionally high levels (such as South Yorkshire, whose low fares policy was abruptly reversed in 1986). A convenient measure of bus use, where absolute ridership data is available, is 'trips per capita per annum' (i.e. total trips per year divided by the catchment population served by the operator). For example, Exeter and Bristol had rather low per capita ridership figures before the minibus/midibus conversions of the mid-1980s, which brought them to around, or above, the per capita levels already experienced in other towns of similar population size and density (White 1997b).

Another example of convergence may be seen among the PTEs. A much lower rate of passenger decline is evident in the Glasgow (Strathclyde) area than the six English PTEs. However, the fares were exceptionally high before deregulation in the Glasgow area and some reduction (at least until c. 1993) seems to have occurred through competition in this case. Its per capita trip rate was also lower than in Edinburgh prior to deregulation, despite low car ownership (White and Farrington 1997).

The 'convergence' phonemenon may also be discerned from the data collected through the National Travel Survey (NTS). This is a household sample conducted nationally, in which all household members complete one week travel diaries covering trips by all modes of motorised travel. It has been conducted on a number of occasions (including 1985/86, just before deregulation), and also continuously since 1989. The sample size is not sufficient to give data for named counties or urban areas (except for very large centres, such as London), but is constructed to enable average trip rates for areas defined by density to be obtained.

Between 1985/86 and 1992-94, the NTS shows a somewhat lower reduction in bus boardings than the operator data, probably due to an over-representation of non-working women with children (who tend to be heavier-than-average bus users). Of greater significance is the variation by type of area. 'Rural' areas (settlements of under 3,000 population) retained the same annual trip rate over this period, at 36, and even showed an increase in passenger-km per person by bus (partly attributable to transfer of education trips from separate contract services on to local public bus services). The lowest reduction in per capita trip rate after this was in the smallest urban areas (3,000 to 25,000 population) at 9 per cent, and the greatest in conurbations outside London.

The rural results may seem particularly surprising, given the concern at the time of deregulation about possible effects on such areas, resulting in the introduction of special grants (a transitional grant per bus-mile run, which has now been phased out; and the Rural Transport Development Fund for innovative schemes, which continues). Part of the explanation lies in the fact that fare levels were not generally subsidised (in contrast to the Mets.) and were thus already high, but have changed little. Examination of a rural region, such as Lincolnshire, indicates that many services have continued to operate on a 'commercial' basis. One major factor may be the abolition of national wage agreements in favour of local negotiations. Lower-wage rural regions can thus recruit sufficient staff at lower rates of pay than previously offered, reducing unit costs. Further consideration of rural issues is given in the paper by Amanda Root elsewhere in this volume (note, however, that the survey described was undertaken during the summer holiday period, and thus does not cover the significant role of bus and coach services for education movement).

A clear exception to the convergence phenomenon is Oxford, in which a relatively high per capita trip rate (by National Bus Company standards) already existed prior to deregulation. Precise figures for each of the two main competing operators (Thames Transit, and the 'incumbent' City of Oxford) are not available, but it is likely that ridership has grown by about 35-40 per cent.[6] Among the factors applying are the extensive priorities given to buses in the city (which will encourage bus use, irrespective of specific operator actions), high parking charges - notably in the city centre, 'park & ride' site provision, and the more constructive form which competition has taken (Enoch 1997).

In terms of lessons applicable elsewhere, possibly of greatest relevance is the Trent experience, indicating the benefits which may be produced from raising service quality in factors such as driver training, vehicle standards, customer information, and so on.

Financial performance

The period immediately following deregulation was marked by poor financial performance. While about 85 per cent of bus-kilometres outside London were operated 'commercially' (i.e. without specific route contracts, but receiving concessionary fares compensation as a form of revenue), and most operators remained in business (only one major regional company collapsed), the average profit margin was very poor. Furthermore, this was based on historic depreciation, which is inadequate to provide for fleet replacement at 'normal' lives (around 15 years for full-sized vehicles, 5-10 years for minibuses/midibuses). The poor average loads produced as a result of falling ridership and increased bus-km (see above) resulted in low average revenue per bus-km (despite increases in real fares, notably in the Mets.). Hence, operators were not able to afford fleet replacement at usual levels, especially during the late 1980s/early 1990s.

In the last four years the position has greatly improved, both in terms of profit margins, and fleet replacement. About 3,600 buses and coaches per year are now being delivered. If minibuses (which do not appear in the figures produced by the Society of Motor Manufacturers and Traders, SMMT) are excluded, this is fairly close to the 'normal' replacement lives mentioned above, although a substantial backlog still remains. However, the sales of the smaller van-derived minibuses have dropped by about 30 per cent, possibly because their replacements are midi-sized vehicles (such as the Dennis Dart), which thus fall in the SMMT statistics. Hence the latter may show not only replacement of existing 'large' vehicles but also a trading-up from the smaller sizes.

Table 3.3 shows an estimate of industry profitability between 1992/93 and 1995/96 (an aggregate for all areas outside London). Table numbers shown thus: [1.1], indicate the source table in 'Bus and Coach Statistics Great Britain 1995/96'. Bus-km continued to rise until 1994/95, then falling slightly as competing services have been rationalised in some areas. Real operating cost per bus-km continued to fall. Hence, total operating cost was reduced by about 5 per cent over the period. Passenger revenue rose slightly in real terms to 1994/95 - after falling continuously from deregulation to 1992/93 - but then fell in 1995/96. Passenger volume continued to decline during this period, although at a slower rate after 1993/94 (the drop between 1993/94 and 1994/95 being less than 1 per cent). Revenue rose only as a result of increased real fares offsetting the volume reduction.

In addition to passenger fares, operators also received business income in the form of concessionary fares compensation, of over £300m per year, and that share of 'support' payments via local authorities which is passed on to operators for tendered services. Including these elements gives a fairly stable total operator income. The net surplus, as a percentage of operating cost, rises from 3.8 per cent in 1992/93 to an estimated 9.9 per cent in 1995/96. These estimates may be compared with aggregated data from published operator accounts from work by TAS, giving a similar improvement between 1992/93 and 1994/95 (one would not necessarily expect the two to be identical, due to other income and profit sources such as advertising and coach activity).

Table 3.3 Bus industry profitability 1992/93 to 1995/6

All areas in mainland Britain, excluding London

Year	1992/3	1993/4	1994/5	1995/6
Bus-km run (mill) [1.1]	2185	2242	2293	2270
Op. cost/bus-km, inc. depreciation @95/96 prices (£)[6.1(b)]	0.89	0.84	0.83	0.81
Hence, total operating cost (£m)	1944.5	1883.3	1903.2	1838.7
Passenger trips (mill) [3.1]	3351	3268	3253	3178
Passenger revenue @ 1995/96 prices (£m) (of which, con[a]) [4.2 & 5.3]	1851 (313)	1871 (327)	1902 (337)	1869 (329)
Share of public transport support to operators (£m) 5.2[b]	166.8	152.1	148.7	151.4
Hence, operator income (£m)	2017.8	2023.1	2050.7	2020.4
Net surplus	73.3	139.8	147.6	181.7
Surplus as % of operating cos	3.8%	7.4%	7.7%	9.9%
Do., from TAS	5.3%	6.9%	7.9%	n/a

a) Concessionary fares compensation from local authorities to bus operators.

b) Based on an assumption from previous research that two-thirds of the published 'support' figures is passed to operators (e.g. in 1995/96 the total was £226m).

c) From Table E1 'Pre-tax profit margins by area', *Bus Industry Monitor* 1996, TAS, Preston, August 1996.

Calculated figures are shown to one decimal place. Totals in millions published as such *in Bus & Coach Statistics* are shown in whole units only, source tables for which are shown thus '[5.2]'.

Their results for 1995/96 have not yet been published, but an examination of five major operator annual accounts for this period (or the nearest equivalent) in *Transit* shows an improvement in profit margin from 12.3 per cent to 13.6 per cent.[7] These operators now represent over 60 per cent of the industry's turnover, and have a much better average profit margin than for the industry as a whole. This is associated with an ability to make greater reductions in costs, through bulk purchases of items such as new vehicles, fuel and spares; and perhaps a willingness to take a tougher approach to cutting labour cost than a management/employee-owned business.

The improved profit margin thus raises both return on capital to investors, and enables operators to invest more heavily in fleet replacement, a characteristic for the larger groups (such as Stagecoach and FirstBus) in particular.

However, this encouraging picture must be qualified in that passenger volume has continued to decline. Cost reductions have come about largely through lower operating costs, and to a lesser extent falling bus-km run. Real revenues have resumed their downward trend evident before 1992/93. This is not a sustainable outcome in the longer-term. Further cost reductions may be difficult to attain as the labour market improves, and could affect service quality. Higher real fares make bus use less attractive vis a vis competing modes.

The other issue arising is that of 'goodwill' payments, and values placed on companies by purchasers. As many of the initial MEBO privatisations have 'sold on' to larger groups (such as Stagecoach or FirstBus), increasingly high prices have been paid. For example, the analysis by Robbins (1996) shows that the average ratio of price paid to annual turnover in 1993 was 0.42. This rose to 0.60 in 1994, 0.74 to 1995 and 0.91 in 1996. Even in the case of London, where all of the routes operated by the then LT subsidiary bus companies were subject to re-tendering within five years, a substantially higher price was obtained for the businesses as a whole (£233m) compared with initial estimates of around £100m. While reflecting some degree of optimism in the industry, such increased prices for companies also require a return on the 'investment' thus made by the purchasers. Hence, a higher profit margin may be needed in future for this reason, as well as to provide replacement asset investment, and a reasonable return on existing capital.

Net changes in public spending

The express coach industry has always operated without explicit public expenditure support, and continues to do so. The only significant exceptions are those routes which carry some passengers over distances of less than 15 miles (the dividing line between 'local' and 'express' services since 1986) and are thus eligible for partial fuel duty rebate.

The deregulation and privatisation of the local bus industry produced large reductions in the net support received via local authorities, following the abolition of general support payments for fares and service level support in 1986, and their replacement by competitive tendering for those non-commercial services (about 15 per cent of all bus-km run) which local authorities wished to provide. As mentioned earlier, the reduction has come about largely in the former metropolitan counties of England. Of a reduction from £547m in 1985/86 to £226m in 1995/96 (at 1995/96 prices), £238m (74 per cent of the total) was from this source (1). One must bear in mind, however, that while significant service loss has been prevented by use of powers to provide tendered services, users in the Mets. suffered substantial real fare increases. Larger public expenditure reductions also occurred in London (see above), in this case with somewhat lower real fares increases than in the Mets.

However, concessionary fares compensation has remained broadly stable, at about £400m (including London), and thus represents the largest element of public spending on the industry (although rightly classified by operators as a payment on behalf of users, rather than a subsidy). Free travel has been retained in some large urban areas (notably London, the West Midlands and Merseyside), but in others growing expenditure constraints have resulted in its being replaced by flat rate cash fares (West Yorkshire, South Yorkshire and Edinburgh for example). Outside major conurbations, free travel is rare but most areas are covered by a range of flat fare or half-fare schemes.

Scotland differs from these trends in that while support payments halved - broadly in line with deregulated areas as a whole - the concessionary fare expenditure also fell by about 20 per cent (White and Farrington 1997).

Until 1993, registered local bus services received a full rebate of fuel duty that would otherwise be payable, approximately halving fuel costs. Subsequently, absolute increases in the duty from the November 1993 budget have been imposed on bus services also. The net total rebate per year has remained fairly stable at about £200m since 1985/86 (at 1995/96 prices).

In addition to the reductions in annual public expenditure, gains were also made through the sale of companies to the private sector. Overall proceeds from the privatisation of NBC were over £322m (National Bus Company 1989). Sale of the London Transport operating companies produced £233m (National Audit Office 1995); in addition PTEs and local authorities gained income from sales of their operating companies in many cases.

However, these gains must be qualified in several respects:

a) A substantial part of the NBC privatisation proceeds was in fact the outcome of a surplus on the pension fund, which would have accrued in any case. The issue of whether the state, or the pensioners, should gain from this is currently a matter of dispute, the Trustees of the Bus Employees' Superannuation Fund suing the Department of Transport for return of the fund surplus, now £168m (at July 1997).

b) NBC and SBG succeeded in meeting all of their interest and debt repayment requirements, producing an operating surplus to do so. In selling the businesses, the state made a 'one off' gain, out of which existing capital debts were repaid, but lost the cash flow that resulted. In its last full year of public ownership, 1985, NBC had a turnover of £807m, making an operating surplus (on historic depreciation) of £35m. Some £15m was paid in interest, and £32m of commencing capital debt and other borrowing repaid during that year, leaving outstanding debt at £96m. Had it remained in public ownership, continued payment of interest on capital debt would have continued. For example, a payment of £10m p.a. over 25 years at a discount rate of 8 per cent p.a. would produce an inflow of £107m.

c) A notable feature of reduced operating costs since deregulation has been a reduction in real earnings of bus industry staff. While growth in productivity represents the main element in reduced costs, this factor should also be considered. The average weekly earnings of bus drivers fell 14 per cent in real terms between 1985/86 and 1995/96, and even more so relative to earnings of manual workers as a whole. In addition, working conditions may have worsened through reduced holiday entitlements. unpaid meal breaks, etc.. This change could be seen as a 'transfer payment' from bus industry staff to other sectors rather than a net reduction in resource spending, but even in the narrower terms of public expenditure may not give as clear a reduction as appears at first sight.

There is evidence of a substantial number of bus drivers in London, for example, being eligible for family credit.[8] Thus, some of the savings in public expenditure may be in part offset by increased spending under headings other than 'transport'. This point applies both within and outside London, through raising the level of commercially-viable services, and reducing the net costs to public authorities of tendered services. Work by Johnson (1997) indicates that for an adult with two children, and a non-wage-earning partner, earning £3.50 per hour and with rent of £50 per week, various forms of public expenditure comprise a substantial part of total income, even on a 50-hour week. While this rate is below the bus industry average, especially in London, similar figures are applicable to drivers (especially of minibuses) elsewhere in Britain.

In the event of a minimum wage and/or the 'social chapter' requirements being introduced, some increases in public spending may result (assuming no cuts in services). But the net effect may be less than appears at first sight if current low wages are being subsidised through other forms of public spending.

While it is the case that a net reduction has been made in public expenditure, the extent of this is somewhat exaggerated by simply looking at the gross privatisation proceeds and net annual savings in 'transport' expenditure.

Contrasts with the rail privatisation process

As is described in other papers in this volume, the rail privatisation process is almost complete, with all 25 passenger train operating companies being transferred to the private sector by 1 April this year. The process is closely linked with that in the bus industry, in that about 53 per cent of the franchise payments go to train operating companies which are now wholly, or partially, owned by bus and coach companies. National Express is the largest recipient of franchise income, some 20 per cent of the total, other major companies from the bus and coach sector being Prism Rail (a consortium formed by bus industry executives) at 9 per cent, MTL (9 per cent), Stagecoach (8 per cent), and FirstBus (4 per cent) (Glover 1997). The annual franchise payments to National Express total around £530m in 1997/98 (plus passenger revenues) compared with a turnover for its entire coaching operation of £170m in 1996. Thus, it is now primarily a rail operator, as well as the greater part of the scheduled coach industry.

Given the performance of the bus and coach industry as described above this might seem somewhat surprising, since the ridership trend has been poor, and the skills of the local bus industry in particular seem to lie mainly in cost reduction rather than marketing or improving service quality. Furthermore, the structure of rail costs (in which payments by operating companies to Railtrack and the Rolling Stock Leasing companies typically form over 60 per cent of their total costs) means that substantial growth in passenger volume and revenue will be necessary to achieve the ambitious targets for franchise payment reductions over the next seven years to which most bidders have committed themselves. In contrast to the bus industry, where most effort (at least until recently) was directed toward cost reduction as the means of improving performance, the rail franchisees are thus forced into emphasising a more positive approach to the customer.

While privatised, the rail industry cannot be described as 'deregulated' however. Stricter controls on service standards, quality and pricing have been introduced, notably in the regulation of certain major fare categories ('savers', and season tickets), initially so that average annual increases should not exceed RPI, and from 1999, RPI-1 per cent. Considerable effort has been directed toward protecting existing service levels enjoyed by users, provision of through ticketing (albeit not for cheaper fare categories), and publication of a national timetable. Service quality is closely monitored by OPRAF, especially in respect of cancellation levels.

This contrasts very markedly with the bus industry, in which introduction of the 1985 Act's policies effectively discouraged inter-operator cooperation in scheduling and ticketing. No safeguards in respect of fare increases were provided, and timetable information became very erratic and variable in quality.

Subsequently, some improvement in provision of bus information has occurred, and a degree of inter-operator cooperation (for example, in joint timetabling, and travelcards) is reappearing, albeit condoned rather than encouraged by OFT and MMC policy.

Why are such marked differences found, when the same overall economic philosophy lay behind the changes introduced? One explanation lies in the much greater media coverage of rail than bus issues, associated with the higher social status of rail users. Much stronger pressure was applied to protect users' current facilities. There was also an awareness of the greater importance of interchange and through ticketing, given the greater average length of rail than bus journeys. The rail industry as a whole also displays a more sophisticated management approach (for example, in use of market research work) than the bus sector.

Encouraging trends in ridership and revenues are being reported by a number of franchisees from the first year of privatised operation. The extent to which this is due to their own initiatives is uncertain, since a general upturn in the economy may also be an explanation (paralleling the growth in BR ridership and revenue in the mid-1980s). Nonetheless, it is clear that the negative effects found at the time of bus deregulation (such as service instability, and high fares increases in the Mets.) have been successfully avoided.

The rail case is also relevant to the question of public expenditure. Unlike the bus case, annual operating support has approximately doubled (in round figures, from £1,000m to £2,000m), as a result of the high payments to Railtrack and the ROSCOs. Conversely, gains through selling-off assets have come on a larger scale, and at an earlier stage (notably the ROSCOs in the 1995-96 financial year, and Railtrack the following year). Whereas a 'circular cash flow' was retained within the public sector prior to privatisation (i.e. the large profits generated by operating company payments to ROSCOs and Railtrack were paid back to the state), this was then lost. The maximum net outflow of public money thus occurs in the current financial year (1997-98), but is due to fall back to pre-privatisation levels around 2002.

If the commitments made by franchisees are workable, then a small net gain in public spending will have been made on a discounted cash flow basis, over the first seven years. This assumes, however, that no further gains would have occurred in productivity under continued BR ownership, and that none of the revenue growth would have arisen either. The latter in particular is likely to be due as much to the general state of the economy as actions of particular franchisees.

Suggestions for future policy

The recent change of government, and the scope to analyse the longer-term effects of coach and bus deregulation, provide an opportunity to review future policy options as a means of improving the industry's performance.

The programme outlined in the Queen's Speech in May 1997 does not include any explicit transport measures. This gives an opportunity to examine more thoroughly the transport policy options before new legislation is introduced. However, some legislation already announced will have implications for public transport policy, notably that concerned with competition law, the government of London, and devolution in Scotland and Wales.

Given the markedly different trends found in London and the deregulated regions, one option would be to adopt a London-style tendering system, notably in other conurbations with complex networks in which the benefits of network-wide planning, passenger information and common off-bus ticketing may be most marked. However, not all differences between performance in London and that in the rest of Britain can be attributed to these structural differences (for example, the retention of free concessionary travel in London obviously affects ridership), and a further restructuring of the industry might itself prove disruptive with effects on ridership similar to those found just after deregulation in 1986.

While retaining the option of applying a London-style franchising system to other cities for the medium-term (and introducing powers to permit this), it may be more productive in the shorter-term to look at modifications to the existing system to improve its performance.

As mentioned above, the application of competition law through the roles of the OFT and MMC in the bus industry has tended to discourage sensible coordination in terms of ticketing and service planning. At the same time, however, the effectiveness of the competition authorities in dealing with highly aggressive competition is also open to question. Enquiries may take a year or more to complete, and while recommendations (such as divestment of acquired undertakings) may be enforced, there are no direct legal penalties (such as fines) on the 'guilty' party, only requirements in respect of future behaviour. For example, a case in which one operator ran just ahead of another at unrealistically low fares, could severely damage the operator affected, yet any action would not follow for some time.

The number of references made to the OFT (often by operators complaining about each others' behaviour) has diminished, as competition has reduced. However, bus industry enquiries continue to form a substantial part of the MMC's output, despite its small size in relation to other industries for which the MMC also has responsibility.

In other cases, the actions taken may only indirectly change conditions. For example, as a result of the FirstBus/Strathclyde merger, which gives FirstBus a dominant role in much of central Scotland, the MMC has required FirstBus to sell off one of its Glasgow depots, and the Midland subsidiary (which operates around Falkirk). This would only indirectly encourage competition (which has in any case emerged from Stagecoach more recently), yet would disrupt the comprehensive Glasgow network now offered, on which a common off-bus travelcard is available. While the PTE offers a card valid on all operators, most sales appear to be for the cheaper card offered only on FirstBus subsidiary companies' services (MMC 1997, Tables 3.1 and 4.10), whose comprehensiveness would be lost if the network were fragmented. Likewise, it is difficult to see what direct benefit is provided by compelling Stagecoach to sell off its

Huntingdon depot and Milton Keynes operations following its takeover of Cambus.

There is a danger that tightening up competition policy enforcement might aggravate the problems through lack of coordination already found outside London, while at the same time London would continue to enjoy the benefits of an integrated system.

There is also a need to harmonise the rail franchising process and competition policy, given the uncertainty created by the reference of the National Express Central Trains and Scotrail franchises to the MMC in May, some time after the transfer of operations had occurred. While the Scottish case could be seen as one in which modes normally competing successfully (rail and coach) have been placed in the same ownership, the Central Trains case (related to NE's ownership of Travel West Midlands, the dominant bus operator in the West Midlands conurbation) is one in which sensible opportunities for common ticketing and network planning have been provided as a result.

One approach could be to change the emphasis of competition policy from the pursuit of competition *per se*, to one in which long-run user benefits were the main criterion. On this basis, for example, competition which produced (sustainable) lower fares could be seen as beneficial, whereas that which simply duplicates an existing operator's service at the same fares (thus lowering its viability, while giving few user benefits) could be seen as harmful. Where extensive mergers are taking place, with associated benefits in coordinated timetabling and ticketing, but a risk of long-run fares increases occurring, a better safeguard against monopolistic abuse than forcing divestment of parts of an operation might be to impose some overall control on real fares increases (as in the case of rail franchises).

A particular irony at present is that it may be easier to produce network benefits through one company taking over another completely, than through encouraging joint operations (which would still leave companies free to compete in other respects, such as bidding for tendered services). For example, in the Brighton area, a merger recently took place between Brighton Buses (the former municipal operator, privatised in 1993 through a management/employee buy-out), and Brighton Hove & District (the former NBC company, now part of the Go-Ahead Group). This enables rationalisation of services to provide a network of high-frequency 'metro' links, and a lower-priced travelcard than the existing inter-operator card.

In the London case, the proposed creation of a London-wide elected authority would restore a degree of democratic control in a situation in which many decisions have been made solely by central government (the principal exception being the funding of concessionary fares through a consortium of the boroughs). A strategic approach to issues such as 'park and ride' provision would enable a more comprehensive view to be taken than that of individual boroughs.

Although the London bus tendering system has worked well, a greater degree of flexibility could be useful. Under the London Regional Transport Act of 1984 the responsibility for provision of services lies with LT itself (rather than bus operating companies), which has been intrepreted to mean that detailed service planning and specification also remains an LT responsibility: for example, the determination of the exact route to be followed, and the timetable, together with local public consultation on proposed changes (Chartered Institute of Transport 1994). The shift toward net cost tendering (rather than gross cost) results in the operator taking the revenue risk for any changes that are made (as mentioned earlier), while having little flexibility to intiate such changes or modify services in the light of operating experience.

Minor modifications to the legal role of LT, in the process of the new London Act, should make it possible to retain the benefits of strategic network planning, comprehensive passenger information, and off-bus ticketing, while introducing greater flexibility at local level.

Public transport initiatives at the local level have been constrained by limits on public expenditure. While the shift to the 'package' TPP system had encouraged a more comprehensive approach to road and public transport schemes, the very limited total funding for such measures severely limits their scope. Even modest schemes, such as short guided busways, may be subject to considerable delay in funding. Greater devolution of powers to a London authority, and likewise in Wales and Scotland, might ease such funding restrictions, although presumably still subject to some constraints through national economic policy.

A final policy issue, although not directly linked with legislation, is the degree of data disclosure. As indicated above, analysis of bus industry performance is hindered by the lack of operator-level data on ridership and physical outputs. In the case of the privatised rail industry, a good deal of data is available on each of the franchises (notably through press statements issued by OPRAF on the occasion of each award). While the rail statistics bulletin now issued by the Department of Transport, Environment and the regions gives only aggregate figures for the whole rail network, OPRAF's annual report gives data on an operator-by-operator basis (for example, of total train-km, passenger trips and passenger-km). One hopes that this will be continued on a annual basis, to enable an explicit assessment of system performance. Hopefully, the same practice could be extended to the bus sector, in which overall performance of the industry can be described, but variations around it only in very fragmentary manner.

References

Chartered Institute of Transport (1994) *Whither the Clapham Omnibus?* Report of the bus working party on future policy options for buses in London.

Department of Transport (1996) *Bus and Coach Statistics Great Britain 1995/96,* Appendix 4, Stationary Office, London.

M. Enoch (1997) 'Oxford and Darlington - the Mess and Success of Bus Deregulation', *Proceedings of the Chartered Institute of Transport in the UK,* vol. 6, no. 1 (March), pp. 29-47.

J. Glover, 'The Franchised Railway' *Modern Railways*, May 1997, pp. 273-277.

P. Johnson (1997) 'A 10p tax rate will not banish the poverty trap', *Independent* 30 June.

D. Kennedy (1996) 'London Bus Tendering: A Welfare Balance' *Transport Policy*, vol. 2 no. 4, pp. 234-249.

London Transport (1997) Annual Report 1996/97.

Monopolies and Mergers Commission, *FirstBus plc and S B Holdings Limited: A Report on the Merger Situation,* Cm3531, Stationery Office, London.

National Audit Office (1995) 'The Sale of London Transport's Bus Operating Companies', HC-29, HMSO, London.

National Bus Company (1989) Report and Accounts for Year to 31 March 1988.

D.K. Robbins (1996) 'Performance Trends in the Bus and Coach Industry', Appendix, Lloyds Bowmaker Corporate Finance Division, Bournemouth.

D. Thompson and A. Whitfield (1995) 'Express Coaching: Privatization, Imcumbent Advantage, and the Competitive Process', in M. Bishop, J. Kay and C. Mayer, OUP, Oxford.

P.R. White (1997a) 'The Experience of Bus and Coach Deregulation in Britain and other *Countries', International Journal of Transport Economics*, vol. XXIV no. 1, February, pp. 35-52.

P.R. White (1997b) 'What Conclusions Can Be Drawn About Bus Deregulation in Britain?' *Transport Reviews,* vol. 17 no. 1, January-March, pp. 1-16.

P. White, P. Dennis and N. Tyler (1995) 'Analysis of Recent Trends in Bus and Coach Safety in Britain' *Safety Science,* vol. 19, pp. 99-107.

P. White and J. Farrington (1997) 'Bus and Coach Deregulation and Privatisation in Great Britain, with Particular Reference to Scotland', Paper at Royal Geographical Society - Institute of British Geographers' Annual Conference, University of Exeter.

Notes

1. Bus & Coach Statistics Great Britain 1995/96, Table 5.2.
2. Ibid, Table 6.2(b).
3. National Travel Survey 1993-95, Table 2.3.
4. Bus & Coach Statistics Great Britain 1995/96, Table 6.2(b).
5. Bus & Coach Statistics Great Britain 1995/96, Table 5.2.
6. Based on reported passenger figures under NBC ownership immediately prior to privatisation for City of Oxford, and South Midland (the latter being taken over by Thames Transit), total ridership figures for Thames Transit quoted in the technical press, and a working assumption that City of Oxford ridership is about the same as before deregulation. Keith Moffatt of the Go-Ahead Group (owners of City of Oxford) has recently quoted a 50 per cent growth ('Modern Railways' July 1997, p. 415), but absolute figures are not given.
7. An unweighted average for profit (after interest, before tax) of Cowie, Stagecoach, FirstBus, National Express (its bus operations in the West Midlands), and the Go-Ahead Group. Together, these groups represent over half of total industry turnover. Note that they tend to have somewhat higher profit margins than smaller operators, and hence than the industry as a whole.
8. I am indebted to Ms Juliet Solomon, an independent transport researcher, for drawing this to my attention.

CHAPTER 4

Competition and local bus services

Mike Parr

Introduction

This paper examines the use of competitive markets to provide local bus services,[1] focusing on experience since the *1985 Transport Act* and other allied legislation became effective. The present government's consultation paper on transport, issued last year as part of its fundamental review, recognised the benefits that competition has brought, but raised a number of potential major options for change.[2] This paper does not develop or explicitly analyse these here, as this would require an *ex ante* assessment of alternative arrangements that do not presently exist on a wide scale. Instead the paper assesses what is the current economic organisation of local bus services.[3] However in doing that it inevitably addresses whether there is a case for substantive change.

Nevertheless, the forthcoming White Paper on transport policy seems likely to emphasise the important role that public transport, and especially local buses, can play in encouraging a switch from car use, perhaps most particularly in urban areas. Certainly this has been the thrust of ministerial statements and is an implicit message of the consultation paper. Quite how this is to be accomplished remains to be seen.

Background

Local bus services remain an important part of passenger transport, though it is well known that such services have been in long-term decline. In 1994-95 nearly four and a half billion journeys were made by local bus, but this is only about half the number made 25 years previously. Indeed going back still further in time, the number of bus and coach journeys was even higher with bus and coach travel being the dominant form of passenger transport from the 1930s until the mid 1950s.[4] Now, of course, it is the private car that dominates.[5]

Nevertheless, local bus markets continue to be significant in money terms, with a national market value of about £2$^1/_2$ billion in Great Britain in 1996-97.[6] The real value of this market has been largely constant over the last decade as real price increases have compensated for falling passenger numbers.

At a more microeconomic level and on average, about 80 bus journeys per head of population are made each year at an annual cost of £40, so the average household spends around £100 on bus travel each year. In practice, local buses are used more by women, children and the lower socio-economic groups, who do not own or have access to a car. 33 per cent of local bus journeys are made for the purpose of shopping, whilst 14 per cent are for education, i.e. transport to and from schools, colleges and so on.[7]

Notwithstanding how local bus transport is, or should be, integrated with other transport modes to take account of externality or equity considerations and impacts, it is clearly important that the provision of local bus services is done efficiently and responsively to passenger demand. The *1980* and *1985 Transport Acts* changed the basis of policy towards bus markets by emphasising the feasibility and desirability of introducing competition so as to achieve that end. Perhaps the fullest statement of the previous government's underlying presumptions and expectations as regards local bus services is to be found in the 1984 White Paper.[8] There it was stated that 'the bus market is a highly contestable one'[9] (and so was deemed suitable for competitive market provision).

This was a view repeated, for example, in 1990, when the Department of Transport gave evidence to the Committee of Public Accounts in its examination of the sale of the National Bus Company (NBC). The Committee reported that, 'The Department (of Transport) pointed out that, as the costs of entry into the bus business were not generally high, the threat of such access ought to be effective in maintaining the benefits of competition'.[10]

Local bus services were thus seen as a good example of a contestable market, which required little active intervention to ensure that economic efficiency was achieved.

As is well known, the 1985 Transport Act then gave effect to this view. It provided, in Part I, for the abolition of road service licensing and the deregulation of local bus services outside London from October 1986 (though this only came fully into effect in 1987). Part II dealt with the special London regime.[11] Part III dealt with the privatisation of the NBC. Part IV was concerned with the establishment of the municipal bus operations as public transport companies and their division into smaller companies. Part V dealt with the introduction of competitive tendering for socially necessary services.

Thus, from October 1986 the Passenger Transport Executives and municipal bus companies were converted to public companies. Private bus operators were now only required to have a licence to operate a public service vehicle and could introduce competing local bus services simply by giving 42 days notice to the Traffic Commissioner for the relevant area. This was in contrast to the much more directional role the Commissioners had previously played. The latter continued, however, to supervise safety standards.

Accompanying these changes was a considerable degree of company restructuring. Local bus service provision was not therefore seen as perfectly contestable. The market organisation did require some restructuring towards a more atomistic ideal.

As regards the NBC, there was some debate about how much restructuring was necessary,[12] but over the July 1986 to May 1988 period there were 62 separate sales of 72 subsidiaries (some were sold together). These were made through trade sales and management buy-outs. No single buyer could purchase more than three of the 72 subsidiaries or acquire companies operating in adjacent areas.[13]

What emerged was a ‘patchwork quilt’ of local bus companies in England and Wales, with the Department of Transport ensuring that companies in adjacent areas[14] were not sold together. This was on the grounds that, even though local bus markets were highly contestable, the most likely entrants were companies already active geographically close to each other.[15] It was therefore felt important that these potential competitors were independent of each other. The break up of the NBC ensured that no single new company accounted for much more than a percentage point or two of the national market.

The same approach was adopted, following the 1989 Transport (Scotland) Act, when the Scottish Bus Group (SBG), part of the Scottish Transport Group (STG), was sold off over the period August 1990 to October 1991. The SBG was broken up, with ten separate sales and similar rules about acquisition, with no single buyer being allowed to acquire more than two companies.

In 1992 the government introduced incentives to local authorities to privatise their bus companies. Since then a majority have been sold to management and employee buy-outs, though a small number remain in the public sector. According to the MMC only 17 local authority bus companies remain.[16]

In London, even though a special franchising regime was set up, London Buses was subdivided into 11 operating subsidiaries in 1989 and these were sold to private operators, either third parties or their own management, over the January 1994 - January 1995 period. Again similar rules on acquisition were made. Purchasers were not allowed to buy companies in adjacent areas and no one was allowed to acquire one or more subsidiaries if the effect was to create a 25 per cent market share.[17]

Thus the 1985 Transport Act, the 1989 Transport (Scotland) Act and the subsequent privatisations of NBC and SBG, coupled with the establishment of the municipal fleets as fully commercial companies operating at arms length from the local authorities concerned, left local bus provision in Great Britain to the forces of on the road competition.[18] Operators could choose their own routes, frequencies and fare levels for commercial services, though there was a registration and notification period for starting and stopping services so as to avoid undue service instability. The much smaller number of socially necessary services,[19] which would not be provided voluntarily by the market, were also to be provided by private operators. They were to be purchased by local authorities after competitive tendering. In parallel the national market was 'deconcentrated' to provide a fair and sustainable,[20] or at least starting, competitive base. It was believed that contestability would ensure that market power did not become a problem. As a backstop, the industry was made subject to the normal competition rules.

Market change

The industry created above has undergone very considerable change in the decade since privatisation and deregulation took effect. Three aspects of that change are worth examination; market structure changes; the involvement of the competition authorities; and a review of market performance, comparing out-turns with the expectations of the 1984 White Paper.

Market structure

Structural change was relatively limited in the first half of the 1987-1997 period. Partly the NBC, SBG and municipal fleets were still in the process of being sold off. Once set up, the new company managements were also preoccupied with coming to terms with their particular geographic areas of operation and the new competitive regime. However, structural change away from the privatisation structural blueprint accelerated in the second half of the period, with a sharp rise in merger activity. This has dominated over organic change via competitive success and failure in the market, which is generally a slower process when there are very many local markets rather than a single national one.

At national level a number of large, and publicly quoted, operators have emerged. The House of Commons Transport Committee's Report in November 1995 reported a share of the top four companies of $32^1/_2$ per cent.[21] The Monopolies and Mergers Commission (MMC) at about the same time reported a figure of 38 per cent, rising to 42 per cent as a result of the Badgerline/GRT merger in May 1995.

The latest MMC report on the situation after the merger between Cowie and British Bus reports that the top four's share had risen to almost 50 per cent, as Table 4.1 shows.

The changes occurring at local level are rather more difficult to determine. Firstly, the TAS directory, on which the above figures are based, is compiled only at regional level and local markets are more numerous than this, being at sub-regional level. Secondly, the definition of a local market is not straightforward.[22] Even if it were, no-one has catalogued and examined all the local markets in Great Britain, which probably run at least into the hundreds. Thus there is no readily available data source on local markets and the MMC in its enquiries has always had to collect data anew in each local market case.

Nevertheless, because local markets are at a greater degree of disaggregation than the national market, the average level of concentration, unweighted across all local markets, is likely to be higher than the national level of concentration.[23] Certainly it is the case in some local markets that only one company operates bus services. This might remain consistent with a contestable view of local markets, if that one operator were still constrained by potential competitors, whether contiguous or not. However, the considerable increase in national concentration and the uneven strength of the major companies across regions suggests that there is limited contiguity of the majors and that they do not actively constrain each other.[24]

Table 4.1 National percentage market shares of the leading companies, August 1996

Company	1991	1992	1993	1994	1995	1996
First Bus	6.3	6.2	6.8	12.8	12.8	19.8
Stagecoach	4.9	4.9	6.9	13.4	13.4	16.1
Go-Ahead	1.7	1.7	1.7	4.3	4.3	6.2
National Express	6.0	5.9	5.9	7.7	7.7	5.2
MTL	0.0	0.0	2.1	3.2	3.2	3.3
British Bus	3.4	3.4	3.9	7.9	9.7	-
Cowie (before merger)	0.6	0.6	0.6	3.5	3.5	-
Cowie (after merger)	-	-	-	-	-	14.9

Source : MMC(1997), op cit., Table 4.1, based on TAS Publications and Events Ltd., Bus Industry Monitor 1996.

What is without doubt is that the market structure put in place in the NBC, SBG, municipal and London privatisations has changed beyond recognition. From being a very fragmented and unconcentrated national market with no one operator having more than a few percentage points, an above average degree of aggregate market concentration has emerged.[25] Whether this was expected by the Department of Transport and its advisers seems unlikely. Nevertheless, such change might not matter if the main companies operated in all markets or were at least contiguous. There could still be strong on-road and potential competition in all local markets. However, this does not appear to be the case for the main operators.

Even then, competition might still not be adversely affected if the smaller companies operating in local markets were effective in competitively constraining the larger companies (and vice versa), i.e. if the market was highly, or workably, contestable. Certainly, Thompson and Whitfield[26] conclude, in examining long distance bus services, that the very dominant national position of National Express does not give rise to concern, with market performance not diverging significantly from the contestable bench-mark.

Involvement of the competition authorities

Given the above structural change, the competition authorities have inevitably been much exercised with the bus industry. The behaviour of bus companies has also occasioned action. Utton [27] argues that the industry has received more attention from the UK antitrust authorities than any other since the formation of the Monopolies and Restrictive Practices Commission in 1948.

Initially, the involvement was by the Office of Fair Trading (OFT) in scrutinising a large number of restrictive trade practices (RTP) agreements, which were submitted for registration after the industry became subject to the competition laws. The bulk of agreements were submitted between September 1986 and March 1987 when about 200 were submitted (it being an offence to have an unregistered agreement). These were, in the main, between ex-NBC companies and the larger municipal and metropolitan operators. They covered a large number of arrangements in place, ranging from agreements to provide information, agreements to use certain bus-stops, through agreements on routes and timetables, and on to agreements about fares and travelcard schemes.[28]

Given the degree of coordination and regulated monopoly provision which characterised the industry before the 1985 Transport Act it was not surprising that the successor companies wished to continue with practices that were once deemed normal operating arrangements. A large number of the agreements contained restrictions on fares and timetables and in October 1989 the OFT indicated to the companies that it would take the offending agreements, 115 in total, to the Restrictive Practices Court. The bulk of the agreements were then abandoned by the companies concerned. The remainder were modified to remove the anti-competitive restrictions. Two cases went to Court; G K Kinch/Midland Fox and Western National/Plymouth Citybus, both price fixing and market sharing agreements. Once this initial batch of agreements was dealt with, RTP activity shrank dramatically as the companies understood rather better what was permissible. New agreements submitted, and variations to existing RTP agreements, are now believed to be at a much lower level.

At the same time the OFT began to receive a steady stream of complaints from consumers and competitors about anti-competitive behaviour. The vast majority of complaints were eventually found to be unsubstantiated or unjustified. However the level of complaint about the industry did not diminish as Table 4.2 below shows.

Table 4.2 Complaints about bus operators to the OFT, 1987-1994

Subject	'87	'88	'89	'90	'91	'92	'93	'94	Total
Bus stations	9	11	7	4	7	6	10	2	56
Subsid-ised services	7	13	15	6	4	16	7	1	69
Predat-ion	23	24	19	23	39	52	51	36	267
Other	18	10	12	18	18	16	40	17	149
Total	57	58	53	51	68	90	108	56	541

Source: Table 1, p.6 of Director General of Fair Trading (DGFT), *The Consequences of Bus Deregulation outside London : a memorandum to the Transport Committee*, April 1995.

A number of complaints led to investigations by the OFT under s.3 of the 1980 Competition Act, all bar one about predatory behaviour,[29] either pricing or route running. The other case, which established a precedent used in a number of later cases, was about the conditions of access to a bus station given by the owner to a rival bus operator. Table 4.3 below shows reports made by the OFT.

The 1994 Deregulation and Contracting Out Act repealed the requirement on the OFT to make a formal Competition Act report before negotiating undertakings to remedy any anti-competitive situation or making a reference to the MMC. Since January 1995 there have therefore been no formal OFT reports on anti-competitive cases.

Table 4.3 Competition act reports made by the OFT, 1986 - 1995

Case	OFT Report	Main Practice Involved	Decision of the DGFT
Southern Vectis Omnibus Co.	1988	Refusal of access to bus terminal	Anti-competitive
West Yorkshire Road Car Co.	1989	Predatory behaviour	Not anti-competitive
Highland Scottish Omnibuses	1989	Predatory behaviour	Anti-competitive *
South Yorkshire Transport	1989	Predatory behaviour	Anti-competitive
Kingston upon Hull City Transport	1990	Predatory behaviour	Not anti-competitive
Southdown	1992	Predatory behaviour	Anti-competitive *
Thamesway	1993	Predatory behaviour	Not anti-competitive
Fife Scottish	1994	Predatory behaviour	Anti-competitive
United Automobile Service	1995	Predatory behaviour	Anti-competitive

Note: * subsequently referred to the MMC, which reported separately in 1990 and 1993 respectively.

However, the OFT was increasingly involved in scrutinising mergers, which became more frequent over the 1990s. It is these mergers which account predominantly for the market structure changes outlined before. Table 4.4 below shows the load from 1990 - 1994.

Table 4.4 Bus mergers and proposed mergers considered by the OFT, 1990-1994

	1990	1991	1992	1993	1994
Total cases considered	19	10	3	23	68
Referred to the MMC	5	0	0	2	3

Source: DGFT, *op cit.*, p. 9.

Over the whole period since 1986 to the end of 1997, a total of 17 merger references have been made to the MMC. Table 4.5 below lists the individual cases.

In a number of cases where an adverse public interest finding was made the merger was not allowed to proceed and full divestment was required so the pre-merger situation was restored. In other adverse finding cases, such as the recent First Bus/SB Holdings case, the Secretary of State has allowed the merger to proceed subject to satisfactory undertakings. Such undertakings apply to the activities of the companies in the reference area or parts of it. In the First Bus/SB Holdings case some divestment was the condition, but normally undertakings have been of a behavioural kind, covering the services to be run and prices charged after the merger. The undertakings are frequently time limited.[30]

Table 4.5 Mergers considered by the MMC, 1986-1997

Case	MMC report	Combined market share in reference area	MMC conclusion
Badgerline Holdings and Midland West Holdings	1989	82	Against the public interest
Stagecoach Holdings and Portsmouth City Bus	1990	42	Against the public interest
South Yorkshire Transport (Sheffield & District, Michael Groves, Sheafline, Hallam/SUT)	1990	50	Against the public interest
Western Travel and G & G Coaches (Leamington)	1990	34	Not against the public interest
Stagecoach Holdings and Formia	1990	36	Against the public interest
Caldaire Holdings and Blue Bird Securities	1991	48	Against the public interest
Stagecoach Holdings and Lancaster City Transport	1993	51	Against the public interest
National Express Group and Saltire Holdings	1994	80	Not against the public interest (this was a coach merger not a local bus case)
Stagecoach Holdings and 20% stake in Mainline Partnership	1995	56	Against the public interest
SB Holdings and Kelvin Central Buses	1995	66	Not against the public interest
Stagecoach Holdings and 20% stake in SB Holdings	1995	61	Against the public interest

Case	MMC report	Combined market share in reference area	MMC conclusion
Stagecoach Holdings and Ayrshire Bus Owners (A1 Service)	1995	33	Against the public interest
Stagecoach Holdings and Chesterfield Transport (1989) Ltd	1996	63	Not against the public interest
The Go-Ahead Group and OK Motor Services	1996	35	Not against the public interest
British Bus and Arrowline (Travel)	1996	12	Not against the public interest
First Bus and SB Holdings	1997	54	Against the public interest
Cowie Group and British Bus	1997	15 in GB, 26 in London	Not against the public interest

The OFT has also made two scale monopoly references to the MMC covering areas that are more substantial than a narrow local market. In both cases adverse findings were made by the MMC. Table 4.6 below details the cases.

The 1995 report found that there was little evidence of active competition between the largest operators, which are now the second and fourth largest in Great Britain, in the reference area.

Taken together this level of scrutiny by the competition authorities might seem to justify Utton's view. There have been two RTP court cases, nine OFT Competition Act reports, two MMC reports following up OFT Competition Act reports, 17 MMC merger reports, two MMC monopoly reports and a number of cases of judicial review.[31] This has led to quasi-permanent regulation of some bus companies in some local markets, where undertakings are in place. This level of involvement of the general competition authorities over the decade has on occasion prompted calls for a specialist regulator to be set up.[32]

Table 4.6 Monopoly cases considered by the MMC, 1986-1997

Case	MMC Report	Issues	MMC conclusion
Supply of bus services in Mid & West Kent	1993	Abuse of dominant position (predation) and vertical integration (discriminatory access to a garage) by Maidstone & District (95% market share)	Against the public interest. The MMC also urged that a review of competition in the bus industry should be undertaken. This repeated a call first made in the 1990 Stagecoach/ Portsmouth merger report
Supply of bus services in the North East of England	1995	Abuse (predation) by Stagecoach Holdings and The Go-Ahead Group (both 33% market shares); the former, through its Busways subsidiary having lead to the collapse of Darlington Transport Company	Against the public interest

Market performance

However, it does not matter in welfare terms what the structural changes have been, or whether the industry has been the subject of much intervention by the competition authorities.[33] Many other industries that have been deregulated and privatised have seen similar levels of structural change and investigation. Indeed in retrospect it was probably inevitable that the break up of the industry would be followed by some reagglomeration.[34] What does matter is whether the impact of the changes has been beneficial in terms of the market performance.

The interim evaluation by Tyson,[35] made in January 1992, has seen the broad trends identified there continue. Table 4.7 below lists some key aspects of aggregate industry performance since 1985. This is for Great

Britain as a whole and there is considerable variation in experience when the data is considered at a less aggregate level.

The number of passenger journeys has continued to decline broadly in line with long-term trends, though there is limited evidence that the decline has to a degree been arrested.

In contrast the number of bus kilometres run by operators, i.e. services and/or their frequency, has increased by around a quarter since deregulation. This is against the long-term trend of a steady decline of around 2 per cent per annum from the mid-1950s to the early 1980s.

Real fare levels have increased by a quarter over the period.

Direct public transport support has more than halved, though concessionary fare reimbursement not counted in the above has increased in importance.

Real operating costs per vehicle kilometre have fallen by nearly a half over the period, though the fall per passenger journey is much less, reflecting the greater vehicle kilometres run and the consequent fall in the average load of each bus.

Interpreting and explaining these facts is more controversial, as deregulation has worked as expected in some areas but not in others.
Demand has not risen. This is partly the result of rising real fare levels, which was not fully anticipated in the White Paper. Some increase in fares was inevitable on those routes that were previously cross-subsidised, but the elimination of cost inefficiency should have restrained this. The more general rise in fares is partly because the public sector bus operators in existence before deregulation were not charging economic prices for services as a whole. [36] Certainly this seems to have been true of the metropolitan areas and London. In those circumstances it is inevitable that, in the private sector, as capital assets are replaced so prices overall would have to rise to provide a proper return. But of more importance to the declining demand appear to be other demand side factors, and in particular, continued growth in car ownership.

Whilst the price trends have perhaps been contrary to expectations, costs and subsidy payments have behaved as expected. Costs have been reduced considerably. This is largely due to more efficient working practices, timetabling, lower wages and driver only buses. The value for money of subsidy has also improved. Competitive tendering is widely credited with lowered costs.[37] An interesting feature of subsidy is that the pattern of subsidy has changed. Ten years ago revenue support of routes and route systems for socially necessary services, was substantially larger than concessionary fare reimbursements - the direct subsidy of individual groups of consumers. Today that position is reversed. This suggests that

Table 4.7 Selected local bus service measures, Great Britain, 1985/86 - 1995/96

Year	Passenger journeys, million	Vehicle kms, million	Real fare index, 1992=100	Real public transport support £m.[38]	Real operating costs, pence per vehicle km[39]	Real operating costs, pence per passenger journey[40]
1985/86	5.6	2.1	87	769	158	58
1986/87	5.3	2.2	93	634	141	57
1987/88	5.3	2.3	94	484	124	55
1988/89	5.2	2.4	94	453	118	54
1989/90	5.1	2.4	94	373	112	54
1990/91	4.9	2.4	94	382	108	54
1991/92	4.7	2.5	95	441	106	56
1992/93	4.5	2.5	98	427	101	57
1993/94	4.4	2.6	101	288	95	56
1994/95	4.4	2.6	106	278	92	55
1995/96	4.4	2.6	107	268	89	53

Source: GSS, *Busdata - 1997 edition*, *op cit.*

subsidy is now much better targeted, though in practice this means that services that were previously loss-making and cross-subsidised are now probably either higher priced or run less frequently.

The rise in real prices might be cause for concern if, combined with the increase in market concentration, there was evidence of excessive profit levels. However, although data on bus industry profitability is published by TAS in its Bus Industry Monitor this does not straddle the whole period. Pre-tax operating margins have increased from a low point in 1990/91, so that by 1994/95 they were around 8 per cent. The historical return on capital was then estimated to be around 40 per cent. However, this is a poor measure of real economic returns and definitive views on the level of profits cannot be made, though it does seem that the larger operators are more profitable than the average.

As regards the dynamic performance of the industry, there have been major changes in the vehicle stock, and significant improvements in vehicle and labour productivity. Between 1985/96 and the end of 1995/96 double deck vehicles declined from 37 to 26 per cent as a proportion of the PSV bus fleet. Over the same period single deckers with fewer than 36 seats more than doubled from 14 to 33 per cent, with most of the increase coming in the midibus category. As expected new types of bus service with more appropriate buses have been introduced and there is some evidence that investment levels are now rising.

Competition issues arising

It is clear that aggregate market concentration, and local market concentration, have increased substantially over the last decade. This might not matter if barriers to entry were zero or so low as to allow the impact of potential competition to constrain the actions of any dominant firm or group of firms. A key competition issue, therefore, is the extent to which the 1984 White Paper's presumption of a high degree of market contestability has proved to be the case.

This is an issue on which a wide variety of views have been expressed. Cowie and FitzRoy[41] believe that contestability theory is discredited and quote approvingly from Dasgupta and Stiglitz[42] that the theory is 'well funded but not well founded' (as it was developed as part of IBM's antitrust defence strategy in the US in the 1970s). In contrast, Ridyard[43] finds it a useful general framework for analysing the ease of entry.

The view taken on this issue, has a potentially much wider policy implication for the transport sector as a whole, for it is often claimed that contestability is a characteristic of most transport markets. In rail, shipping and airlines (as well as road transport) it is conventionally thought that there are well developed second-hand markets for the main non-infrastructure assets used. Entering such markets does not therefore require expenditure to be sunk irretrievably. Indeed, by definition, cars, lorries, buses, planes and so on are transportable to new locations and uses without substantial cost. Because they are not tied to a specific use they are not sunk and expenditure can be recovered on exiting a market. Before addressing this main issue, however, there is a related issue to consider.

One of the 1984 White Paper's other presumptions was that economies of scale were minor, so that the implied optimum size of firm, based on cost considerations alone, was small. Consequently the industry was restructured into small operators and, additionally, it was felt that new small scale entry was possible. The growth of large operators in the last decade must, on this view, be motivated by non-cost, and possibly market power, considerations.

However, there is increasing evidence that there are economies of scale associated with the purchasing of new vehicles, fuel, spares and insurance costs. If larger firms have newer vehicles then running costs per mile may also be lower. It also seems to be the case that larger firms enjoy certain financial economies. Their cost of capital is below that of small firms.[44] However, there do not appear to be any comparable cost economies of scope associated with multiple route running, unless demand on a route is so thin that vehicles and staff must be shared across routes.

Aside from economies on the supply side, there may also be economies of scale on the demand side, because of discounted return, multi-journey and network ticketing.

Thus, the White Paper presumption that local bus provision is a constant cost activity may well be incorrect. As yet there are no empirically sound estimates of the economies involved. Irrespective of the size of any economies of scale, this does not, in principle, change the contestability of the market as economies of scale, per se, are not a barrier to entry. If capital markets are perfect then new companies or entrants from other markets should be able to obtain the necessary capital to set up at the required scale.[45] Any imperfection in capital markets would alter this and might cause an absolute cost disadvantage for a small scale entrant.

Whilst economies of scale do not of themselves represent an entry barrier, they do alter the scale at which entry needs to occur. If there are significant economies of scale and the size of these is such that they can only be achieved by operating across a number of local markets, then entry would have to be simultaneous multiple market entry. Entry to a single market would not be feasible as the new entrant would be at a cost disadvantage to the existing firm. Simultaneous multiple market entry seems unlikely if there are informational costs associated with local market conditions.

Single market entry from a below optimum size firm can also be met selectively and will not be successful unless the entrant has other cost or quality advantages. Nevertheless, even if the optimum size of firm is quite large one might be able to rely on the threat of entry from other existing firms to constrain an incumbent. In extremis, even if the market is a natural monopoly it may still be that the threat of entry constrains the monopolist, and this was the startling implication of contestability theory. This brings us back to the main competition question of the contestability of bus markets.

Contestability analysis made its first appearance in the US in the late 1970s and early 1980s[46] and very quickly had an impact on policy, particularly in the deregulation of civil aviation and a number of merger decisions therein.[47] The status of the theory remains controversial.

At a theoretical level it has been shown that small deviations from the key assumption of zero sunk costs and lags in the response of incumbents to entry can cause the predicted competitive market outcome to disappear.[48] If true, this makes the theory non-robust. Schwartz,[49] in particular, emphasised that only market data can establish whether price and quantity responses are sufficiently rapid. This has led the theory and policy debate towards the examination of the manipulation of entry conditions and post-entry expectations of potential entrants by incumbents through a variety of types of strategic behaviour.

If the impact of workable contestable markets can only be appreciated through investigation of actual markets, then the views of competition authorities who have investigated a large number of local bus markets are of some importance.

The OFT points to two reasons for believing that potential competition is not a strong disciplining force in local bus markets.[50] First, the development of a reputation for aggressive responses, particularly through predatory output or price changes, is said to influence entrants. As detailed earlier, there have been a large number of predatory allegations in the industry. Second, there is a reluctance of the major operators to engage in direct competition with other large rivals by invading their territories. The OFT concludes that, 'potential competition may not be as potent a force as was assumed when the industry was deregulated.'

The first of the OFT's reasons is theoretically weak. Reputation alone is unlikely to deter rational entry unless backed either by a real difference in cost efficiency or a deeper financial pocket than the entrant. The latter will be the case when the entrant is small-scale. Thus it is the second reason, tacit collusion not to compete by the larger firms, which is more compelling. One should also add in the fact that bus companies can respond quickly to entry with price changes. New registration of services takes a little longer. As already noted, speed of response in the market is an important factor and critical to the robustness of contestability theory.

The MMC has also rejected the view that local bus markets are perfectly contestable. It has rejected a number of mergers, which would not have mattered if potential competition was as powerful as contestability implied. If bus markets were perfectly contestable then the emergence of high market shares in a local market would not be cause for concern. The MMC does seem to have come to the view that reductions in actual competition can be more important than the potential competitive threat.

It has identified a number of barriers to entry or growth:[51]

- access to bus stations;
- the availability of depots and maintenance facilities;
- the arrangement of passenger information;
- pre-paid ticket schemes;
- the reputation of incumbent operators for aggressive behaviour in defence of their territories.

These are not all to be found in every local market, but have been found individually and sometimes in combination in cases.

Access to existing bus stations owned by an incumbent can be important to a new entrant, but, following the OFT's 1988 Southern Vectis test case, a precedent for non-discriminatory access has been established[52] and it appears to be a very sporadic problem.

Large depots are no longer generally needed by bus operators, only secure parking space, and maintenance can be contracted out, so the second factor has similarly only occasionally been found to be an issue.

Passenger information is more important. A new entrant may need to incur sunk costs on information provision and advertising to promote its services (which an incumbent may not have had to do). Any such need is probably less on high frequency services. It has also been argued that passengers get much of their information from sources other than timetables, and particularly from using the service. There may then be additional costs of running services at an initial loss to establish a reputation. However, this factor presumes a degree of consumer loyalty to incumbents, and conscious choice of bus operator, which does not seem characteristic of local bus users.

The fourth factor is more debatable. The MMC have argued that network tickets of a large incumbent constitute a barrier to entry as consumers prefer through-ticketing. Certainly through ticketing is an important feature of international airline markets, where passengers fly through hub airports on to other destinations. Whether through ticketing is as important in local bus markets is much more open to doubt. Certainly, a new entrant, entering at sufficient scale, could replicate the network and not suffer the disadvantage.

The fifth factor has already been commented upon. As prospective local bus operators have to register services 42 days in advance, incumbents have time in which to plan a response. Here the relevant contestability condition is clearly violated. Price responses can be immediate and so occur before entry, though any quantity response will be slower as incumbents also have to give 42 days notice for service and timing changes, except in a number of restricted circumstances.[53]

Nevertheless small scale operations remain prevalent across Great Britain[54] and the MMC believes that new entry, at small scale, 'has (historically) been a frequent occurrence in most parts of the country, and reflects the fact that barriers to entry are not significant at this level'.[55]

Thus on the key question of the contestability of local bus markets it does seem as if the reports of the competition authorities cast doubt on perfect contestability but not on a reasonable degree of contestability. Potential competition from small scale entrants still seems an important factor, though probably a declining one in the long term. Assuming that present trends in concentration continue, it is legitimate to question whether competition, and cross local market entry, between the major companies will be strong in the future. Having reached the limits of expansion by merger, the top companies can now only further expand by head to head competition so this makes stronger competition more likely. Tacit collusion rather than active rivalry remains a possibility. However, this feature is no different from a host of other equally important product and service markets which are similarly concentrated.

Many bus lobbyists have also pointed to a variety of market outcomes which are said to be examples of malign competition. Firstly, it has been noted that although competition has led to more (wasteful) bus kilometres being run, it has not led to a growth of passenger kilometres. This is a confusing criticism of competition, as operators have run more buses, but consumers have chosen not to use them. The reasons for this failure to attract more custom seem more connected to the fundamental attractions of rival transport modes rather than the supply side organisation of local buses.

Problems of passenger information are sometimes also said to be more acute under competition. I doubt this is such a problem as is sometimes maintained (because most bus journeys are regular rather than taken entirely off the cuff.) If centralised information points are needed then it should be relatively straightforward for local authorities to undertake more provision.

Wasteful 'bunching' of buses is also sometimes said to occur. This mainly seems a consequence of demand itself being thicker at some points of the day rather than at others. There can be periods of short term 'bus wars', but these do seem to be temporary rather than permanent features of local markets. Head to head competition is often part of a struggle for control of a route. Certainly the MMC investigations do not find these matters to be major issues. Often criticisms about bunching are linked to the idea that services where demand is low, such as at slack times of the day, ought to be provided even if they are not economic. This can only occur if cross subsidy between different consumers is allowed to exist.[56]

There is then the argument that service regularity is important, and frequent service changes harm the industry as a whole as travel by bus becomes more uncertain. This really relates to the registration rules rather than to the desirability of competition. The government has already recognised that there is a balance to be struck between stability and flexibility.[57] Changed registration rules could deal with this.

Finally, there is the finding that investigations by the competition authorities are not always sufficiently timely or effective in remedying competitive defects.[58] However here the defect is one primarily of the competition legislation rather than the use of competition in local bus markets. The new Competition Bill, mirroring the provisions of the European Commission's Article 85 and 86 approach, should go some way to rectifying that.

Conclusions

A number of conclusions can be drawn from this review of local bus market deregulation.

First, the fairly constant decline in the popularity of local bus transport over the last 40 years, when a variety of bus policies have been followed, has continued. This background represented an unpromising set of circumstances in which to introduce competition. The continued long run decline in bus usage seems caused primarily by factors on the demand side, such as rising income levels, increasing car ownership and changing land use patterns, rather than any remaining inefficiencies on the supply side. As income levels increase so the value of time spent travelling and the conditions of travel become more important. People prefer faster and less time sensitive consumptive modes. The use of cars for local travel can be rational, even when it increases congestion and gives no substantial time saving, when consumers do not bear the full social costs of such decisions and see quality differences in the transport modes. Bus deregulation and privatisation could only affect the supply side. Whilst this might change the nature of the product and its price on offer, it is only though other policies that the more fundamental factors on the demand side can be affected.[59]

Second, this confusion of supply and demand side factors needs to be recognised in proposing any change to the present competitive organisation. Further changes to the supply side, whether minor or major, seem unlikely alone to reverse the long term decline of bus usage and encourage modal shift.

Third, some of the key underlying presumptions of bus deregulation have not been borne out. There may be economies of scale on the supply and demand sides so that small scale operation and entry is increasingly less likely. The fair and sustainable industry structure created a decade ago has changed dramatically to become a concentrated national market structure with higher local market concentration.

Fourth, and notwithstanding the third conclusion, some of the expectations of the 1984 White Paper have come true with deregulation bringing a number of clear and significant benefits. Cost efficiency has increased substantially. If strong product market competition has been a weak force giving this, then it may be that capital market competition and the incentives created by management ownership have had an effect. In the provision of socially necessary services via competitive tender, costs have similarly been reduced.

Fifth, and on balance, competition remains feasible and desirable, even though the market is not perfectly contestable. Some routes and urban areas can and do support multiple operators. In general, and given the industry changes though, one would expect rather more concentrated provision to become the norm. It will then be particularly important to maintain actual competition where it exists. An active competition policy to police any exploitation of a dominant position is needed as well as guarding against tacit collusion between the majors.[60] In this respect local bus markets appear to be no more different from many other areas of companies are abusing their position. The forthcoming reform of competition law should guard against this.

Notes

1. Local bus services, or stage carriage services, are normally distinguished in the literature from long distance, or express coaching, and other bus services, such as contract hire. Historically these distinctions were more meaningful as they each had substantially different regulatory requirements on them.
2. Department of Environment, Transport and the Regions (DETR), *Developing an Integrated Transport Policy: An Invitation to Contribute*, August 1997. There, at paras 22-23, a number of options are flagged, ranging from limited enhancement of the current arrangements to wider reliance on (monopoly) franchising or the 'quality partnership' approach.
3. A number of other reviews have been made, such as; C. Nash, 'British Bus Deregulation', *Economic Journal,* 103 (419), July 1993, pp.1042-49, I. Savage, 'Deregulation and Privatisation of Britain's Local Bus Industry', *Journal of Regulatory Economics*, 5, 2, June 1993, pp.143-58 and, S Glaister, 'Deregulation and privatisation: British Experience', in, G. De Rus, C. Nash, (eds.) (1997), *Recent Developments in Transport Economics* (Avebury, Aldershot). See also the House of Commons Select Committee on Transport, *The Consequences of Bus Deregulation*, HC 1995/96, 54-1.
4. It was in 1957 that car and van transport overtook bus and coach as the dominant mode, as measured by the share of passenger miles. See Table 9.1., in GSS (1996), *Transport Statistics, Great Britain 1995* (HMSO, London).
5. See, DETR (1997), *Developing an Integrated Transport Policy: Factual Background*, or, for more detail, GSS, *ibid.*
6. This figure includes concessionary fare reimbursements from local authorities, but excludes revenue support grant and fuel duty rebate.
7. The other main uses over 1992-94 were commuting (20 per cent) and leisure (20 per cent), see Table 12.5, in, GSS (1996), *Social Trends 1996* (HMSO, London). For more recent and extensive analysis see Part 4 : The National Travel Survey, in, GSS (1997), *Busdata- 1997 edition; A Compendium of Bus, Coach and Taxi Statistics* (DETR, London).
8. Department of Transport, Scottish Office and Welsh Office (1984), *Buses* (HMSO, London, Cmnd.9300).

9. *Ibid.*, para 20, p.52, though the criteria by which 'high' contestability was adduced were not set out.
10. Committee of Public Accounts (1991), Ninth Report, Session 1990-91, *Sale of the National Bus Company* (HMSO, London) para. 13, p. vii. However, the Department made some small changes to the regulatory regime in February 1995 so as to allow Traffic Commissioners to restrict the circumstances in which duplicate buses could be run. Previously operators had been allowed to run such buses without registration. This flexibility had allegedly been used in an anti-competitive way by incumbents against new entrants.
11. The past and present position in London is best explained in the Monopolies and Mergers Commission (MMC) report of March last year; MMC (1997), *Cowie Group plc and British Bus Group plc : A report on the merger situation* (HMSO, London, Cm.3578), as Cowie and British Bus both operated in London.
12. See J. Vickers and G. Yarrow (1988), *Privatization : An Economic Analysis* (MIT. London,), pp. 366 - 384.
13. Though this desire was breached twice.
14. Now called, in the regulatory literature, contiguous companies.
15. This view on the supply side substitution possibilities, part of the key step of market identification, was taken on by the competition authorities. The acquisition of a bus company the operations of which are not overlapping or contiguous has not normally been a candidate for reference to the Monopolies and Mergers Commission. This view also seems to have transferred to the mixed road/rail mergers now underway.
16. MMC (1997), *op cit.*, para.3.5.
17. For further details, see MMC (1997), *ibid.*
18. Northern Ireland and London to one side.
19. Often called non-commercial or contract services, as opposed to the commercial or non-contract services. Today these account, outside London, for under 20 per cent of bus kilometres run.
20. This phrase comes from s.48 of the *1985 Transport Act* which made this, the promotion of fair and sustained competition, the main objective for the NBC in splitting itself up.
21. Then Stagecoach, FirstBus, British Bus and West Midlands Travel (owned by National Express). Market share estimates can be made on either an output (bus kilometres) or value (turnover) basis. The latter is the most common and is used here.
22. It is beyond the scope of this paper to consider the way local bus markets have been defined in competition cases, temporally, geographically and by service.

23. This can be grasped intuitively by considering a single small operator, with 1 per cent of the national market, but operating exclusively in a single local market where it is the sole operator. In that local market the concentration is 100 per cent. Inevitably the share of the big four must be higher in the remaining 99 per cent of local markets than when averaged across all markets.
24. This is also the view of NERA in, *The Effectiveness of Undertakings in the Bus Industry*, OFT Research Paper 14, December 1997.
25. The 1996 national bus market C_5 ratio of 62 per cent compares to an average 1992 C_5 ratio in 3 digit manufacturing industries of 40 per cent.
26. D. Thompson, A. Whitfield (1995), Ch. 1, Express Coaching: Privatisation, Incumbent Advantage, and the Competitive Process', in, M. Bishop, J. Kay, C. Mayer (eds.), *The Regulatory Challenge* (Oxford University Press, Oxford) p. 41.
27. M. A. Utton (1995), *Market Dominance and Antitrust Policy* (Edward Elgar, Aldershot) p.129.
28. For a fuller description of the main types of registrable agreement, see OFT (1995), *Restrictive Trade Practices in the Bus Industry* (HMSO, London).
29. These bus predation cases seem to have helped the OFT develop its methodology for analysing predation. See G. Myers, *Predatory Behaviour in UK Competition Policy*, OFT Research Paper no.5, November 1994, where two of the three case studies are bus cases. The methodology was carried forward in the 1995 United Automobile Service report, where a range of price elasticity values were used in the incremental loss calculation.
30. For a critical examination of the undertakings imposed in bus cases see NERA, *op cit.*.
31. Including the key case brought by South Yorkshire Transport which resolved what meaning was to be put on 'a substantial part of the UK', as this test, in s.64 of the *1973 Fair Trading Act*, has to be met for a merger to be referred to the MMC.
32. The MMC raised this issue in the 1993 monopoly report.
33. If their actions are guided by some sort of cost benefit analysis.
34. Certainly before the 1930 Road Traffic Act, which ushered in 50 years of heavily regulated monopoly provision, there were three dominant bus groupings: the Thomas Tilling Group, British Electric Traction and Scottish Motor Traction.
35. W. J. Tyson (1992), *Bus Deregulation Five Years On : Report to the Association of Metropolitan Authorities and the Passenger Transport Executive Group* (AMA), which was part of a series of monitoring studies undertaken for the AMA and PTEG.

36. This has two aspects to it. Firstly, whether the NBC and other public sector operators were meeting the rate of return targets set by the Treasury before privatisation. Many were not, but this was largely compensated for in the sale price of assets achieved. However, assets have eventually to be replaced and the due rate earned. Secondly there is the issue of whether the private sector rate of return, or cost of capital, is higher. The unweighted average equity *betas* for the top five bus companies, all quoted (FirstBus, Stagecoach, Cowie, National Express and Go-Ahead), from the London Business School Risk Measurement Service, January-March 1997, is 0.83. The market capitalisation weighted average is 0.85. This shows the companies are a little below average risk in relation to the market as a whole.
37. See, for example, D. Kennedy, 'London Bus Tendering: The Impact on Costs', *International Review of Applied Economics*, 9, 3, 1995, pp. 305-17, and the Transport Committee's report.
38. Revenue support only. This excludes concessionary fare reimbursement and fuel duty rebate and is at 1995/96 prices. There is no requirement on Local Authorities to set up concessionary fare schemes and there appears to be considerable variation between authorities.
39. Including depreciation at 1995/96 prices.
40. Including depreciation at 1995/96 prices.
41. C. G. Cowie, F. FitzRoy, *Deregulation and the Demand for Scottish Bus Transport*, University of St. Andrews, Department of Economics, Discussion Paper No.9508, July 1995., p.2.
42. P. Dasgupta, J. Stiglitz, 'Potential Competition, Actual Competition and Economic Welfare', *European Economic Review*, 32, 1988, pp. 569 - 77.
43. D. Ridyard, 'Contestability Theory and its Practical Impact on Competition Policy Decisions', *The Business Economist*, 26, 2, 1995, pp. 1-11.
44. Though this has not been subject to close examination.
45. The important distinction in entry barriers is between absolute and strategic barriers, see, London Economics, *Barriers to Entry and Exit in UK Competition Policy*, OFT Research Paper no.2, 1994.
46. W. J. Baumol, J. Panzar, R. Willig (1982), *Contestable Markets and the Theory of Industry Structure* (Harcourt Brace Jovanovich, New York), and, W. J. Baumol, 'Contestable Markets; an Uprising in the Theory of Industry Structure', *American Economic Review*, 72, March 1982, pp.1-15.
47. E. Bailey, 'Contestability and the Design of Regulatory and Antitrust Policy', *American Economic Review*, May 1981, pp.178-183.

48. See for example, W. G. Shepherd, 'Contestability versus Competition', *American Economic Review,* 74, 4, September 1984, pp.572- 587.
49. M. Schwartz, 'The Nature and Scope of Contestability Theory', *US Department of Justice Antitrust Division, Economic Analysis Group Paper*, 1986, subsequently published in *Oxford Economic Papers.*
50. DGFT, *op cit.*, p.11-12.
51. MMC, *FirstBus plc and S B Holdings Limited: A report on the merger situation* (HMSO, London, Cm.3531, January 1997) paras 2.46 and 4.45.
52. Though actually calculating non-discriminatory access prices is not easy.
53. Details are to be found in Appendix 3 of, *Local Bus Service Registration: A Guide to the Arrangements for Providing Local Bus Services* (Booklet PSV 353A (Rev 2/96)).
54. According to figures from the *Bus Industry Monitor*, quoted by the MMC (1997), in para 4.5, the national market share of independent operators has in fact risen from 12 per cent in 1989 to over 14 per cent in 1996.
55. *Ibid,* para 4.12.
56. Economists are generally against cross subsidy as it results in economic inefficiency, with demand too high on the subsidised routes. There is also an equity dimension to cross subsidy.
57. Department of Transport (1993), *The Operation of Local Bus Services outside London*, (HMSO, London).
58. NERA, op cit.
59. There appears some dispute in the literature on the appropriate short- and long-run price elasticities, and the long run relationship between car ownership, bus use, and land use changes, all key demand side factors.
60. Though there is little that can be done to force oligopolistic companies to compete.

CHAPTER 5

Rural transport after the deregulation of buses

Amanda Root[1]

Synopsis

This paper describes travel patterns, energy use and attitudes to travel in two rural villages. The cultural factors that are influencing travel and options for policy innovation post transport privatisation and deregulation are assessed. The importance of understanding both newly emerging travel patterns and households as mediators of lifestyle and travel demands are stressed, and the environmental and policy implications are explored.

Abstract

The following account is based on a quantitative and qualitative study of travel by residents in two rural villages in South Oxfordshire.

The findings suggest that lifestyles are now largely based upon the opportunities arising out of the use of cars which is tantamount to the 'privatisation' of transport (i.e. a reliance upon a household-based provision of the means of travel). This kind of privatisation is exemplified in reducing usage of public transport and increasing usage of cars. The question of what people want, the lifestyles they lead and the options they perceive should be paramount in any discussion of new transport planning. Household resources determine high and low energy use, but are themselves constrained by cultural factors and infrastructural provision. Thus it is argued that policy implications should involve innovative approaches, and their success will depend largely upon whether they can adapt to the changing travel needs of rural populations.

Introduction

Both the villages in this study, Chalgrove and Cholsey, have expanded dramatically in population in the last thirty years. Both now contain about 3,000 residents and approximately 1,000 households. The survey and interviews revealed that central to many of the respondents' perceptions is the sense of migration, attendant new houses, jobs and other opportunities. These largely positive attitudes to transience and the malleable nature of 'community' inform attitudes to travel and its provision. Transport infrastructure, (which has been altered by deregulation) in addition to household resources, will be discussed as an important determinant of travel needs and behaviour.

The two villages used in this study are in South Oxfordshire. Both villages have major roads going through them, or on their outskirts. The mode of transport used the most in both villages is the private car: 96 per cent of households in these villages owned cars. The villages have different types of public transport infrastructure. Chalgrove has a limited bus service that operates approximately hourly during morning and evening peak periods, and two hourly at other times in the day. The buses, which go through Chalgrove to Oxford in one direction and a town called Watlington in the other, stop at about seven o'clock at night. Cholsey has a railway station, with trains about every half an hour until late at night and an approximately hourly bus service to Oxford, via a local market town called Wallingford and to Reading.

Much of the housing in Cholsey (population 3,428) was built between the two world wars by the local councils, so the housing estates are substantial, and somewhat more uniform in appearance than those of Chalgrove (population 2,832). Neither Chalgrove nor Cholsey have the 'exclusive' or picturesque charm of some nearby villages and towns, but both places are perceived as desirable as indicated by the fact that house prices are as high or higher than those for comparable properties in parts of the City of Oxford (Root et al 1996a).

Both villages are sought after as residential locations and have broadly similar socio-economic profiles (Root et al 1996a). Not only is the City of Oxford, with its prosperous 'sun belt' economy within about fifteen miles for each village, but Reading, London and other towns and cities are within reach as commuter destinations. Official levels of unemployment are consistently below UK and Oxfordshire averages, at about 4 per cent for both villages[2] (op. cit.).

Why focus on rural areas?

The growth of rural populations and their travel highlights the growing importance of rural transport. Three hundred people a day are leaving the UK's major cities to live in the countryside and 4.4 million new houses are projected to be needed by 2016, approximately a quarter of which are likely to be in rural areas (Ramesh,1997). Traffic on rural roads is predicted to double or even treble by 2025 (CPRE, 1996).

Chalgrove and Cholsey respondents are of interest because they typify a pattern of 'semi-rural' or 'peri-urban' lifestyles that are becoming more and more common in parts of the UK and elsewhere in Western Europe (Champion, 1989, Root et al, 1996a). Residents in these areas are often more car-dependent than others (Curtis, 1996, RAC, 1995) and drive approximately 1.5 times further than urban dwellers (Table 5.1).

Table 5.1 Share of population and car mileage by type of area

Settlement types	>250k	>25k	>3k	Rural
% of population	40	31	18	11
% of car mileage	32	30	21	17
Ratio of mileage over population	0.80	0.96	1.16	1.55

Source: Stokes, 1995.

In this study, due to the existence of an earlier survey, it was possible to find out that the residents of Chalgrove and Cholsey travelled approximately twice as far in 1996 as they did in 1978, increases in travel that are similar to national average increases of distance travelled in this period (Root et al 1996b).

The doubling of distance travelled does, however, pose serious threats to the physical and social environment (Whitelegg, 1993). It is part of the purpose of this paper to explore, using those in the study as examples of wider trends, the roles of lifestyles and changing employment patterns in determining levels of travel and potential for service innovation and change.

What of the picture for bus operators? The scenario is not rosy. The majority of travel by rural residents is by car, so much so that the recent Rural White Paper commented that car travel should be recognised as the 'norm' of rural transport (Department of the Environment et al, 1995). Bus usage by rural residents is only about 75 per cent that of urban residents, at about 200 miles per year per person (Stokes, 1995). Since deregulation bus passenger distance has declined in all areas although bus kilometres have increased (White, 1996).

The Rural Development Commission (RDC) plays an important role in stimulating and supporting bus, taxi, taxi-bus, volunteer-driver and other forms of community transport in rural areas. It spent £745,000 on such schemes in 1994/95 (RDC, 1996a). Largely through the Rural Transport Development Fund and often in conjunction with local authorities, the RDC seeks to enable the mobility of those who would otherwise find it difficult to take part in the activities most of us take for granted - having access to jobs, being able to get to hospitals or doctors when necessary or having a choice of shops. But even these subsidies are often caught up with the tide of demand from travel: although there is evidence that schemes were set up to compensate for the lack of conventional public transport in rural areas, community and voluntary transport schemes were more likely where conventional public transport was also well provided (RDC, 1996b).

The increase in bus kilometres might be largely attributable to the introduction of lower-cost 'hopper' minibuses on more routes, but overall the picture is a depressing one from the perspective of environmentalists, the bus industry and others who wish to see public transport usage increase. (In some areas similar trends are shaping railway use, but here the focus is on buses). What, if anything, can be done to reverse this trend?

Methodology

One of the methodological objectives of this study of Chalgrove and Cholsey was to examine the travel patterns and focus in particular on the needs of young adults. A team of six interviewers visited households identified from the Electoral Register and sometimes by local contacts, as likely to have residents in the 16-29 age group. (This banding was chosen to coincide with *National Travel Survey* categories.) Interviewers visited in August 1995, and if the household did contain a 16-29 year old person and if those concerned were willing, each member of the household was left a form to complete, which consisted of a day's travel diary and a questionnaire. The questionnaire contained 35 questions about employment, costs of transport and who in the household pays them, control of household finances, attitudes to cars and other forms of transport and opinions about public spending priorities, including transport. The questionnaire was reproduced in Root et al. (1996a).

Table 5.2 Sample by age group in Chalgrove and Cholsey

	Age					
Village	**Under 12**	**12-15**	**16-29**	**30-59**	**60+**	**Total**
Chalgrove*	1	7	50	82	3	143
Cholsey	4	20	38	69	3	134

* two missing cases from Chalgrove

All the travel diaries were put onto a database, in all 1,692 individual journeys. Two hundred and seventy nine people filled in travel diaries and questionnaires. Information was gathered fairly evenly from both villages: from Chalgrove 145 people (52 per cent of the sample), and from Cholsey 134 people (48 per cent of the sample) returned and completed questionnaires (Table 5.2). The sample was almost equally divided between the sexes. Information about the socio-economic groups of individuals from the two villages is given in Table 5.3.

Like all self completed questionnaires, the questions had different response rates. There is always the concern that those who elected not to answer a question may bias the results in an unknown way. In addition, each respondent will have put his own interpretation on the questions which need not correspond to the intended meaning, thus again making interpretation of the data ambiguous.

The data in the travel diaries relate to one weekday's travel in the summer. The implications of the reported travel patterns should be seen as 'pointers' rather than firm evidence. However, it will be noted that many of the findings in the study are similar to those reported elsewhere which suggests that despite the limitations the data may have captured the main features of the rural villages.

The classification of trips by mode has been simplified to concentrate on those which occurred most often. All travel by car has been included in the 'car' class, so, for example, hitch-hiking and 'friend's car' were grouped as travel by car except for the purposes of assessing effects of car ownership when only owned cars and company cars were included.

Table 5.3 Socio-economic groups in the sample in Chalgrove and Cholsey

	Village	
Socio-economic group	Chalgrove	Cholsey
Number of employed people		
Professional & managerial	11 *(16)*	25 *(34)*
Other non-manual & skilled manual workers	46 *(67)*	40 *(55)*
Semi- and unskilled workers	12 *(17)*	8 *(11)*
Number of unemployed people		
Student	11	12
State Benefit/ Pension	6	6
Housewife	5	3
Other (including no response)	54	40
TOTAL	145	134

Note: Percentages are given in brackets

Deregulation and its impact on Chalgrove

It is not the role of this chapter to cite in detail the regulatory and legislative changes that have impacted on bus services. However, it is important to highlight the specific changes that took place to rural bus services, using Chalgrove as an example. (No similar alterations to transport services took place in Cholsey.)

The company that had been running the bus service in Chalgrove withdrew from doing so, giving the legal minimum notice. Oxfordshire County Council instituted a wholly subsidised temporary service while putting the route to tender. Just as the tendering process was nearly complete, another bus company decided that it would run a service, albeit one that was reduced from previous levels. Chalgrove was left with an approximately two-hourly service to Oxford in one direction and Watlington in the other. The last bus runs in the early evening. There was a level of complaint about the new service: for instance, residents asserted that the changes were not publicised and that even weeks after the alterations groups of people were to be seen waiting at the bus stops at the times when the former services ran.

These changes happened in the context of the 1985 Transport Act which enables local authorities to coordinate and secure services that are not commercially viable, but which are deemed 'socially necessary'. Oxfordshire County Council is subsidising approximately 80 per cent of bus routes in Oxfordshire, but because these are the less frequently running routes, such subsidy amounts to only about 15 per cent of bus mileage (Helling, 1997). Parish Councils also have the authority to support some transport services, such as community buses, postbuses and contract taxis, but their scope is limited by lack of resources. It may be important to begin to target these transport subsidies better, but, at the moment, the diversity of transport needs in different kinds of rural areas is still defying more effective targeting (Cloke, 1985).

Current travel patterns

The miles travelled by different modes in the two villages are shown in Table 5.4. The journeys are the totals of distance covered by all those who filled in their travel diaries and those who travelled on the 'travel day'. Data are added together for all journeys.

Cholsey residents have access to better public transport services and travel further by 'environmentally friendlier' modes (bus, cycle or train), more than twice as far than in Chalgrove (10.5 miles/person and 4.0 miles/person, respectively).

But the opposite is the case for aggregated 'car modes': Chalgrove residents travel 20 per cent further by car than Cholsey residents (28.6 miles/person in Chalgrove compared with 23.8 miles/person in Cholsey). More miles were travelled in company cars in Chalgrove than Cholsey,[3] and similarly more miles were travelled by those accepting lifts with friends. Lift-giving is affected by gender. Just over twice the number of lifts to household members were given by women (27 by women and 12 by men). Women gave nearly three times as many lifts as men to non-household members (11 by women and 4 by men).

Table 5.4 Miles by mode of transport

Mode	Chalgrove	Cholsey
Car	3,192	2,601
Company car	606	447
Friend's car	353	135
Bus	198	383
Walk	172	179
Minibus	123	13
Cycle	77	154
Hitchhike	49	---
Motorcycle	45	---
School bus[4]	8	---
Taxi	8	7
Train	---	669
Underground	---	10
Total miles	4,830	4,598
No. of people in sample	145	134
Average no. of miles travelled/person	33	34

Notes: -- no recorded journeys; numbers have been rounded

Table 5.5, which lists the modal averages of distances travelled shows that people go furthest by train, car and bus. The average journey lengths by car and bus indicate the predominance of journeys to Wallingford from Cholsey (two or three miles) and to Oxford from Chalgrove (eight or nine miles).

Table 5.5 Length (miles) per single journey by mode of transport

	Chalgrove		Cholsey	
	Average	Max.	Average	Max.
Bicycle	1.0	11.0	0.5	15.0
Bus	9.0	14.0	2.0	138.0
Car	8.0	169.0	3.0	135.0
Taxi	4.0	4.0	2.0	3.0
Train	n/a	n/a	11.0	50.0
Van	3.0	84.0	3.0	46.0
Walk	0.2	8.0	0.5	4.5

There is some evidence to suggest that the percentage of those who used both modes of transport from each village (3 per cent in Chalgrove and 8 per cent in Cholsey) is different.[5]

Energy use

The impact of the findings on energy use has been estimated in Table 5.6. These figures include journey length with mode and passengers as the main variables.[6]

The environmental gain made (2,300 MJ per day, 16,000 MJ per week, 839,500 MJ per year) amongst those sampled in Cholsey is probably as a result of the existence of rail use. The energy use in Chalgrove is 0.0191 MJ/mile/person/day and 0.0179 MJ/mile/person/day in Cholsey. If the 679 miles travelled on trains had been undertaken by car, then the energy use for Cholsey would rise to 0.0204 MJ/mile/person/day.

The data relates to individual journeys. Although all modes of transport were used in estimating transport energy use, the dominant contributor to energy consumption is the car. Thus energy use for transport is, in practical terms, synonymous with energy used by cars in this survey.

Table 5.6 Estimated energy use and mode of transport

Mode	Estimated energy use MJ/passenger mile[7]	Chalgrove (MJ/day)	Cholsey (MJ/day)
Bus	0.83	164	318
Car	3.21	10,245	8,349
Company car	3.21	1,944	1,435
Cycle	0.10	8	15
Friend's car	1.60	564	15
Hitchhike	1.60	78	N/A
Minibus	1.15	141	15
Motorcycle	3.13	141	N/A
School bus[8]	0.83	7	N/A
Taxi[9]	1.15	9	8
Train	0.89	N/A	595
Underground	1.08	N/A	11
Walk	0.25	43	45
TOTAL: MJ per day		**13,345**	**11,006**
Average miles per person		33	34
No. of people in sample		145	134

High and low energy users

Household energy consumption was estimated following Root *et al.* (1996). Forty per cent of the total energy consumption associated with travel was consumed by 25 per cent of households, while 25 per cent of lowest energy using households used only 10 per cent of the total energy (Figure 5.1). Thus, the quarter of the households that travel the most are responsible for four times the energy use - and probably pollution - of the quarter that travels the least. If our sample had included pensioners, the range would probably have been extended. Even so, this is a substantial variation between households with similar age profiles in two similar villages in South Oxfordshire.

Figure 5.1 Estimated reported energy consumption by household in two Oxfordshire villages, 1995.

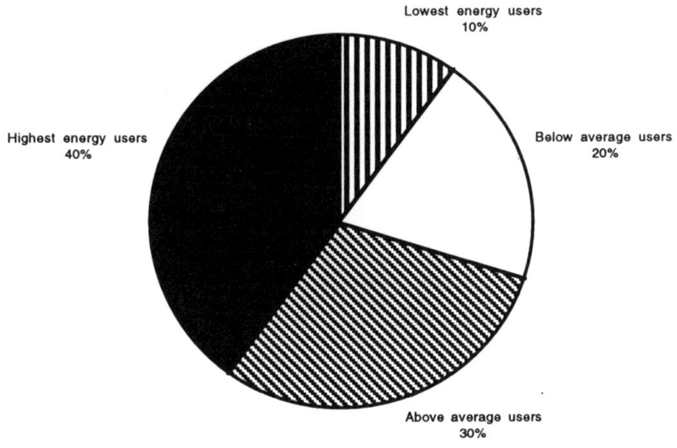

The segments each contain 25 per cent of the households, and their size shows the proportion of the total energy for travel expended.

Looking at the extreme ends of the distribution, however, the decile of households which used least energy were defined as 'low' energy users and the 10 per cent of households that used most energy were defined as 'high' energy users. A summary of the characteristics of the two contrasting decile household types is given in Table 5.7.

Table 5.7 Description of responding rural households in the high and low energy groups in two Oxfordshire villages, 1995

	Low energy group	High energy group
Respondents per household	2.2	3.5
Total reported miles travelled (all households)/day	269	2,722
Total energy use (all households)/day (MJ)	400	7,700
Percentage of respondents earning over £20,000	0	17
Percentage of respondents in full-time work	44	59
Modal age group of respondents	30-59	30-59
Percentage of households with at least two cars	90	100
Number of households	10	10

Thus, high energy households were economically more active, better off and all owned cars. They travelled over ten times as far as the low-energy users, but, because of the different modes used, the ratio of energy used was nearly double this (1:19).

These differences in car ownership, age and earning power resulted in different methods of travel and the frequency and distance travelled (Table 5.8). The high energy users are travelling further by all methods, except by bus and cycle, than the low energy users. The low energy users do more walking trips than by car, though, of course, for substantially different distances. In both groups, half of all journeys are by car.

Trips to work and at work were a greater proportion of trips for the high energy households (31 per cent, 1,583 miles) than the low energy households (10 per cent, 35 miles). Conversely, shopping trips represented 11 per cent of the journeys made by high energy households and 29 per cent by low energy households. Additionally, no managers resided in low-energy using households; the association of distance and income was observed nationally by Stokes (1995).

Although high energy households used buses less than low energy households, overall they travelled more miles per person by public transport (mainly train). This reflects the level of economic activity of these households rather than usage for environmental reasons.

Table 5.8 Daily use of selected travel modes by respondents of low and high energy use rural households in two Oxfordshire villages, 1995

	Total number of trips		Total distance travelled (miles)	
Mode	**Low energy**	**High energy**	**Low energy**	**High energy**
Car	29	99	79	1,818
Walk	59	70	37	95
Bus	11	5	87	28
Pedal cycle	14	11	11	4
Train	2	12	28	256

Understanding choice

An analysis was conducted on answers given to the question 'was a car available for this journey?' when non-car modes were used. Scores of 100 indicate that a car was available every time a non-car journey was made (Table 5.9). As could be expected, professionals and managers consistently have higher levels of choice than those in other socio-economic groups. Interestingly, however, only the unskilled manual group experienced no choice when using their bicycles. All the other groups had varying levels of opportunities to use cars, but chose not to do so. One question that is key to this issue is why did these people make these choices? Are they choosing this way because they have no practical travel alternatives? Do parking restrictions, for example near inner-city workplaces, make car use inconvenient? If they had cars, why were they not using them? If we knew the answer to this question we would be well on the way towards constructing a viable, profitable and sustainable public transport system. Unfortunately we do not know enough to answer this question with any certainty. All we have is 'snapshots' of views from attitudinal surveys and qualitative research. Some of these are now assessed.

Table 5.9 Level of choice of transport mode for journeys by occupation in two Oxfordshire villages, 1995. (Percentage of trips which could have been made by car but the traveller used another mode)

Occupation	Bicycle	Foot	Train	Bus	Overall
Managerial	-	100	100	-	100
Professional	100	100	100	67	96
Skilled manual	93	82	100	-	85
State benefit/ pension	100	77	-	-	81
Skilled white collar	100	77	62.5	78	78
Housewife	-	63	-	50	62
Unskilled manual	0	46	-	-	39

The qualitative work that was undertaken showed that some of the poorer residents, who could be described as 'travel poor' did feel trapped in Chalgrove and some expressed the desire to move, but said they could not because of the higher house prices in Oxford. The question of choice is a difficult issue, because choice operates at many levels and has various meanings for different people, but it was clear from discussion groups that there were many who felt Chalgrove lacked amenities, like shops, and they did not have the money to travel to them or to move away from the village.

There was a difference in men's and women's responses to the idea that their quality of life suffered from the time spent travelling (37 per cent of men and 25 per cent of women). This difference may reflect the fact that men travelled, on average, 31 miles per day by car and women 17 miles.

Seventy-eight per cent of respondents said it was important to conserve fossil fuels. Fifty per cent thought that the quality of rural life was threatened by car use and 55 per cent agreed the health risks associated with car pollution required action to reduce car use. However, most respondents did not associate health risks from car pollution with the quality of life in the countryside (only 30 per cent of respondents linked these two aspects).

Community life

The following comment illustrates some of the strong feelings about lack of travel opportunities from a group of young people in one of the villages studied. The boys perceive that lack of facilities leads them to crime:

> Spike: "We need transport or we need facilities. Or all we do is get in trouble. True or false?"
> [Clamour of voices]
> AR: "How do you think transport would help?"
> Spike: "Get us out of trouble, won't it? We'd be doing things. We might get pissed an' that, but..."
> Andy: "What's stopping you going to Oxford and beating someone up? That [transport] ain't going to keep you out of trouble is it?"
> Spike: "Yeah! (emphatically) But what's stopping us here... smoking drugs and that? If we went to Oxford..."
> Andy: "There's nothing stopping you smoking drugs, man..."
> Spike: "Yeah, but if we'd got transport you're going to be doing other things ain't you?"

This example is used to show the implications of the relative absence of mobility. But the converse is also true: other research has found that community life is adversely affected by traffic levels. For instance, the number of social interactions on a street is inversely related to the amount of traffic using it (Whitelegg, 1993). The implications of better public transport usage stretch far beyond a reduction in CO_2 or increasing health due to more exercise, to, probably, more social cohesion and less social exclusion.

Key groups and cultural change

The analysis above begins to make visible some of the groups that have limited choice and so will need public transport services. These are:

- the elderly
- housewives
- the young
- the unemployed.

But it is also clear that the members of these groups are not going to have travel needs like the previous generation who, to a large extent, were able to get permanent, full-time jobs (Funken and Cooper, 1995). This new group, for whom the employment market is likely to offer part-time temporary jobs, are, like Spike and his friends, likely to want to travel throughout the day and night to a variety of destinations. Trips to night-clubs, to see friends, to new workplaces etc. are likely to replace the more 'standardised' lifestyle of a previous generation who could often obtain regular work more locally and whose lifestyles were usually based in the village (Root et al, 1996a).

It is the argument here that deregulation creates new opportunities for the provision of different sorts of transport services, but that these need to be created on the basis of understanding current choices and desires for travel. Following Giddens (1991) lifestyles are defined as:

> (...) routinised practices, the routines incorporated in habits of dress, eating, modes of acting and favoured milieus for encountering others: but the routines are reflexively open to change in the light of the mobile nature of self-identity.

Lifestyles are constructed in particular 'sectors' - time/space slices of an individual's activities (Turrentine, 1994). Aspects of this time/space division were found in relation to time use, for example. Time was equally precious to high and low energy users and it was the waste of time that was resented most in relation to public transport (Root et al, 1996b). In this study, with the dominance of the car, there is considerable scope for these lifestyle sectors to be 'de-localised'; separated across time and space.

Those respondents without a car experienced difficulties. A third of all respondents said that there had been a job they had applied for, but did not take, because of the difficulties of getting to the workplace. Those who, in the focus groups, had to rely on buses talked with anger and frustration of the difficulties and problems it imposed on their lifestyles.

Opportunities for change are constrained by the spatial location of the villages in relation to amenities and workplaces. Scope for modal switches or travel reduction are also reduced by dominant lifestyle factors. In Chalgrove and Cholsey it is not possible to live a 'normal' lifestyle without access to a car. The service frequency of approximately two hourly buses means that residents are highly constrained in terms of planning outings. Some respondents reported huge inconvenience if they missed the bus - sometimes due to the smaller Chalgrove bus getting hidden behind a bigger bus at the bus stops in central Oxford. It was also reported that the bus times no longer suited the lifestyle of some user groups: for example, the alteration meant that it was no longer possible to catch a bus for shopping in Oxford immediately after taking children to school and return in time for the children's lunch-hour. Similarly, the infrequency of the service caused problems. Travel to neighbouring Watlington for an errand such as accompanying an elderly relative on an optician's visit can take almost all day if buses are only available at two-hourly intervals. The constraints of the bus service stopping in the early evening also means that travel to and from Oxford for a film or for a meal is impossible in the evening.

Despite adequate local shops and other facilities, and some opportunities for local employment (Root et al, 1996b) most people in the village go elsewhere for their paid work, shopping and leisure activities (ibid.). For those with access to a car, living such a de-localised lifestyle causes few difficulties. 'We have no alternative' said one teenage girl of her car use, a statement that was substantiated by the infrequency of buses in Chalgrove.

Dissatisfaction with lifestyle

There is also an issue about those who had complete choice to use a car, but choose not to do so - particularly striking are the group that choose to use the train. Do travel operators know why they made these choices? I believe that we know too little to be able to understand such behaviour and build on it to make public transport in rural areas more attractive and hence commercially viable.

The attitudes of respondents from high and low energy households were compared (Table 5.10). Attitudes of household members reflected some quantitative differences between the households but health and fuel conservation were considered important by both groups.

Socio-economic differences may help to explain some of the differences in attitudes. The high energy users were, perhaps predictably, more likely to agree that the car is essential for work and less prepared to pay extra for their motoring. They recognise that they use the car more partly as a result of pressure on their time. They were more aware of the damage the car does to the rural environment and of the health risks associated with car use. These high energy users were more responsive to the idea of car sharing, would like money spent on public transport rather than on roads and are more likely to believe that their quality of life would be improved by less travel. These travellers are not very happy with their lifestyle and are aware of its negative impacts on them and the environment. The impression given is of a group feeling 'trapped' into using a car.

Table 5.10 Attitudes expressed by the respondents of low and high energy use households in two Oxfordshire villages, (percentage of those responding in each type of household)

	Low energy users	High energy users
Agreed that lack of time encourages car ownership	50	93
Agreed that health risks require less car use	79	81
Agreed on importance of conserving fossil fuels	74	81
Want money spent on public transport rather than roads	53	75
Agreed that a car is essential for work	27	72
Unwilling to pay more for motoring costs	50	55
Unwilling to use a car sharing scheme	44	37
Agreed that a car is essential for shopping etc.	53	52
Quality of rural life is threatened by car use	22	45
Quality of life would improve with less travel	11	34
Modal time to wait for bus (minutes)	10	10
Modal time to wait for trains (minutes)	10	10
Modal time to wait in traffic (minutes)	Have to wait; no choice	Have to wait; no choice

The respondents in low energy using households were equally likely to believe that the car is essential for shopping and would tolerate similar travel delays as high energy using households, confirming that using a car for some activities is of equal importance for both high and low energy users.

It would appear that desires to live in particular areas and aspirations and actualities towards particular lifestyles can operate independently of, and as an inhibitor on otherwise 'rational' responses to dislike of high levels of car use. The wish to live in the country and work in cities is one such example of an area where lifestyles will potentially contradict change in transport goals.

The evidence presented in this paper suggests that high energy users are more sceptical about the car than low energy users, and more willing to try car-sharing schemes and to agree that money should be spent on public transport rather than roads. Such attitudes are unsurprising given the propensity for high-energy users to belong to the groups rich in economic and 'cultural' capital (i.e. social status evidenced through qualifications, taste, etc.) that often innovate and lead social change (Bourdieu, 1979).

The upward trends in distance travelled are showing no sign of slowing. It is possible to speculate that although this survey revealed substantial differences in travel between men and women, a large part of such differences will be reduced as there is greater equality in employment opportunities and if responsibility for children and other dependants becomes shared more equally between the sexes. There is likely to be, therefore, even more pressure on the environment and on community life through greater use of private cars, and small scope for improvements through attitudinal change alone.

Conclusions

The picture that has been painted is one of rural transport, at least in the accessible part of the South East studied, where there has been a polarisation: one third of the population exists in what can be called 'travel poverty' which means that, in the wake of the deregulation and privatisation of transport, they do not have access to the travel opportunities that they would like. At the other end of the spectrum, there are those whose travel uses most energy and they too are dissatisfied. They are three times as likely to say that their quality of life would improve with less travel than those who travel the least.

The private sector is often praised for its ability to innovate. Since deregulation we have a situation where some groups of people are choosing to use public transport, but many are not. It seems clear that some creative and new solutions need to be found because 'more of the same' will not necessarily mean that people from either of these groups will again use buses or trains in large numbers. The simple solution which might be characterised as 'throwing buses at the problem' will not do. Obviously solutions will depend upon locality, region, culture and economic development, but we can never achieve any sort of 'blanket coverage' of the countryside with public transport. We never had it in the first place (Adams, 1996) and, anyway, we have had more than 40 years of development that has rested on the availability of the private car, so land-use development now militates against the possibility of comprehensive public transport.

We need imaginative and innovative solutions that tailor the public transport of the twenty first century to the needs of the lifestyles of those examined above and of the generation yet to come. Both carrots and sticks are needed to encourage the rural 'travel rich' and the 'travel poor' out of their cars and into more public transport. These could include:

- easy to understand environmental and energy labelling for cars and public transport, so that the public has accurate information about the environmental impact of their travel choices;
- mandatory targets to be included in the Road Traffic Reduction Bill;
- more 'partnership' work between local authorities and transport operators to improve access, 'packaging' of bus and train franchises and potential for rail re-openings;
- requirements for national travel information service including a free telephone helpline and requirements for transport operators to publicise services;
- encouragement of the development of more unconventional public transport e.g. taxi-bus services and the establishment of more car-pooling and car-sharing schemes in rural areas;
- reorganisation of fuel subsidies for community buses to encourage more sharing of often under-used community transport vehicles;
- encouragement of green commuter plans, including an assessment of tax-free travelcards to be made available through employers;
- land-use and transport planning that aims to be more sustainable, e.g. the creation of more 'bus lanes' on motorways; the development of 'public transport corridors' where appropriate and an overhaul of the processes by which new roads are evaluated against other options;

- reconsideration of petrol price rises as recommended by the Royal Commission on Environmental Pollution but with tradable allowances for fuel for average rural mileage at current prices, adjusted for inflation.

The importance of harnessing the imagination and creativity of both public and private companies to meet the changing transport needs of rural populations is becoming, almost daily, more obvious.

References

J. Adams (1996) 'Can Technology Save Us?' *World Transport Policy and Practice*, Vol. 2, No. 3, MCB University Press, Bradford.

P. Bourdieu (1979) *Outline of a Theory of Practice,* Cambridge Studies in Social Anthropology.

D. Banister (1992) in M. J Breheny, *Sustainable Development and Urban Form,* Pion, London.

A.G. Champion (ed.) (1989) Counterurbanisation - the Changing Pace and Nature of Population Decentralisation, Edward Arnold, London.

P.J. Cloke and C.C. Park (1985) *Rural Resource Management,* Croon Helm, London.

CPRE (1996) *CPRE's Traffic Trauma Map: a forecast of traffic growth on rural roads from 1994 to 2025*, CPRE, London.

C. Curtis (1996) 'Can Strategic Planning Contribute to a Reduction in Car-Based Travel?' *Transport Policy*, 3, (1/2) pp. 55—65.

Department of the Environment, Ministry of Agriculture, Fisheries and Food (1995), *Rural England,* HMSO, London.

C. Funken and P. Cooper (eds.) (1995) *Old and New Poverty: The Challenge for Reform,* Rivers Oram Press, London.

A. Giddens (1991) Modernity and Self-Identity: Self and Society in the Late Modern Age. Stanford University Press, Stanford.

D. Helling (1997) *Personal Communication*, Oxfordshire County Council, Oxford.

P. Hughes, *Personal Transport and the Greenhouse Effect*, Earthscan, London.

C. Oppenheim (1993) *Poverty: the Facts*, CPAG Ltd, London.

RAC (1995) *Car Dependence*, Royal Automobile Club, London.

R. Ramesh (1997) 'Plan to Make drivers Pay for Congestion', *The Independent*, 18 June 1997.

A. Root, B. Boardman and W.J. Fielding (1996a) *Rural Travel and Transport Corridors - Interim Report,* Energy and Environment Programme, Environmental Change Unit, University of Oxford.

A. Root, B. Boardman and W.J. Fielding (1996b) *The Costs of Rural Travel - Final Report.* Energy and Environment Programme, Environmental Change Unit, University of Oxford.

Rural Development Commission (1996a) *Rural Development Commission Annual Report 1995/96*, RDC, London.

Rural Development Commission (1996b) *Research Findings: Community and Voluntary Transport in Rural England*, RDC, London.

G. Stokes (1995) *A Note on Income Inequalities in Car Driver Mileage,*. ESRC Transport Studies Unit, University of Oxford.

G. Stokes (1995) Assessing the Effects of New Transport Policies on Rural Residents, TSU Paper 836, University of Oxford, Oxford.

T. Turrentine (1994) *Lifestyles and Life Politics: Towards a Green Car Market,* Institute of Transportation Studies, University of California, Davis, CA.

P. White (1996) 'The Future of Public Transport 10 Year After the Transport Act of 1985,' *Proceedings*, Chartered Institute of Transport, Vol. 5, No. 2, June 1996, London.

J. Whitelegg (1993) *Transport for a Sustainable Future: the Case for Europe,* Belhaven Press, London and New York.

Notes

1. Thanks are due to Professor Gordon Clark, Acting Director of the Transport Studies Unit, and Dr Brenda Boardman, Programme Leader of the Energy and Environment Team at the Environmental Change Unit, Oxford University, who have guided the work discussed above. I would also like to thank William Fielding, who worked with me and other members of the Energy and Environment Team, Environmental Change Unit, for their ideas and support. Without the help of Sylvia Boyce and Ann Heath of the Transport Studies Unit this paper would have contained many more mistakes than it currently does. All remaining omissions and errors are, however, the author's responsibility.
2. In order to register as unemployed, evidence of availability for work including evidence of child-care arrangements has to be produced. Those who are married with a spouse who is employed do not receive

benefits, hence as many as a third of those who are unemployed do not register as such (Oppenheim 1993)

3. More people with company cars might move to Chalgrove, as they do not need public transport.
4. An atypical figure, as most of the research was carried out in August, when schools were shut.
5. Statistical significance p=0.068.
6. Some other factors, such as fuel economy, are omitted.
7. Estimations based on figures by Banister in Breheny, M. J. (1992: 165). Car occupancy is assumed to be one for car drivers, but 2.0 for car passengers. Buses and trains are assumed to be 33% full electric trains. Both diesel suburban and InterCity trains use more energy. Similar figures but give a lower MJ figure for cars, are given in Hughes (1993).
8. An unusually low figure, as most of the research was carried out in August, when schools were shut.
9. 'Other private' is the category used, as taxis are assumed to be diesel fuelled.

Part 3

Railways

CHAPTER 6

The privatisation of British Rail

Richard Davies[1]

Introduction

This chapter describes the way in which the privatisation of British Rail was carried out. The project was the largest transformation ever carried out on a state-owned industry in the UK. Over 100 separate, autonomous companies were formed from a single, vertically integrated corporation first created fifty years ago. Its scale was comparable to the restructuring of enterprises in Eastern Europe at the start of the 1990s. The chapter considers:

- British Rail as it stood, in 1993/94, prior to restructuring,
- the way in which it was broken up,
- the change in the way the industry was financed as a consequence,
- the privatisation of the companies formed out of that restructuring, and
- the implications of the way that it was privatised for future transport policy.

Why was British Rail privatised?

Privatisation, as a tool of government policy, serves a variety of objectives: provision of better and cheaper services, promotion of competition and the realisation of sales proceeds normally being the most prominent. The debate about whether value for money was achieved in this privatisation must take place later, as a clearer market develops in railway assets and services. For now, the remarkable feature of this privatisation is the way in which, largely because of regulatory[2] decisions taken early on, very little about the day-to-day running of the railway network has changed. There are new liveries and new branding, but a very similar service pattern is being operated to that in 1993/94, the same fares structure exists and the same trains are in use. The bulk of the changes that privatisation will bring, including reduced subsidy levels, new trains, new service patterns and different fares structures, are largely yet to come. The effect of reduced subsidy levels, if achieved, is likely to be the most far-reaching aspect of privatisation. Such reductions call into question many aspects of the current regulation of the sector and, in particular, the franchising process.

British Rail in 1993/94

The British Railways Board was created by the 1962 Transport Act as the successor to the rail and shipping activities of the British Transport Commission, itself set up to hold the various interests in shipping, railways, hotels and road transport that were nationalised on 1 January 1948. The Board operated passenger and freight services[3] in Great Britain and was almost entirely vertically integrated: that is to say, it owned its own trains, infrastructure and carried out almost all track and train maintenance itself.

Previous restructuring initiatives

The Board had been reorganised successively over the years and, indeed, had already experienced some privatisation:

Disposal of non-core activities

- British Transport Hotels were privatised in the early 1980s,
- its shipping division, Sealink, was sold to Sea Containers in 1984,
- its train manufacturing subsidiary, British Rail Engineering Ltd., was privatised in 1988 along with Travellers Fare, which supplied catering services on stations and on some trains,
- its overseas consultancy division, Transmark, and Meldon Quarry, producing stone for track ballast, were privatised in 1993, and
- extensive sales of surplus lands occurred.
-

Joint ventures with the private sector

- some freight locomotives (the Class 59) and many wagons were bought by private sector owners for operation by BRB,

- in the early 1990s it entered into a commercial joint venture with a privatised company, BAA, to build the Heathrow Express service between Paddington and Heathrow Airport,
- it transferred local services in both Newcastle and Manchester to the local Passenger Transport Executives for conversion into metro systems: operation of the latter is undertaken by the private sector,
- it contracted with John Laing to operate Ashford International passenger station for Channel Tunnel services, the cost of the station being borne by the contractor, and
- it explored in some detail the means by which the Channel Tunnel Rail Link project might be financed in the private sector and its subsidiary, Union Railways Ltd., helped organise a competition to select a promoter in 1993-1995.

All told, the scale of the Board's activities had reduced over the years and it had developed considerable experience of working with the private sector. The Board did so both because of the funding constraints that it faced and in order to realise the kind of efficiency improvements that can often be obtained by public-private partnerships. These developments largely predated the government's Private Finance Initiative.[4]

The Board had also participated in several studies over the years as to how the rail network might be privatised and had, in parallel to these, reorganised its operations along business lines. The process of 'sectorisation' began in the early 1980s and led to the creation of a relatively simplistic internal market between:

- the producers of rail services in the regions and functions, and
- the specifiers, who marketed those services and set commercial policies for them.

A high-level attempt at creating an access charging regime was also put in place at this time. Sectorisation created organisations that became well known to the travelling public, such as Network SouthEast, Regional Railways and InterCity.

In 1992-94, the Board implemented its Organising for Quality initiative, a major restructuring which completely re-orientated its activities along business lines, and finally broke the functional and geographic orientations that had existed since nationalisation in 1948.

Financial structure of BRB in 1993/94

The Board's operations in 1993/94 were substantial and its financing complex. Its turnover from passenger and freight operations, before subsidy, was about £3.1bn, two-thirds of which came from passenger services, and its operating expenditure was £3.6bn. The Board's financing regime was intended to ensure that passenger services would not cumulate substantial ongoing deficits: revenue grants were made to cover forecast deficits and capital grants were provided to replace assets used on grant-supported services. In principle, both types of grant were made in relation to a base set by services that could be provided on a fully-commercial basis.

The InterCity business unit moved out of grant in the late 1980s. Freight services were not subsidised per se by Government grant, but the internal pricing regime for freight track access ensured that its services paid only marginal costs (incrementally to passenger services). Of the passenger businesses, both Network SouthEast and Regional Railways were eligible for grants from central government and local Passenger Transport Executives (see below): revenue grants to cover operating losses and capital grants towards the cost of asset renewals. In fact, in 1993/94, Network SouthEast covered its operating costs directly and did not receive any revenue account subsidy, but Regional Railways required £460m of revenue grants and a further £85m of such grants were provided to the Board to cover corporate costs.[5] Both NSE and Regional Railways received capital grants, totalling £450m in 1993/94. All told, the sum of revenue and capital grants received amounted to just under £1bn.[6]

In addition, the Board was completing a major programme of investment to provide passenger and freight services through the Channel Tunnel. These were intended to be operated on a commercial basis and were funded by a combination of internal earnings from elsewhere in British Rail and loans from the Government. At 31 March 1994, the total external debt of BRB was £2.4bn, the vast majority of which arose from investment in the Channel Tunnel and debt from losses on freight operations.[7]

The initial restructuring

To prepare the industry for sale, it was first restructured. The restructuring took place under state control and the Government's hope was to establish as much of a financial track record using the new structure as possible in the state sector before subsequent sale. In the event, this was not always possible because of the length of time needed to restructure.

The path chosen by the Government, as is well documented elsewhere, was to separate infrastructure ownership and operation from service provision. This reversed the vertical integration created by the Organising for Quality initiative. In this model, network operation has substantial natural monopoly power associated with it and consequently should be tightly regulated. Service provision, however, is essentially a competitive activity and the lower the barriers to entry can be made, the more competitive it will be.

This split, which had been pioneered in other network industries such as telecommunications, gas and electricity, was taken to its logical extreme in the case of railways by means of full legal separation of infrastructure (Railtrack) from both passenger and freight train operations, each of which initially stayed in British Rail's control.[8] Railtrack's licence bars the company from participation in train operations.

The closest parallel to this structure is the National Grid in the electricity sector and now Transco in the gas sector. The separation was taken to the ultimate extreme in the case of the railways as a result of Railtrack's decision to 'outsource' infrastructure maintenance and renewal activities. These services are procured by Railtrack from independent suppliers, a step designed to maximise the amount of competition available in the sector. This structure meant that Railtrack, on Vesting Day on 1 April 1994, took over only about 10,000 of the 120,000 workforce that British Rail had. These were signalling and operations staff, together with managerial, commercial and property specialists. At Vesting Day, Railtrack was left purchasing about £1.2bn worth of services from British Rail whereas British Rail procured over £2bn of track access services from Railtrack for its various rail businesses. This was part of the famed 'money-go-round' that bemused politicians, journalists and analysts alike during the restructuring's early days.

Immediately after the Official Opening of the Channel Tunnel on 6 May 1994, ownership of BR's international passenger subsidiary, European Passenger Services (EPS) was transferred directly to the Government, in readiness for its eventual privatisation in 1996.

Figure 6.1 illustrates the initial split.

Figure 6.1 The initial split of British Rail on 31st March 1994 (all activities remained in the public sector in 1994/95)

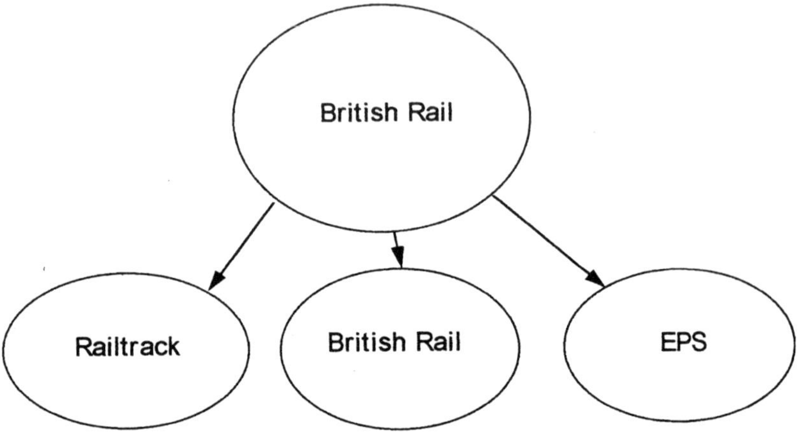

Further changes took place after Vesting Day. Progressively, British Rail was broken up into:

- 25 passenger train operating companies (TOCs),
- 6 freight companies (3 bulk freight companies, plus Rail Express System (principally handling Post Office traffic), Freightliner and Railfreight Distribution (principally Channel Tunnel freight traffic),
- 13 infrastructure maintenance units (IMUs) and track renewal companies (TRCs),
- 3 rolling stock leasing companies (ROSCOs), and
- a number of engineering consultancy and design organisations and other support companies.

For reasons of space, this chapter primarily concerns the privatisation of the passenger train operating companies.

Figure 6.2 The secondary split, implemented between April 1994 and December 1996.

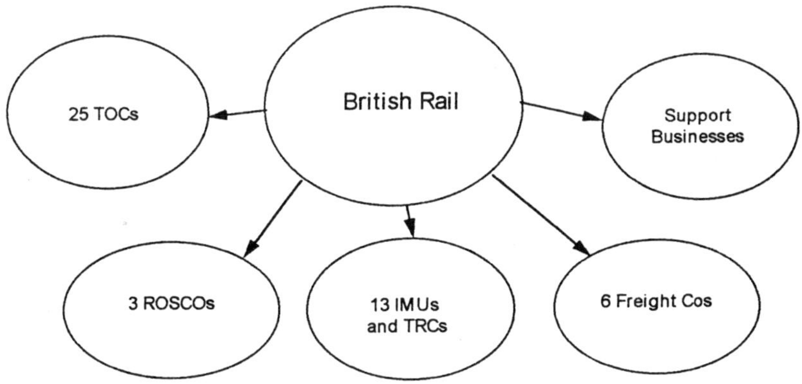

Changes to the financing regime

Even though each of these companies continued, initially, to be owned by the state sector, this split necessitated major changes to the financing regime for the rail sector. The Treasury's requirement was that, despite the restructuring, the total net funding that it gave the rail sector should not increase, except where specific one-off restructuring costs (such as the cost of systems and legal advice in relation to the preparation of contracts) could be identified. This had important ramifications, discussed below.

The first challenge was to create pricing regimes for use across the new commercial boundaries that this restructuring had brought. These principally consisted of:

- track access charges for both passenger and freight services,
- rolling stock leasing charges, since almost all passenger rolling stock was now owned by the ROSCOs,
- charges for infrastructure maintenance and renewal,
- prices for supply of services, such as provision of ticket office facilities and train crew, between the 25 TOCs, and finally
- the pricing of public service contracts put in place between British Rail and the Franchising Director, to ensure continuity of passenger service provision during privatisation. These contracts ceased upon privatisation, but included a profit element for the first time.

Pricing based on replacement costs

After a considerable period of debate, the use of consultants and a limited amount of financial modelling, the charging regimes applied followed the classic replacement-cost principles formulated elsewhere in the nationalised sector and which are intended to mimic the presumed long-run equilibrium that would apply in a perfectly competitive market. The total revenue collected was specified to be the sum of:

- depreciation calculated according to a Modern Equivalent Asset Value (MEAV) valuation,
- a rate of return, calculated using an appropriate rate reflecting the risk involved in holding the assets applied to an appropriate MEAV valuation,
- operating expenses, and
- an allowance for the cost of operating a performance regime.

This latter is a novelty for such pricing regimes: the Railtrack pricing structures include a performance regime which is intended to incentivise the asset owners to improve performance.[9]

In the case of rolling stock, discounts were given to older rolling stock to reflect the cost saving and revenue improvement that would come from replacement with new rolling stock. No similar adjustment was made in the Railtrack regime. Apart from this, no direct account was taken of the earning power before subsidy of the asset in these calculations: it was essentially a cost-based approach.

Such an approach implicitly assumed the following:

- pricing should simulate the outcome of a competitive market for provision of the asset, assuming that one could have been created,
- subsidy would exactly match the difference between cost of rail service provision and revenues earned, and
- subsidy would continue to be provided at this level for the foreseeable future, meaning that the extent and level of railway services would continue at current levels for the foreseeable future.

It was a corollary of this approach that there would be no reduction in rail services at all for the foreseeable future and that no account should be taken of any temporary scarcity value of certain asset types nor of the relative abundance of others.

In addition, the totals were indexed in a way designed to encourage efficiency using the familiar RPI - x formulation. Railtrack's fixed charges are indexed at RPI - 2 per cent, whilst the capital element of rolling stock lease charges are fixed in cash terms and the non-capital element is indexed at PPI[10] - 3 per cent.

Difficulties in implementing this approach

This formulation posed a number of difficulties in practice to implement. The assessment of MEAV values is inevitably problematic: there is great uncertainty in practice as to how certain assets would be replaced and inevitably a number of subjective judgements are required. This is particularly true for assessing the MEAV value of historic railway stations which are subject to listing and other planning controls.

In addition, it is possible that the costs BR faced in replacing assets would not be those that would apply in a more competitive supply market. BR faced, historically, limitations on the goods and services that it could source from overseas, but such restrictions would not apply in the future.

Table 6.1 Total passenger track access charges and ROSCO rentals, in 1995/96

	1995/96 Year Charges £m
Railtrack charge	
Railtrack fixed access charge	1800
Traction electricity	100
Variable track access charge	50
Subtotal	*1950*
ROSCO lease charge:	
Capital portion	435
Non-capital portion (largely heavy maintenance costs)	380
Subtotal	*815*
Grand Total	2765

Notes: Railtrack fixed charges include station and depot fixed rental charges, together with operating costs of independent stations. ROSCO charges excludes new rolling stock in the course of delivery (Class 365/5), the rentals for which were not fixed until the start of the 1995/96 year.

Table 6.1 indicates the total charges arrived at by this means. For information, the total replacement cost of infrastructure was assessed to be about £20bn and that for rolling stock to be between £8-9bn, the latter figure making some allowance for cost savings that would be available in the private sector.[11]

In addition, the TOCs between them incurred approximately £1.4bn in costs to operate, market and retail train services. In round terms, the apparent cost base of the passenger sector increased by just over £1bn and with it the subsidy requirement doubled overnight, from £1bn in 1993/94 to just over £2bn in 1994/95.

Structure of charges

Somewhat less thought was given to the structure of charges for track access and leases than to their respective levels. The decision was taken that use of these assets should be subject to long-term, fixed price contracts and that the amounts paid under these contracts should not be tied directly to measures such as passenger revenue or the number of trains operated. In other words, they are essentially take-or-pay contracts. This decision was taken so as to make the sale of the Railtrack and the ROSCOs easier.

Railtrack's access charges

The track charging regime adopted was complicated and of limited relevance in practice. This was because, in practice, franchises were sold with take-or-pay commitments to Railtrack in order to underwrite its cashflow and also had the benefit of full indemnification against the effects of the Regulator's review of access charges in 2001. Indeed, it is for this latter reason that there is no particular reason why the same regime must be followed during the 2001 review and it is to be hoped that a number of modifications are made to it.[12]

The regime, in highly simplified form, has two components:

- a variable charge, including the cost of traction electricity, comprising about 8 per cent of the total, and reflecting the immediate damage to track caused by each type of train on the network,
- a fixed charge designed to reflect the cost of other maintenance and operations costs, asset replacement and a rate of return. This charge was, in turn, computed using a two stage process as follows:
    - in the first stage, each TOC on a given route section was charged the marginal cost of infrastructure required to support its services, against an assumption that sufficient assets would be provided for all other operator's services. Thus an operator that triggered a need for a four-track railway rather than a two-track railway would be charged the cost of the extra two tracks, and

- in the second stage, a 'residual' was computed as the difference between the sum of the charges in the first stage and the total required revenue from consideration of the total asset base on that section, zonally and at national level. This residual was then apportioned to operators on a basis that has not been directly made public, but appears to have been a combination of train or vehicle-mileage operated and passenger revenue generated by each TOC.

In practice, the first stage calculations seem to have identified a relatively small amount of the total infrastructure to different users (perhaps about 25 per cent) indicating the substantial extent to which operators make common use of infrastructure. Thus a considerable amount remained to be allocated under the 'residual'.

In addition, simplified costings were generally used so regional and traffic-related differences in Railtrack's operating costs appear not to have affected final quoted access charges to any great extent.

The track access agreements do contain, at the Regulator's insistence, some provision for a TOC to reduce its access rights and hence its access charges, but Railtrack is compelled only to reduce its access charge to the extent that it avoids costs as a consequence. Since, in the economic framework that underlies Railtrack's access prices, its costs are largely fixed in relation to output, such amounts will be very small in practice.

Structure of charges made by other companies

Similar arrangements applied in the case of a number of other sales, including those for track maintenance and renewal, and train maintenance. Given the importance of creating a long-term cashflow, the structure of charges was less crucial. Looking forwards, however, it becomes more important since it is in practice arguably the structure as much as the level of charges that dictates economic behaviour and shapes the allocation of commercial and technical risks in the sector.

In the case of the ROSCOs, to the extent that long-term borrowings are secured against new rolling stock purchases it is largely inevitable that rolling stock leases will largely be of fixed form.[13] About 85 per cent of Railtrack's access income comes from fixed charges.

Passenger Transport Executives

PTEs, which were created by the 1968 Transport Act, are responsible for transport provision in those cities in England and Scotland that were formerly metropolitan counties.[14] They contracted with British Rail and other transport operators to provide local bus, metro and rail transport in each of these cities. PTEs were historically charged for their services on a marginal cost basis: the cost of providing a national system was calculated first and then local services were charged only a marginal cost above the cost of the national system. The Government decided that, in future, PTEs should face the full cost of the services that they provided and so charging for PTE services was, on 1 April 1994, switched over to the basis of an allocation of full costs. The consequences of this approach were spectacular: PTE revenue grant paid to BR rose from £110m in 1993/94 to some £340m in 1994/95. The increase was so substantial that, in practice, PTEs could not afford it and the Government decided to create a special mechanism, the Metropolitan Railway Grant paid to the PTEs via local authorities, to subsidise the difference. Greater Manchester PTE withdrew support altogether on 1 April 1996, in part because of this substantial increase (funding was then taken over directly by the Franchising Director). As franchising proceeded, the Government agreed to changes to the pricing regime faced by PTE services including the removal of any requirement to contribute to Railtrack's fixed access charge; Greater Manchester PTE subsequently resumed funding responsibility in April 1997 and the burden of these charges fell directly upon Government through subsidy payments to the franchisee.

Thus, from an initial emphasis on exposing the 'true' cost of providing PTE services, policy seems progressively to have moved on to focusing on finding ways for PTEs to continue to provide finance for local passenger services, it being taken as read that such service levels could not be reduced. Indeed, PTEs have a good track record in catalysing funding from a variety of sources, including urban regeneration and European regional development funds, which has historically helped increase investment levels above those that central Government could have afforded on its own. All parties now appear to recognise the danger of losing this catalytic role in future rail investment.

Why did subsidy increase so much?

The immediate consequence of the creation of these price structures was a substantial increase in the grant requirement for passenger services.

The increase in grant requirement can be attributed to the following:

- a capital charge based on full replacement cost of assets now applied. A presumption of replacement of the entire rail network was thereby created even though there was no contractual or statutory obligation on any party to carry out that replacement. Railtrack's licence obliges it to continue to maintain the network, but not to replace all of its assets with modern equivalents.[15] The previous grant regime can be characterised as a pay-as-you-go environment and did not, in its latter years, provide adequate funds to maintain steady-state conditions. The new regime would, at least in principle, do that,

- the capital charge allowed for a return on assets, to reflect the opportunity cost of capital tied up in railway investment. A return on assets was needed fundamentally in order to create some value to privatise. Since the industry as a whole was substantially loss-making before grant income, it would not have been possible to privatise many of its assets without assurances about grant income. Without such a return, the assets were valueless,
- profits charged by infrastructure maintenance companies and other suppliers, such as BR Telecommunications (subsequently sold to Racal), and
- a substantial sum of £200m per annum, which was included in Government's payments to British Rail to simulate the profit margins that might be needed when franchises were sold.[16]

The decision to set prices according to costs and not value, together with the ability on the part of Government to set negative financing limits to absorb the extra cash 'created' by the new charging regime, meant that the role of subsidy altered. The issue was not that of determining how much rail service could be provided within a fixed subsidy budget, but that of working out the 'true cost' of rail service, presumably in the expectation that this knowledge would prove so unpalatable to taxpayers that they would demand some rebalancing of the transport budget. If the subsidy number was above the target number contained within the overall financial limit of the industry, Government would simply demand more cash back from companies still in the state sector by changing their financing limits. Subsidy, rather than service levels, thus became the dependent variable in the equation of railway financing. Given the public sector's quest for 'value for money', this was a strange situation.

Consequences of the subsidy increase

As we have seen, the subsidy increase was intended to provide either TOCs or asset owners with sufficient cashflow in future to provide for replacement costs of railway assets together with a commercial return on them. In order to ensure that the total underlying expenditure of the rail sector stayed in line with previous planning estimates, however, the Government restricted the use of these funds for investment. It set financing limits on Railtrack and the ROSCOs to claw back the extra funds paid in subsidy by means of negative external financing limits. It is an indication of the topsy-turvy nature of nationalised industry financing and budgetary control that, to achieve this, it found the substantial debt incurred by BRB prior to 1 April 1994 of positive use to enforce these limits. Almost exactly half of the debt at 31 March 1994, £1.2bn, was allocated to Railtrack: the interest and principal due on it by Railtrack to government was the means by which Railtrack's negative financing limit was enforced. (The balance of the debt was substantially written-off prior to Railtrack's sale in May 1996, at which point the company escaped from nationalised industry financing controls to leave it with just £235m of long-term debt.) The allocation of debt between BRB and Railtrack was almost entirely done on the basis of the amounts needed to enforce Railtrack's negative financing limits rather than on the basis of where the assets that the debt had finance were allocated.

There were other distributional implications of the subsidy increase. In the first instance, all InterCity routes except Gatwick Express became loss-making before subsidy. This reversed long-standing government policy that InterCity services should not be subsidised since they operate in a competitive environment in which there were unlikely to be overriding benefits from provision of more railway service than the market required.[17]

Table 6.2 indicates the range across all 25 TOCs in the ratio of required subsidy to passenger income in 1995/96 that was created by the charging regime. The ratio can be thought of as the additional contribution in subsidy that the government makes per pound of passenger income provided from the farebox. It indicates a broad spread from over 4.0 (Merseyrail Electrics) to -0.2 (Gatwick Express). The former Regional Railways TOCs are grouped together at high ratios, with none less than 1.9. The former NSE TOCs are grouped together with ratios in the range of 0.3 to 1.0.

Sales process

The diversity of businesses created from the privatisation of British Rail necessitated a diversity of means of privatisation. Most privatisations were achieved by trade-sales and were carried out by British Rail itself. The largest sales, those of Railtrack, the ROSCOs and EPS were carried out by Government, the first by flotation and the rest by trade-sales. The 25 TOCs were sold by means of franchising, described below.[18]

Substantial sales sums were achieved, totalling approximately £4.5bn. Table 6.3 describes the sums achieved.

Table 6.2 Variation in ratio of subsidy to passenger income between TOCs, 1995/96

TOC	Subsidy/ Passenger Income
Merseyrail Electrics	4.5
North West Regional Rlys	3.9
Cardiff Railways	3.6
Regional Railways North East	3.2
Central Trains	2.9
Island Line	2.9
ScotRail	2.8
South Wales and West	1.8
InterCity Cross Country	1.2
Anglia Railways	1.1
North London Railways	0.9
Thames Trains	0.7
Chiltern Railways	0.6
LTS Rail	0.6
W Anglia and Gt Northern	0.6
South Eastern	0.5
South Central	0.5
InterCity West Coast	0.4
Great Western	0.3
Great Eastern	0.3
Great North Eastern Rly	0.3
Midland Main Line	0.3
South West Trains	0.2
Thameslink	0.2
Gatwick Express	-0.1

Notes: Anglia is here classified as an InterCity TOC since it earns the majority of income from InterCity services: it acquired a number of regional services on 1 April 1994.

Table 6.3 Principal sales proceeds from BR privatisation

Unit sold	Sales proceeds £m	Profit in 1994/95 £m	Ratio of sales proceeds to 1994/95 profit
BRML (6 heavy maintenance sites)	15		
13 infrastructure maintenance and renewal companies	169	34	5
30 support organisations	65	n.a.	
3 heavy freight companies	225	n.a.	
Freightliner	5*	n.a.	
3 rolling stock leasing companies	1800	270	7
Railtrack	2000	269**	7
BR Telecommunications	130	31	4

Source: BRB accounts and Railtrack prospectus

Notes

Profits are last reported results in the public sector and are stated, where possible, after interest and tax. The capital structure of some companies changed at privatisation and pro-forma profit estimates have not been disclosed in all cases.

**This company benefits from both a restructuring grant and a track access grant, designed to reduce the amount it has to pay for track access.*

*** Railtrack is 1995/96 profit forecast, post-capital reconstruction.*

The value for money achieved by a number of these sales including the infrastructure units, the ROSCOs and Railtrack was at the time of writing being examined by the National Audit Office.

Introduction of competition in the passenger market

The theoretical purpose behind the split of infrastructure operation from service provision, as described earlier, was to promote the maximum degree of competition within the value chain by holding out the possibility of open access competition in due course. It is an obvious danger of such a policy, however, that it merely concentrates the total amount of monopoly power that existed previously into one entity, in this case Railtrack. Such concentration necessitates tighter regulation than would otherwise have applied.[19]

Objectives of competition

Competition was directed at achievement of two goals:

- more efficient (ie. lower cost) service provision, and
- innovative service provision (such as an increased number of direct services, an increase in the number of classes of service on-board trains and higher service frequencies).

The essential policy choice faced at this stage was one that had rarely been directly encountered thus far in the UK's privatisation programme: whether it is preferable to promote competition in the market (i.e. the provision of train services themselves) or for the market (i.e. the right to operate train services).

Competition in the market

Competition between train operators, however, poses substantial problems of co-ordination and market organisation. In principle, the Railways Act permits service providers to run services where and whenever they wish on the rail network, subject to obtaining access rights from Railtrack and the approval of the Rail Regulator to the grant of those rights. However, such an approach could lead to a chaotic situation developing:

- rules would be needed to allocate capacity at terminal stations and key junctions: many freight and some lightly loaded passenger services would be threatened by a completely free market in such capacity since they do not cover their direct costs,
- off-peak services cannot be assumed to cover their costs, even at the margin, encouraging operators to reduce services that many customers still regard as 'essential', and,
- the regular interval timetable that has gradually evolved to cover almost the entire country and is popular amongst users would be

threatened by services that 'cream skim' the most lucrative slots. As an illustration of what happens in competitive markets, airline services on competitive routes typically bunch together in a struggle to obtain the most traffic at peak periods rather than being spread out during the day.

After a consultation exercise in 1994, the Regulator announced that he was persuaded that he was necessary to 'moderate' competition on the rail network. His policy meant that substantially no new inter-operator competition would be permitted on the network until April 2002. This permitted the privatisation of passenger services to proceed substantially on the basis of competition for the market, under which pre-packaged tranches of services would be put out to tender.

Competition for the market

Central to the Government's privatisation plan was the concept of a 'franchise', being the right to operate a specified set of passenger services in exchange for a subsidy payment (or a premium payable to the Government). This was, in many ways, the most revolutionary step in the restructuring programme because it involved a novel business proposition: that of assuming the obligation for between 5 and 15 years to operate rail services using existing assets and staff, in exchange for subsidy. This was franchising.

The franchise proposition

The franchise proposition can be summarised as follows:

- the services of each train operating company (TOC) in turn were put up for franchising. Each TOC had access rights and rolling stock leases, together with contracts for all other material purchases and sales, to permit all of its services to be operated initially,
- each TOC must always supply services described in its Passenger Service Requirement, set after public consultation by the Franchising Director,
- each TOC can supply additional services, subject to obtaining sufficient access rights and rolling stock to operate it,
- each TOC was franchised complete with staff including management team. Staff were entitled to some protections arising from the Railways Act and employment legislation designed to protect employees during privatisation and corporate sales,

- substantial revenue and cost risk is borne by the TOC. Exceptions were made to permit some Passenger Transport Executives to retain revenue risk (together with the right to set fare levels) in their areas. Subsidy payments would be indexed at RPI and the risk arising from the Regulator's review of access charges was also kept with the Franchising Director,
- certain fare types are regulated, which has the effect of restricting fare increases and limiting changes to the wider structure of fares,
- the cost of relieving overcrowding in Greater London commuter services is initially shared between the TOC and the Franchising Director. Above specified levels of overcrowding, the cost falls solely on the Franchising Director,
- for some franchises, there is a commitment to provide replacement rolling stock,
- the franchise is awarded for between 5 and 15 years. Most franchises are for 7 years' duration: long periods were only awarded when substantial rolling stock replacement was required,
- a performance bond must be put up, which is analogous to a refundable deposit that a tenant pays when renting a house,
- limitations are placed on the ability of the TOC to mortgage assets or the benefits of contracts that it has: in the event of insolvency, a special regime will apply, termed a Railway Administration Order, designed to ensure that rail services continue to be provided. Consequently, any ability of creditors to seize assets is constrained,
- the benefits of the franchise agreement cannot be transferred to a third party without the approval of the Franchising Director. The Franchising Director can thus control any secondary market in franchises and can, at least in theory, ensure that any surplus value generated is returned to the taxpayer or the customer by means of better services, and
- bidders bid for the right to operate services. The rights are encompassed in a Franchise Agreement between the TOC and the Franchising Director. The bid requiring the lowest subsidy would normally win the competition.[20]

The nature of the franchise is such that it is largely a fixed cost business: typically half to two-thirds of its cost base is contractually committed under its access agreement and rolling stock leases. The balance of costs are for provision of train crew, conductors, carrying out light maintenance on rolling stock, and provision of retail staff at stations. In practice, much of these are effectively fixed as well insofar as they are required in order to provide the Passenger Service Requirements.

Effectiveness of these two forms of competition

The effectiveness of these two approaches can be compared. Undoubtedly, the route of franchising was quick and simple to implement compared with setting up administrative and market mechanisms to resolve the 'free-for-all' that might have developed in a more open market. It allowed the private sector to acquire TOCs as ongoing concerns, complete with senior management, lowering the barriers to entry to the rail market that would otherwise arise from the need to set up rail operations from scratch. Competition in the market, whilst dynamic and innovative and hence most likely to lead to creation of new services that better match market requirements, would have been difficult to organise in practice and would have made it harder still to offer network benefits such as through ticketing and connecting timetables. The franchising route undoubtedly was also very effective at committing franchises to cost reductions and service improvements which reduce the subsidy requirement from the public sector.

Subsidy commitments

Figure 6.3 indicates a typical own cost/subsidy profile for a franchisee, making allowance for an ongoing inflation rate of 2.5 per cent per annum. It can be seen that the subsidy reduction results from the balance between revenue and reduction of the TOC's own costs in real terms. The typical franchise consequently is not without risk: its profit is subject to considerable gearing between a revenue growth rate applied to total passenger revenue and a cost reduction target which can only, at best, apply to about a third of costs.

An important question for any kind of franchise competition of this kind is that of the allocation of risk between the parties. The nature of the franchise agreement is such that the maximum financial loss that a franchisee could sustain is the performance bond, which was normally set at about 10 per cent of franchise turnover. However, the franchisee would inevitably suffer an indirect effect on its reputation as well, of particular relevance where the franchisee is a listed company.

Figure 6.3 Typical own cost/subsidy profile in cash terms for a seven year franchise

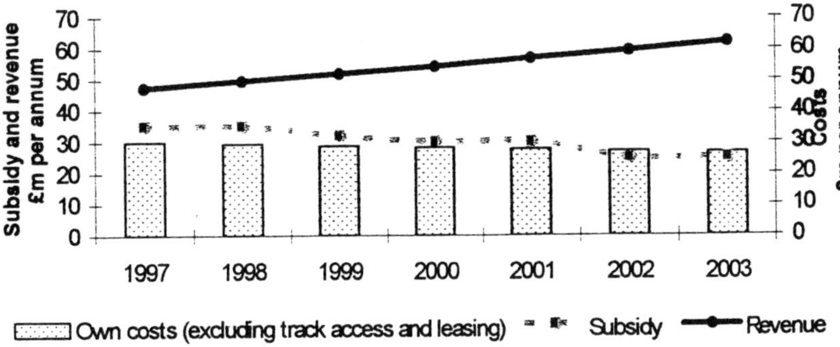

Speed of franchising

Despite initial scepticism by would-be bidders, the first two franchises were awarded in February 1996 (in respect of Great Western and South West Trains). Bidders were soon attracted to the market: most were bus companies (the bus sector was itself undergoing rapid consolidation at the time as the end-game of bus deregulation of 1986 played itself out). Very competitive bidding conditions developed, apparently buoyed by share price increases of the initial franchisees. The then Conservative Government also applied extreme pressure to ensure that the process of franchising was completed prior to the 1997 General Election, presumably in the hope that rail privatisation could be made irreversible.

By March 1997, all franchisees had been selected. and Table 5.4 indicates the types of bidders that won franchises.

Management buy-outs were successful in only two TOCs (Chiltern and Great Western) and in each case consortia were formed with other backers. There were six successful management buy-ins, five of which were through two buy-in vehicles (Prism and GB Railways). If it was the case that the Franchising Director's awards were made primarily on the basis of lowest subsidy requirement, this pattern indicates that incumbent management were more cautious in their bidding than the new entrants. New entrants from the bus industry would have been more directly aware than rail management of the magnitude and realisation of cost savings from bus deregulation, but rail management were, arguably, better aware of the commercial challenge associated with generating increased off-peak use of rail services in a competitive transport market.

Table 6.4 Successful franchise bidders

	Number of franchises awarded
Bus and coach companies, in whole or in part (includes one with minority participation by management)	12
MBOs, majority controlled by management	2
Virgin Rail Group	2
Sea Containers	1
Management buy-ins (of which 5 were backed by floatation on London's Alternative Investment Market (AIM))	6
Compagnie Generale des Eaux	2
Total	**25**

Source: OPRAF franchise awards

Bidding became very competitive, with later franchises being let on the basis of considerable subsidy reduction. This gives two advantages to the Franchising Director:

- a declining subsidy requirement which was a contractual commitment of the franchises, insulating the Franchising Director from the vagaries of the economic cycle (although it should be recalled that should franchisees fail to deliver this aspect, they would lose the franchise and losses would be capped at the size of the performance bond, leaving the obligation to continue to provide services, and hence subsidy, with the Franchising Director).
- the amount of cash needed in future years has become more easy to forecast. The actual cash requirement, the variable analysed closely by the Treasury during public spending rounds, is set by adding forecast inflation to the subsidy sums plus a forecast payment in respect of the performance regime in place between TOCs and the Franchising Director.

Figure 6.4 indicates the substantial subsidy reduction expected in coming years. The drop in 2004/05 exists because a number of franchises are up for renewal then and it is not possible to say what subsidy might be needed for them. The franchises with longer lengths, by and large, require less subsidy. The dotted lines indicates a simple extrapolation, for illustrative purposes, of the trend between 1996/97 and 2003/04, but is certainly not a forecast!

Figure 6.4 Total subsidy commitment in 1996/97 prices for all 25 TOCs

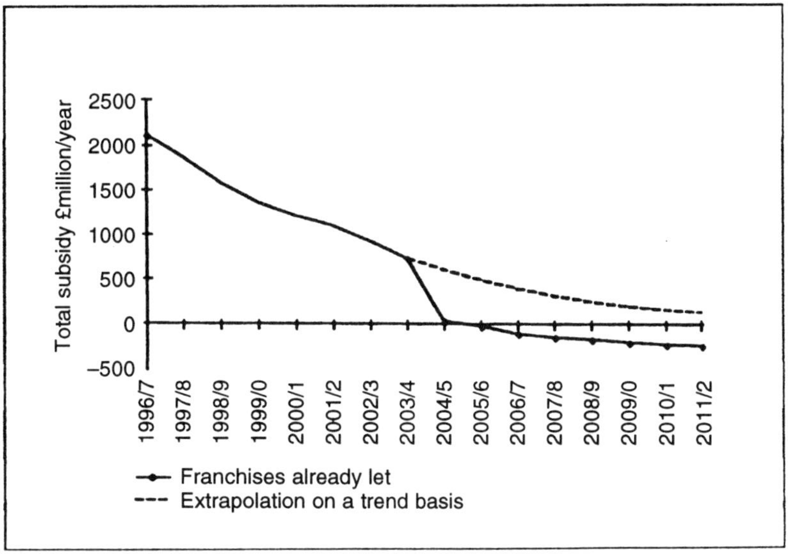

On the basis of franchisees' commitments, total subsidy is expected to fall by about two-thirds between 1996/97 and 2003/04. Indeed, it might be concluded that, if these trends continued and the effect of opening up of competition in the market is minimal, the Franchising Director's function might itself become a profit-making activity within 15 years or so. The Railways Act requires premia earned from franchises to be treated essentially in the same way as tax income to the Government, requiring the Franchising Director to bid for the full value of loss-making services in government spending rounds. Thus, as a matter of Government budgeting, the Franchising Director's 'profits' cannot be used automatically to cancel out his 'losses'. The degree of cross-subsidy within the Franchising

Director's support operations will inevitably become a matter of controversy in future years.

Changes underway and plans for the future

The first franchises have now been in the private sector for over two years and the most recent almost a year. What progress has been made in that time? Aside from new liveries and branding, there has been remarkably little change to service patterns and frequencies on the system. This is largely a consequence of two decisions taken early on in the franchise process:

- that BR's existing timetable should be taken as the base for franchising[21], and
- that fare regulation should be applied to key fares where a TOC might have some degree of market power or which provided 'network benefits' to customers by allowing them to make journeys involving many TOCs easily.

The first decision led to a requirement to essentially 'freeze' the train plan[22] when access rights were granted and franchises designed in 1994/95. The consequence of this is that the franchised TOCs have only limited rights to any additional or different train plan to that which they have inherited from the state sector. There have been some modest frequency increases on certain routes: about 5 per cent more trains were offered in the 1997/98 timetable compared with the previous year although most of these were on short trips so the increase in train-miles was considerably smaller. Anglia Railways, for example, operates to Ipswich twice per hour rather than once an hour previously. Connex South Central has introduced a new service between Gatwick Airport and Rugby via Kensington Olympia. Wales and West Railway has been able to link up services it previously offered to provide an increase in the number of direct services.[23] A number of other changes are in the offing for May 1998's timetable.

The second decision largely explains the lack of change to the fare structure, the complexity and high level of which have come in for considerable criticism over the years. The existence of certain types of fares[24] and their pricing is constrained by regulation: these constraints in turn set limits on other fares. A limited number of new fares have been introduced since privatisation:

- Virgin West Coast is offering a new range of fares, named Shuttle Advance and Virgin Value, which must be booked a day ahead and are only valid on off-peak services. Virgin, however, wishes to increase the price of immediate purchase wherever it can do so,

- Midland Main Line offer a promotional fare for small groups of four or more bookable a day ahead and, in January 1998, restructured their service to offer two types of service for business customers,
- Scotrail offers an enhanced range of combined travel plus sleeper supplement on its Anglo-Scottish sleeper services,
- Some of the benefits of the Network Railcard in the South East area have, however, been removed and its price has been increased.

Initial results

Passenger mileage continues to grow rapidly: last year it expanded at about 6 per cent, due to a combination of marketing initiatives, growth in employment levels, improved revenue collection, changes to staff travel rules and increases in GDP. Similar levels of growth occurred during the late 1980s during a comparable period of economic prosperity. All TOCs that are quoted or are subsidiaries of quoted companies[25] have reported operating profits from the initial period following privatisation, although a number have taken restructuring provisions at the same time to cover the cost of making the transition to leaner ways of working.

The pace of change will inevitably increase in coming years. New rolling stock and station refurbishments should increase the attractiveness of rail services to leisure and other off-peak users of the system. More creative use of yield management techniques, tied to a wider range of fares, seems likely on some principal InterCity routes following airline practice.

Achievement of subsidy commitments

High-level analysis of the franchise bids suggests that there are two routes by which the subsidy targets can be met: revenue growth and cost reduction, or a combination of the two.[26]

If the bids are achieved solely by revenue growth, then simple arithmetic based on the fact that the ROSCO and track access elements of the TOC's costs are subject to RPI - x type price controls suggest that an average of between 4 and 6 per cent growth in revenue in real terms must be achieved each year for the next seven years. If, however, TOCs can make a one-off reduction of 30 per cent of their own added costs excluding leasing and track access and then reduce their own costs by 2 per cent each year in real terms, the required revenue growth is about 2.5 per cent in real terms. Such a one-off reduction is at the extreme of expectation for a sector still subject to a considerable quantity of inevitable economic and safety regulation. Clearly, a large number of combinations of these variables is possible: a higher rate of cost cutting means that a lower revenue target need be achieved. Some of this increase can be achieved by real price increases, where this is permitted under the price control arrangements.[27] Figure 6.5 compares the effect of a 4.5 per cent growth rate and a 2.5 per cent growth rate in passenger numbers compared with the historic trend since 1986/87.

Figure 6.5 Possible implications for passenger volume of subsidy commitments

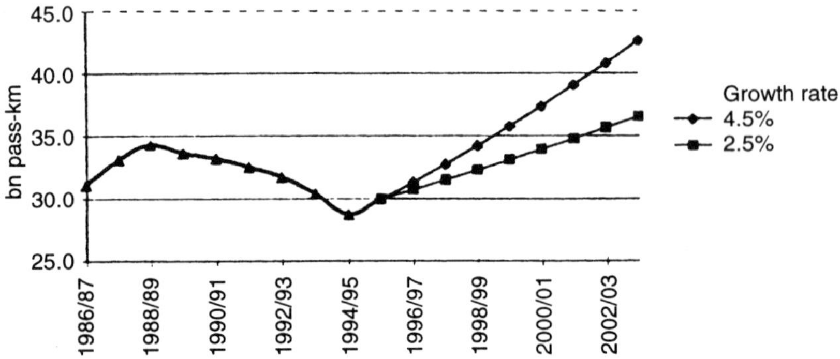

What is clear, however, is that, in any scenario, these numbers are high by historic trends. Even at the peak of the last business cycle, in 1988/89, volume growth on BR was no better than 6.4 per cent, much of this driven by growth in Central London employment and increases in GDP generally. Franchisees, however, will probably benefit from the following factors:

- changes to fares structures, designed to increase revenue from peak traffic (where permitted under regulation) and stimulate additional off-peak traffic by more precise means of yield management,
- access to capital to carry out marketing and refurbishment investment rapidly,
- increasing traffic congestion, together with increases in real fuel prices will encourage some substitution of rail for car, and
- in many cases, highly motivated managements that are strongly incentivised by equity participation, share options or bonuses to achieve the revenue targets that have been set for them or they have set themselves.

Implication of the subsidy commitments

An immediate implication of the subsidy commitments is an increase in passenger volumes. Parts of the rail network and some fleets of rolling stock already operate at close to capacity, posing problems of cost and quality to TOCs if new demand is generated at peak times. The search for off-peak traffic to fill an operation whose size is dictated by peak requirements is, of course, the obvious strategy for any transport operator.

The gradual move to premia payments

Eight of the franchises are expected to make premium payments (i.e. negative subsidy) to Government by their end. Given the charging regime in use, outlined above, such premia indicate that these TOCs believe that they are fully capable of generating commercial returns on the replacement cost of the assets they use. Given the degree of cross-subsidy implicit within each franchise (i.e. the weaker routes are supported by the stronger routes, and off-peak services supported by peak services), the number of individual service groups within TOCs expected to generate commercial returns above replacement cost must be higher still. This will be a major achievement, particularly to those who recall successive waves of anguish over railway economics over the past forty years[28] and one which is not achieved in all quarters of the private sector either.

Its achievement poses a number of challenges for railway policy in the future:

- when these franchises end, who will obtain the surpluses over fully-allocated replacement cost that these premia represent?
- will the surpluses attract open access operators and how much will be competed away by them in seeking access to the market[29]?
- do such services 'need' to be franchised in future, or will the market provide them of its own accord?

Who will capture the surpluses?

In the first instance, it is the TOC that obtains these surpluses. The surpluses could be transferred back to the Franchising Director by means of a new franchise competition: an efficient competition for the right to operate such services should transfer back all of the surplus to the public sector. In this scenario, the public sector would essentially be selling concessions, as distinct from obligations, to operate services. This already happens in the case of Gatwick Express, which has been premium paying since May 1996.

Such services could be left entirely to the market to provide. If concessions are not sold off, the cross-subsidy inherent in the franchising system would be lost with the inevitable consequence that the total subsidy requirement would increase. Whilst the efficiency of that cross-subsidy is obviously debatable to theoreticians, it does have obvious practical benefits. These may be particularly hard to resist in a situation in which not all of the subsidy commitments by franchisees are met and the public sector needs to find increased funding to maintain service levels. That said, the consequence of leaving the market to provide such services should be, in the long term, to transfer the bulk of the surpluses to consumers as competition amongst service providers opens up. Either way, it seems, the general public will eventually benefit: the question is whether they would prefer to do so as taxpayers or as consumers.

The surpluses could also be captured by Railtrack. Regulators are normally persuaded by arguments of economic efficiency that pricing network services both above fully-allocated cost and in a discriminatory manner (between, in this case, franchised and non-franchised services) is to be avoided. However, the Rail Regulator also has a duty to take account of the subsidy budget when setting policies and he will also be aware that capacity on the rail network is constrained in a number of places and that some of capacity rationing pricing has efficiency advantages. In the most general sense, these arguments suggest that the Rail Regulator could allow a degree of cross-subsidy within the rail businesses in order to reduce the subsidy budget. In other words, it would appear to permit him to capture the surpluses by means of Railtrack access charges and recycle them to 'poorer' TOCs in order to reduce their access charges below cost. Existing regulatory controls on income from property lettings at stations, freight and open access services already have this effect.

Perhaps more perniciously, given the lack of regulation of the ROSCOs, the surpluses could also be captured by the rolling stock lessors. If a competitive rolling stock leasing market existed, prices would reflect costs and not the economic value to the TOC generated by the use of those assets. In circumstances where the surpluses are generated on routes with limited choice of rolling stock, it is possible that lessors may capture surpluses once the pricing in the existing lease contracts comes to an end. Lease lengths do not always match franchise lengths. Thus, lessors may attempt to increase lease prices during the final years of franchises both in recognition of the balance of economic power between franchisee and lessor and in order to set a market for an incoming franchisee.

The role that subsidy might play in future

That said, it remains the case that many observers believe the subsidy commitments to be ambitious, but not commercially impossible given the managerial freedoms, including freer access to capital, that privatisation has brought. The recent announcement of the possibility of renationalisation of the Eurostar services and heavy restructuring of London and Continental Railways, set up to build the Channel Tunnel Rail Link, indicates the risks run by companies reliant on ambitious revenue forecasts.

If surpluses are not achieved, the question of how the deficits are to be distributed in future becomes more pertinent and all of the arguments described above apply in reverse. Although the existing price structure for track access and rolling stock reflects long-run average costs on a replacement cost basis, there can be no guarantee that the Rail Regulator, the Franchising Director and, in the background, the Government will continue to follow this course. Government may not be persuaded, any more than their predecessors were, that the theoretical advantages that come from pricing at such levels justify paying out more than is strictly needed to keep the rail system going on a day-to-day basis.[30] Indeed, the interesting possibility exists that, in his forthcoming price review, the Regulator could back charging on a replacement cost basis, to preserve the equality of treatment of Railtrack with other network utilities, whereas the Franchising Director might have insufficient funds to support such levels. In this case, the Regulator's price controls would, presumably, become ceilings, and Railtrack via the TOCs would have to negotiate charges less than this amount with the Franchising Director. In other words, monopsony power would prevail over monopoly power.

In such a negotiation, there can only be three dimensions to the debate:

- route closures,
- number and quality of services to be retained on each route, and
- access prices to be set below replacement cost levels.

Since route closures have largely been replaced in recent years by modest route re-openings, choices about rail subsidies have largely become questions of the amount of rail service to be provided on uneconomic routes and the quality of rolling stock and infrastructure provided there. The possibility of pricing below replacement cost levels merely mimics the action taken by BRB to make the books balance in difficult years in the past: it reduced the number of investment schemes and, more occasionally, took 'maintenance holidays'. Lowering Railtrack's prices in the first instance reduces its profitability: if such lowering made Railtrack more reluctant to invest in assets whose life would extend beyond the life of franchise agreements, this could conflict with its licence obligations and the aspirations of its customers. The key point remains, however, that even with all the change wrought by the Railways Act, the Government still is not required to provide any long-term assurance about the funding that it will provide or about the extent and quantum of rail services.[31] The predictability of subsidy and, behind that, the specification of the social and commercial objectives that the rail sector should fulfill in the long term, remains as uncertain as ever. This uncertainty, which has been in place at least since the 1962 Transport Act, is likely to continue to be the most predictable feature of rail policy in future. And, within the variety of public agencies that regulate the rail sector, it seems clear that it is the Franchising Director, in his role as procurement agent on behalf of government, who is likely to have the biggest influence on the development of that policy. His task, whether or not his role is absorbed in a Strategic Rail Authority, is a difficult one. He must hold franchisees to their existing commitments, ensure that the taxpayer and commuters get a fair share of any spoils that might arise from any takeovers and define the future shape and nature of the rail network. The Franchising Director must now move on from the euphoria generated by the sale of franchises and face the more humdrum task of making the numbers balance for the future.

Notes

1. Richard Davies is currently a Senior Consultant with KPMG. Until the end of 1996, he was Financial Planning Manager, Privatisation, at British Railways Board's Group Headquarters and participated in several different aspects of BR's restructuring including the new financing regime, and in the negotiation of contracts to enable Train Operating Companies (TOCs) to be privatised. Any views expressed in this paper are solely those of the author and not necessarily of any other party. Data in this paper is correct as at February 1998.
2. 'Regulation' here covers both the role of the Rail Regulator and the Franchising Director, who have regulatory oversight of different aspects of rail services.
3. British Rail's last train operated on 20 November 1997, state operation of railways thus failing narrowly to survive until its fiftieth birthday.
4. Even as early as 1967, the Board entered into an innovative leasing arrangement with English Electric for Class 50 diesel locomotives, under which payment was tied to the performance of the locomotives.

5. Such as the provision of the British Transport Police.
6. Two further forms of grant should be mentioned. The passenger businesses were also credited with 'deferred grant' in 1993/94 for accounting purposes. This grant reflects amounts received in prior years for repair and renewal of some infrastructure assets: the Board changed its accounting policy on 1 April 1992 and capitalised asset renewal whereas these costs had previously been transferred direct to the profit and loss account. No cash was received by the Board in the 1993/94 year in respect of the deferred grants. In addition, the Board received some grants from central government, Passenger Transport Executives, the EU and local authorities in respect of capital investment, amounting to about £100m.
7. The precise split between these is not disclosed in the Board's accounts. However, insofar as BR's total Channel Tunnel investment was of the order of £2bn, and allowing for some interest on debt, it is clear that most of the total of £2.4bn would have related to Channel Tunnel had the debt been identified with the assets that were funded by it.
8. Not all infrastructure transferred to Railtrack: Waterloo International Terminal and North Pole Depot are used by Eurostar services, some dedicated freight facilities together with some redundant railway viaducts and rights of way stayed with British Rail. The Waterloo and City Line in London, together with the Putney Bridge to Wimbledon section of the District Line, were transferred to London Underground.
9. Strictly speaking, the ROSCO leases also contain some performance-related elements, but these are at de minimus levels in relation to total rentals.
10. The producer price index rather than the retail price index was used here, as it was felt to be a better proxy to the inflation to be expected on heavy maintenance costs.
11. Railtrack's route network comprises 10,000 route miles, so the average replacement cost of track, station and depot infrastructure implied by this valuation is £2m per mile and the average access charge is £200,000 per mile per year. This figures could be contrasted with the approximately £3bn required for the Channel Tunnel Rail Link (CTRL), a route mileage of about 68 miles, which implies an average cost of about £45m per mile. The difference largely arises because of the complex nature of the tunnelling and engineering works needed for CTRL, including the provision of new station facilities. This contrast gives some indication of the difficulty involved in

making the CTRL viable.

12. Such as the introduction of a greater degree of variability between charging and infrastructure output, thereby exposing Railtrack to a greater amount of the economic risk of the rail industry, the use of more specific cost allocation mechanisms (particularly in relation to operating costs), and greater transparency as to the allocation of any 'residual' (see below).
13. Indeed, many 'large ticket' leases transfer the risk of changes in interest rates and taxation regime to lessees, thereby further insulating lenders from economic risk.
14. They are the metropolitan areas centred on Birmingham, Glasgow, Leeds, Liverpool, Manchester, Newcastle and Sheffield.
15. Railtrack did agree to a tighter obligation through a licence modification in July 1997, but it still falls short of a commitment to replace all assets with modern equivalents.
16. Once enough company accounts are filed for the privatised TOCs, it will be possible to estimate how realistic this number was!
17. Arguably, however, this principle had already been reversed by the Government's decision in 1992 to award EPS, which was expected to be profitmaking, as a dowry (ie. a cross-subsidy) to the constructor of the Channel Tunnel Rail Link.
18. Strictly speaking each TOC was sold by BRB to a franchisee for a nominal sum (£1) upon privatisation, but came with the benefit of a franchise agreement which only became effective at the point of privatisation: any value that the TOC had relative to the subsidy and other commitments in the franchise agreements thereby accrued to the franchisee. In theory, the efficiency of the franchise competition itself is measured by the extent that such value was reduced to zero. Five of the franchises are owned by AIM-listed companies (Prism Rail Group plc and GB Railways Group plc) set up solely to own franchises. Their capitalisation on AIM, at the commencement of franchising, was considerably in excess of the amount needed for working capital and to provide performance bonds to the Franchising Director. This surplus is a measure of the value not extracted from franchise bidding.
19. This argument takes a long term view that the objective of regulating Railtrack is to restrict its monopoly power. It is not obvious, given the very competitive nature of most transport or rail markets, that an infrastructure provider in fact possesses much monopoly power in such situations.
20. Franchisees, under the Railways Act, were selected by means of a two

step process. A pre-qualification round was used to eliminate those without the necessary managerial and financial standing or which might inhibit competition between operators on shared routes. In the final bidding round, the bidder submitting the lowest bid (or highest premia) in net present value, it has been reported by the trade press, was normally selected. There was, however, no statutory or administrative obligation on the Franchising Director to pick the lowest bidder each time; the details of bids not accepted have never been made public other than some elements of those for the first three franchisees for which limited details were published by the NAO. Neither has OPRAF disclosed weightings that it applied to the assessment criteria that it used.

21. A point that was specifically challenged through the Courts in the run-up to the first franchises being awarded because the franchise competitions required TOCs only to operate the 'Passenger Service Requirement', which is a level of service lower than BR operated. The Requirements were deliberately specified to include only those services which are loss-making (i.e. do not cover their direct costs) because the Franchising Director has always assumed that operators would provide services that do cover their direct costs of their own volition. In many cases, franchisees voluntarily assumed the obligation to operate more than the Requirements in their Franchise Agreements.
22. Limited 'flexing rights' exist on both sides, but essentially the track access agreements are 'take-or-pay' in nature and give TOCs long-term rights to capacity for the duration of their franchises together with the long-term obligation to pay for it.
23. The Regulator's policy on moderation of competition focuses on competition between pairs of stations. One of its consequences is that new entry is possible where existing flows between such pairs are negligible.
24. Ordinary singles, returns, season tickets and savers are regulated: other fare types are indirectly regulated by the ceilings this regulation imposes.
25. These, at 13, represent just over half of the 25 franchises, and are Anglia, Cardiff Railway Company, Central Trains, Gatwick Express, GNER, Great Eastern, LTS Rail, Midland Main Line, Scotrail, Silverlink, South West Trains, Wales and West, and West Anglia and Great Northern. In addition, Great Western and North Western Trains are part-owned by First Group plc.
26. The Passenger Service Requirements forbid TOCs from reducing

service levels on uneconomic routes.

27. For a description of the price capping arrangements, please see the chapter in this book by Stannard.
28. The Beeching studies of 1962 and 1964, 'The Rail Problem' by Richard Pryke and John Dodgson 1974 and the Serpell inquiry of 1983 come to mind.
29. The 'surpluses' ultimately reflect a difference between the cost of providing rail services and their value. In an efficient market and without capacity constraints, fares to customers can ultimately be expected to be competed down to the level set by competitors' costs.
30. Some changes have been brought by the fact that Railtrack is licensed. It is required, in effect, to maintain the operating capability of the rail network in perpetuity but receives no guarantee of its access charges from franchised passenger operations beyond each five year period of price control. The current price control expires on 31 March 2001.
31. Its commitment to the funding for franchise agreements does, however, provide some short term assurance on these matters which did not exist when the sector was in state control. In addition, the closure procedures created under the 1993 Act are certainly more cumbersome than those they replaced increasing the barriers to exit of rail support on the part of the state sector.

CHAPTER 7

Railway regulation: Looking ahead to the millennium

Chris Bolt

Introduction

The passage of the Railways Act 1993 was the trigger for the biggest restructuring that the railway industry in Great Britain has ever seen. The activities of the British Railways Board have been split into some 100 different companies and sold to the private sector through flotation, trade sales or franchising. From the first sale in March 1994, the pace of privatisation and restructuring increased during 1995, with the initial access agreements put in place and the start of the franchising process. 1996 opened with the first major sales, the three Rolling Stock Companies, and also saw the award of most of the franchises and the flotation of Railtrack. By the time the ScotRail franchise started operation under National Express ownership at the end of March 1997, the process of restructuring and privatisation was virtually complete.

Putting in place the contractual matrix of licences and regulated access agreements that underpin the relationships between the different railway operators was the central activity of the Office of the Rail Regulator (ORR) during Phase I. It involved a number of important policy decisions, such as determining the level of Railtrack's access charges and deciding the extent to which, and means by which, competition should be moderated in the initial franchises.

Merely putting the contractual matrix in place and transferring railway operations from the public to the private sector has been a major achievement, which some commentators thought could not be achieved. The franchising process has also delivered commitments to reduce subsidy and improve services to an extent which has exceeded most expectations.

But the railway industry is now moving from this initial phase of setting up the structure to Phase II, making it work. This Phase will present new challenges.

Looking ahead

The central challenge of Phase II, for both the Franchising Director and Rail Regulator, is to ensure delivery of these commitments by franchisees, and by other regulated operators, Railtrack in particular, and to ensure that the new structure operates to bring even greater improvements. This is reflected, for example, in ORR's high level aim which is 'working together with the industry to create a better railway for passengers and freight users'. But all regulators need to recognise that economic regulation is not an end in itself, but a means to an end: their task is to promote the operation of market forces wherever possible, on the basis that competition is generally the best way of promoting the interests of both customers and producers, while protecting the interests of all parties where there are market failures.

As with other regulators, the promotion of competition is one of the statutory duties placed on the Rail Regulator. But there are other important issues for the Regulator to consider, as well as competition, in promoting the interests of customers. Maintaining the benefits for passengers of a national network requires cooperation between operators. While this might not be in the short term commercial interests of individual operators, it is a key element in promoting greater use of the rail network, and will bring benefits to operators and passengers. Achieving the right balance between competition and cooperation will be an important test of the new railway structure.

There are many external influences on the development of the railway industry besides regulation itself. We have seen, for example through the windfall tax, how Government can take decisions that affect the financial position of utilities. The role of railways could also be significantly affected by wider transport policy proposals to be set out in the Integrated Transport White Paper promised by the new Government for Spring 1998. For example, if the perceived imbalance between charges for using the railway compared with the road network is changed, there could be a significant boost for the use of rail by both passengers and freight users. It is, by the same token, important that the framework of railway regulation should be responsive to, and supportive of, objectives in respect of sustainable development and the environment.

Nor must we forget developments in Europe, where last year's EC White Paper on transport clearly identified the need for 'a new kind of railway'. The extent of restructuring that has already taken place in the UK could put British operators in an advantageous position as open access develops, and private sector involvement is sought in operating networks.

Many of these factors are outside the direct control of the transport regulators. But as well as developing the monitoring and enforcement mechanisms necessary to ensure that the public interest is protected in Phase II, there are a number of further developments in the regulatory framework which could have a significant effect on the future development of railways.. Two such developments are the extension of competition and the 'periodic review' of Railtrack's access charges to be completed by 31 July 2000.

Competition or cooperation?

Vertical separation between the monopoly network owner and operator, Railtrack, and train operators was, in part, designed to provide the basis for on-rail competition without the risk of anti-competitive behaviour where the network operator also competes in downstream markets. The design of the initial franchise map already allows for competition between different operators - there are, for example, four operators providing services between London and Birmingham - and it was expected that other 'open access' operators would enter the market to compete with incumbent operators.

In accordance with Guidance from the Secretary of State (which the Regulator was, until 31 December 1996, under a duty to take into account), the Regulator decided following consultation that on-rail competition between passenger operators should be moderated for the period of the initial franchises. While the Regulator confirmed his view that allowing competition to increase over time would be in the interests of customers and operators, he accepted that an early threat of open access competition would increase risks to potential franchisees to such an extent that the success of franchising might be prejudiced. There was also a concern that, because new entrants would target the most profitable flows, the cross-subsidies inherent in the packaging of services into franchises would be unwound, increasing the subsidy payments required to maintain the remaining loss-making services.

Even with competition moderated, there remains a concern that breaking up passenger services between 25 franchised operators will result in a loss of so-called 'network benefits' - the ability to purchase through tickets from any station to any other station, irrespective of which operator provides the service; the ability to obtain accurate information on all train services both at stations and by telephone, and to purchase tickets from any operator on an impartial basis; and, in the event of a claim arising, the ability to obtain redress in a straightforward way, without needing to identify the operator responsible. The licences issued to operators by the Regulator contain provisions to protect the interests of passengers in these areas. The licences also contain requirements to protect the interests of disabled passengers, and to prevent anti-competitive behaviour by operators.

The Regulator has recently published his 'Regulatory Objectives for Passenger Train and Station Operators'. In that statement, he stressed the importance of operators cooperating to provide these network benefits. He also stated clearly his intention to take necessary action to secure their delivery. The recent 'awful' performance of the National Rail Enquiry Service is one area where such enforcement action may need to be taken, if the performance targets established for the service are not met quickly.

There is, however, a fine line between cooperation between operators in line with licence provisions designed to protect users against the possibility of market failure and collusion which is intended to distort competition and can be expected to operate against the public interest. One of the tasks for the Regulator during Phase II will be to monitor the behaviour of operators to ensure that this boundary is not breached.

One further question that arises in considering the development of competition is the structure of the industry itself. The 25 franchises are controlled by 13 separate firms, and further concentration is being predicted. Railways are, of course, already in competition with other modes of transport, and the Monopolies and Mergers Commission has, in its recent report into the award of the Midland Main Line franchise to the coach group National Express, confirmed that the loss of opportunities for competition between rail and these other modes can be expected to operate against the public interest. But at the same time, there is little doubt that the involvement of bus and coach companies in rail franchising has been a major contribution to its success.

In other parts of the industry, there have also been mergers, both horizontally and vertically. Horizontal mergers have, for example, occurred in the infrastructure maintenance and track renewal market, with Jarvis recently proposing to acquire two of the other companies in the sector. The acquisition of the Porterbrook leasing company by Stagecoach, the operator of South West Trains, raised more significant questions, involving as it did reintroduction of a degree of vertical integration. Following consideration by the competition authorities, behavioural undertakings in respect of fair trading and confidentiality were given by the parties in lieu of a reference to the MMC.

So the 'blackboard structure' put in place in 1994 is already changing. Is this a gradual process, or will there be rapid changes? And how far do the contracts which underpin the new railway themselves effectively reintroduce a degree of integration? Fifteen year contracts for track maintenance, or build and maintain deals for new rolling stock will change the nature of commercial incentives as much as mergers. That is why the Regulator, who provides advice to the Director General of Fair Trading on railway mergers, has said that he would want to see substantial public interest benefits before supporting further mergers in the industry.

The Regulator will be consulting on the development of on-rail competition shortly. It is impossible to predict how, and how fast, open access will develop. What is clear is that there are operators waiting to take advantage of the greater freedoms that Stage II of the Regulator's Moderation of Competition policy will bring. We will need to see how such competition develops, and what the benefits for passengers are.

Freight

Rail freight has been declining - in absolute as well as relative terms - for a number of years. The Regulator has an explicit duty to promote the use and development of the railway network for the carriage of goods, to the greatest extent he considers economically practicable. ORR recognises that the demands of freight users are not the same as passengers, and that the framework of access planning and charging needs to take this into account. For example, ORR has recently approved a new five year agreement covering the majority of services operated by English Welsh and Scottish Railway Ltd, Railtrack's major freight customer, which is designed to allow EWS to quote more competitively, and speedily, for new business than was previously possible.

A reversal of the long-term decline of rail freight now seems possible, and Railtrack has published a 'ten point plan' to reflect its commitment to this sector. But an important issue to be considered as part of the Government's transport policy review is whether freight needs a 'champion' to balance the role of the Franchising Director in supporting passenger services. There are circumstances where passenger and freight users are effectively competing for scarce capacity on the network, but where enhancement of that capacity to accommodate additional freight traffic would require public sector financial support. Balancing the interests of passengers and freight users could be one of the objectives for the proposed Strategic Rail Authority.

Railtrack's periodic review: opportunities and challenges

Although Railtrack's access charges are in regulated access agreements rather than its licence (as with other utilities), they are subject to 'periodic review' in much the same way. Under the terms of the access agreements, the Regulator has until 31 July 2000 to set out his policy conclusions on the future level and structure of access charges; Railtrack and operator then have to propose modifications to their agreements to implement those conclusions.

Many of the features of the periodic review - cost of capital, regulatory asset base, future investment needs, operating cost efficiency, dealing with out-performance or under-performance in the first period - are common to all utilities. The Rail Regulator will reflect the emerging regulatory consensus in reaching decisions on these issues.

There are, however, a number of matters unique to railways that will form part of the overall review programme. These include, for example:

- the subsidy which underpins the commercial profitability of the privatised railway industry;
- the mismatch between the length of franchises and the investment horizons for the infrastructure and rolling stock enhancements which operators may be seeking in order to realise their business plans;
- uncertainty about the nature of the process to be adopted for the next round of franchising, and about the level of subsidy that will then be available; and
- the involvement of the Regulator in approving all access agreements.

As in all regulated utilities, monitoring delivery of the capital programmes underpinning decisions on charges is an important, but difficult, challenge for the Regulator. For example, it is easy to monitor inputs; but effective incentive regulation depends on an ability to monitor outputs in terms of appropriate performance measures. Even if shortfalls in performance can be identified, there is a need to decide whether to take immediate action, or 'log up' the shortfall until the next periodic review. Railways will be no different from other utilities in posing these challenges for the Regulator, although the fact that a considerable amount of taxpayers' money is provided to maintain levels of service for the minority of people who use railways perhaps makes the public interest issues even more important.

The recent modification of Railtrack's licence will give the Regulator appropriate enforcement powers, similar to those that already exist for other industries. By requiring Railtrack to have an effective dialogue with existing and potential users of the network in developing its plans for maintaining, renewing and enhancing the network, it should in future be easier to identify capacity bottlenecks, and to determine how best to deal with them.

One important policy issue to be addressed as part of the review is whether the existing structure of access charges - a large fixed charge, representing some 90 per cent of the total charge, with low variable charges for track usage and, where relevant, traction electricity - should be maintained or changed. Already, questions are arising about whether decisions about the use of capacity at the margin are being taken on an appropriate basis. Are train operators being misled about the costs of seeking additional rights because charges do not, for example, directly reflect additional congestion costs? Do freight and passenger services compete on a fair basis for additional capacity? Do low additional access charges give Railtrack the right commercial incentive to enhance capacity?

It is easy to see, on paper, how development of a charging structure which is more cost reflective will help to deal with these issues. The arguments are very similar, for example, to those being considered by Ian Byatt in the water industry. What is less clear is the extent to which a practicable access charging system could be developed which avoids the need for individual negotiation and for case by case scrutiny by the Regulator. One obvious difficulty arises from the distinction between access **rights** (in terms of the number of train paths, maximum journey times etc) and timetable **slots**. The Regulator has to approve all access agreements and, in order to maintain flexibility in the use of the network, has generally sought to avoid approving rights in a form which involve specific timetable commitments. It is therefore only when timetable bids are made, as part of the process for developing the annual Summer and Winter timetables, that issues about the competitive impact of a new service on an incumbent or about the congestion costs imposed can be properly determined.

What we may be looking for, therefore, is a charging system which provides better **initial** signals to both operators and to Railtrack, but which still requires some further negotiation and regulatory scrutiny in individual cases. Certainly, the Regulator would support measures which reduced his involvement on a routine basis. The key question is how far the system can be developed to give commercial incentives to the different operators who work in the public interest. Would, for example, the development of secondary trading for access rights help to achieve this?

The Government's 1993 publication 'Gaining Access to the Railway Network' envisaged that, once the initial franchises were awarded, the system of gaining access would develop along the following lines:

- more service provided on an open access basis, with less provided under franchise agreements;
- operators seeking subsidy for individual loss-making services rather than packages of services; and
- Railtrack, rather than the Franchising Director, taking the lead in 'marketing' access rights and identifying packages of rights to be offered to potential operators.

It is too early to say whether developments in the Regulator's policies on charging and competition will help the industry to move in this direction - or indeed to be certain whether such moves would necessarily be in the public interest. But the need for a more commercial charging structure clearly underpins any moves in this direction.

Conclusion

It is not the purpose of this paper to predict either the outcome of the periodic review, or the extent to which competition will develop over the next few years. What is clear is that regulatory developments over the next few years, both in introducing Stage II of Moderation of Competition and in preparing for and carrying out the periodic review, will have a significant impact on the development of the railway sector in the next millennium.

CHAPTER 8

Liberalised access to the railway network

Philip O'Donnell[1]

Introduction

The development of liberalised access to the British railway network offers both challenges and opportunities to the railway industry. In this paper a number of the key issues are considered. The objectives of the paper are:

- to review the objectives behind the 1993 Railways Act in respect of liberalising access to the railway network and to explain why progress with liberalisation has been limited;
- to explain the mechanism set in place by the Rail Regulator for the introduction of limited liberalisation in 1999 and to raise a number of questions about how Railtrack and train operators are expected to respond and;
- to consider some of the policy issues associated with more radical liberalisation of access, including the implications for access charges, investment, franchising of passenger services, network benefits and integrated transport policy.

The views expressed in this paper are the author's own, and do not necessarily reflect Railtrack policy.

Legislative framework and emerging policy

The 1993 Railways Act places a duty jointly on the Secretary of State for Transport and the Rail Regulator 'to promote competition in the provision of railway services.'[2] Railtrack's Network Licence reinforces the competitive thrust of primary legislation:

> 'except in so far as the Regulator may otherwise consent, the licence holder shall not in its licensed activities, or in carrying out any other function contemplated by this licence, unduly discriminate between particular persons or between any classes or descriptions of person.'[3]

The then Government, in its thinking about restructuring the railway industry at the time of the 1992 United Kingdom General Election, placed emphasis on liberalising access to the railway network, so called 'open access'. In the previous year it had supported the European Commission in introducing complementary measures (the Open Access Directive).[4] At that time access to the British railway network was controlled by British Rail, a statutory monopoly, but a number of parties were expressing interest in gaining access to the network in order to operate their own services. One, Stagecoach, went as far as sponsoring its own service in conjunction with British Rail. Richard Branson and his Virgin Group also expressed interest and there were others who wanted to operate their own freight trains. The Government of the day saw open access (third parties operating over the then BR owned rail network) as a means of introducing both private capital and competition into the railway industry.

The Government's early approach to privatisation of freight provides a further insight into its enthusiasm for liberalisation. The BR freight operation was seen as profitable. Its major revenue and profit source was the movement of raw materials, primarily coal, where rail appeared to enjoy a significant cost advantage compared to road. As ever with railway economics, the treatment of common costs impacts on the assessment of 'true' profitability. However, the point here related to the attitude of the Government - it was prepared to break up the single BR freight operation into competing companies for sale, foregoing the sales value of a monopoly to secure the longer run benefits of competition.

Progress over the last five years

The Government's early enthusiasm for greater competition was not sustained. In the case of freight, having gone to the effort of restructuring the BR freight activity into six free-standing businesses, the Government was eventually to change tack. It agreed to sell the three most profitable businesses (the heavy haul divisions), as one entity. And for passenger services, the Government eventually gave the following guidance to the Rail Regulator:

> 'It is the policy of the Government that competition on routes to be franchised should be moderated to the extent necessary to ensure the successful launch of the first generation of franchises.'[5]

It is important to understand the reasons for the shift in the emphasis of policy. The Government, in order to honour its commitment that the privatisation of the railway network would not result in a substantial withdrawal of passenger services, decided that the services would be franchised with specific service obligations for which the Government would pay if necessary. The Government, through the Franchise Director, has an interest in minimising such payments - an interest which is best pursued by letting franchises by competitive tender, where the sales value of the franchisee is enhanced (the payment to the successful franchise bidder reduced) by limiting direct competition.

There is an obvious tension between the value the Franchising Director can extract for a franchise, and the extent to which the franchise is to be subject to competitive market forces. In the jargon of the time, this was seen as a conflict between competition for the market (bidding for a franchise) and competition in the market (between one franchisee and another or an open access operator). The tension is all the more important because at least in the early years the great majority of franchises involve public support in the form of payment. From the perspective of public finance the concern is therefore that liberalised access and competition is being 'purchased' at the expense of the tax payer.

Much the same concerns of value for money for the tax payer, and achieving 'successful' sales drove the decision to allow a single buyer to acquire most of the British Rail freight operation. However, no competition restrictions have been placed on freight access.

The Rail Regulator's policy statement

Following the guidance from the Government and after consultation, the Rail Regulator determined that in his view while substantial benefits to both train operators and passengers were to be realised from an increase in competition, it would be necessary for competition between passenger train operators to be substantially restricted for an initial period. He gave two reasons:

> 'that the franchising process itself creates considerable potential for increased competition even without allowing new entry by so called 'open access' operators. This reflects the geographical overlap between [franchised] train operators arising from the way the franchise map is drawn.'
> 'the rail industry is still in an early stage of a period of fundamental restructuring. ...[It] is difficult to predict with confidence the effects of allowing unrestricted - or uncontrolled - competition on train operators [or] the Franchising Director [for whose financial position the Regulator is required to have regard by the 1993 Railways Act]....I do not believe it would be wise to expose the parties to the risks of such an experiment.'[6]

The Rail Regulator determined that there should be very tight controls over competitive new entry which would apply until 31 March 1999, with further restrictions to remain in force at least for three further years, until 31 March 2002.

In drawing up controls he applied a number of 'specific principles' as his basis for moderating competition:

> (a) 'train operators should expect to face either early exposure to possible competition or the firm prospect of increased exposure to competition...;
> (b) the potential financial impact of competition on franchisees should be limited and predictable...;
> (c) franchisees should have the opportunity for commercial expansion as well as protection from excessive competitive risk;
> (d) entry of non competing services should not be constrained;
> (e) ...the location of competitive entry should be market led rather than administratively determined;
> (f) new owners of the passenger railway should have some say over where they feel protection is needed, and decisions

should, if possible, be taken in the light of experience of private sector behaviour'.[5]

The Regulator was seeking to ensure that there should be a transitional path to liberalisation and that the prospect of competition should remain.

The Regulator looked at both price and non price mechanisms of moderating competition. He concluded that reliance either on access charges or access deficit charges to moderate competition would be unsatisfactory. He had in mind:

- the risk that output decisions might be distorted by creating marginal prices which were far removed from marginal costs;
- such solutions may require a great deal of data not reliably available to the industry and;
- pricing solutions appeared to create substantial uncertainty for the Franchising Director, potential franchisees and Railtrack conflicting with the Regulator's duty to allow parties to plan their business with a reasonable degree of assurance, and possibly hampering the overall objective of privatising the industry.

Controlled competition in contestable markets

The Regulator therefore adopted a regime embodying negotiated charges for open access combined with administered rights of exclusivity. He believed that by determining in advance the degree of exposure to competition which each franchisee could face, the adverse impact on the franchising process would be minimised (because bidders could allow for the risk) while at the same time, the opportunity for competition would be created.

The mechanism is intended to allow the market to determine as far as possible where competitive entry should occur. The protection given is over the markets which operators serve rather than, for example, the routes on which they operate. Markets are defined in terms of station to station flows, e.g. Paddington to Reading. This has been refined to recognise that some stations in close proximity to major termini could effectively serve the same market. New entry is restricted through contractual control over Railtrack's ability to sell access rights to any operator in excess of those contained within their initial access agreements.

The Regulator's intention is to give franchisees effective exclusivity against new entry for an initial period of four years; and to impose significant restrictions on the extent of new entry for a further three years. In both cases, restrictions are subject to pre-existing access rights agreed between Railtrack and operators providing services on the same flows. In some markets there is a significant level of pre-existing inter-operator competition.

The mechanics of moderated competition

Under the Stage 1 regime (applying until end March 1999), operators are entitled to nominate all of their point-to-point flows provided that they satisfy a materiality test. They then enjoy protection from new entrants, although they do face the risk that an operator with pre-existing access rights might intensify competition.

The Stage 2 regime (applying from April 1999) is intended to introduce some competitive new entry. As with Stage 1, the key building block is the station to station flow. Franchisees are expected to re-nominate flows but will have only partial protection from new entrants. Up to a competitive threshold level set by the Regulator, new entry can occur, and there may be competition for established (franchised) operators' business. Stage 2 is designed to limit the amount of an operator's passenger revenue that is open to competition from another operator. The mechanics of the Stage 2 regime are complex and need to be understood in order to appreciate the economic incentives created for both operators and Railtrack.

The franchised operator nominates a set of flows on which he is seeking protection - these are known as contestable flows. Any flows that are not nominated, are known as free flows, and Railtrack can sell unlimited access on those flows. If Railtrack sells access on a contestable flow, it becomes known as a contested flow and counts towards the competitive threshold. Railtrack may not sell access on contestable flows if the total operator revenue earned on those contested flows plus the revenue on flows which are already competitive equals or exceeds a threshold percentage of the total nominated revenue (expected to be 20 per cent) -the competitive threshold- although Railtrack can still sell additional access on flows that are already contested.

Stage 2: An illustrative example

The pie chart shows how the total fare box revenue of an operator might divide between revenue earned on competitive, contested, contestable and free flows. The diagram shows the operator has nominated £100m out of its total revenue of £150m, and assumes that the threshold of contestable revenue available for sale is set by the Regulator is 20 per cent, equal in this example to £20 million. Given that the operator already faces pre-existing competition on flows worth £10m, and that we are assuming that Railtrack has already sold some additional paths over contestable flows worth an additional £5m, this ıeaves £5m of contestable flows on which Railtrack is able to sell paths to new entrants.

Figure 7.1 Division of train operator revenues - a hypothetical example

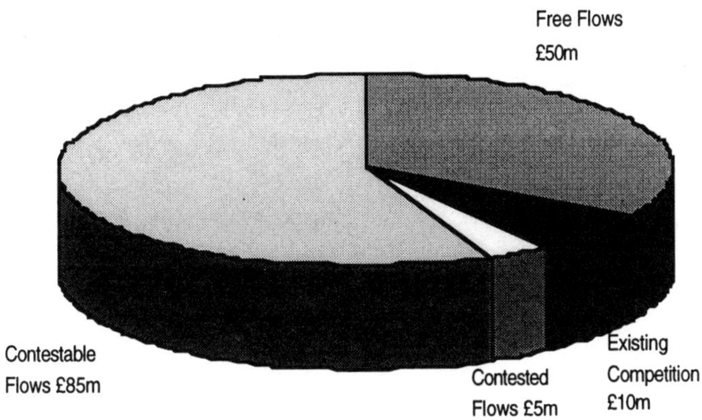

Total revenue £150m
Nominated revenue £100m, Threshold £20m,
Train operator revenue on contestable flows available to be sold £5m.

To illustrate the concepts further a numerical example has been prepared based on a hypothetical railway operation between London and Bristol. Consider the case of the Isis Railway Company. Isis is subject to pre-existing competition between London and Reading and London and Didcot. These flows are therefore contested and count against the 20 per cent threshold.

The table overleaf shows the flows that Isis has nominated. On the Bath to Bristol flow worth £5m, Railtrack has already sold access to a new entrant. Additionally the two flows Paddington to Reading and Paddington to Didcot are subject to pre-existing competition and therefore count towards the threshold. Since only £5m of competitive threshold remains available to be sold, it can be seen that the only remaining opportunity to sell access on contestable flows is on Swindon to Bristol. Clearly, if Railtrack had not sold access on Bath to Bristol, it could have sold access on Paddington to Bath which may have been more commercially attractive.

In practice, the list of nominations would be likely to be hundreds or even thousands of flows long. The purpose of working through this hypothetical example is to better understand both the commercial and wider policy issues.

Table 7.1 Illustrative Nominations - Isis Trains

From	To	£m	Existing competition?
Paddington	Bristol	45	No
Paddington	Swindon	25	No
Paddington	Bath	10	No
Paddington	Reading	8	Yes Contested
Bath	Bristol	5	No Sold Contested
Paddington	Didcot	2	Yes Contested
Swindon	Bristol	5	No
Threshold (20%)		£20m	
Pre-existing competition		£10m	
Sold under Stage 2		£5m	
Balance (available to be sold)		£5m	

Nominating strategies for the incumbent operator

In making nominations, the franchised train operator must bear a number of factors in mind.

Flows on which there is existing competition already count against the competitive threshold. Franchisees should nominate all these flows in order to reduce the number of contestable flows that can be sold and their exposure to new entry. It may be the case that some franchises will already have competition on flows that could account for more than the 20 per cent threshold, in that case Railtrack would not be able to sell any additional access on contestable flows during Stage 2.

Franchisees should be aiming to protect those profitable flows most vulnerable to competition. Franchisees will need to balance the number of economically marginal flows they nominate against the level of competition they are willing to accept on their more valuable flows. A larger number of marginal flows nominated increases the absolute level of the contestable revenue and hence increases the risk of competitive access in terms of total revenue exposed.

Franchisees should not nominate flows if they believe there is little scope for competition either because of the economics of the flow or the availability of spare capacity for new entrants.

In principle franchisees should carry out a business risk analysis examining the trade-off between the security of the inclusion of an incremental flow in its set of nominations and the greater competitive exposure created by enlarging the competitive headroom before the threshold is reached. Such an analysis may call for a very detailed understanding of the economics of the operator's business including the scope for competitive reaction and cost cutting.

There may also be opportunities for game-playing by train operators. For example, two operators owned by the same holding company might buy all the available contestable paths in the other's contestable flows and therefore protect themselves from new entry.

Issues for Railtrack

Stage 2 also raises a number of issues for Railtrack to do with the basis on which it sells track access.

1. **First come, first served?** In selling to an open access operator a path over a contestable flow, Railtrack uses up revenue opportunities on other contestable flows. Does selling on a first come first served basis make either commercial or public policy sense? Should Railtrack be trying pro-actively to identify and market the best opportunities defined in terms of attractiveness to competitors and beneficial impact on its own profitability?
2. **Pricing:** Track access charges for additional open access are negotiated between Railtrack and its customers, and approved by the Rail Regulator. If Railtrack believes that the price offered by train operators for paths (based upon their expected additional profit) does not match the value that Railtrack assigns to the paths due to the opportunity cost of contestable flows foregone, does Railtrack refuse the sale?

The ground rules Railtrack is expected to apply or be required to apply under Stage 2 are as yet unclear but the potential for conflict with customers as a result of the economic incentives is evident. As has proved to be the case with other utilities, the attempt to liberalise access appears to require complex rules and apparently greater regulatory intervention. The impact in terms of new entry, competitions and net customer benefit is uncertain.

Year 2002, fully liberalised access?

Within the commercial agreements which the Regulator has already approved, there is only limited freedom to vary the terms of Moderation of Competition before 2002. However, after that, a move to more radical liberalisation of the access regime is possible. This is likely to be subject to consultation well in advance of 2002 and the industry parties need to take the prospect of increased competition post 2002 into account in developing their medium term commercial strategies. At this juncture, the key issues appear to be the following.

Franchising and the budget constraint

When the Regulator formulated his policy on Moderation of Competition there was major uncertainty about on what terms or indeed whether, franchises could be sold. Given the subsequent success of the franchising process, what is the case for a continuation of moderation of competition beyond 2002?

On the basis of the published franchise bids, the former Inter-City operations are making payments to the Franchising Director by 2002 (i.e. those franchisees are able to pay economic rents for their franchises). What is the policy justification for rents if they are the result of moderation of competition? Should the users of the former Inter-City services be expected to pay rents which competition would be expected to eliminate? What if rents are being used by the Franchising Director to support other rail services? Is cross subsidisation justified at the cost of restricting output in the more profitable markets and possibly increasing road congestion? Does this make sense in the context of an integrated transport policy? However, if such cross-subsidy is disallowed is the tax-payer willing to meet the bill?

Investment in rolling stock

The restructuring of the railway industry was initially accompanied by a famine in new rolling stock orders. However, the new franchisees have committed to procure significant quantities of new vehicles. In some instances the Franchising Director has deemed it necessary to encourage the process by awarding longer franchises, up to fifteen years and with extended moderation of competition to which the Regulator has agreed. To what extent is moderation of competition necessary to secure new rolling stock investment?

The answer depends on train procurers' confidence in the robustness of the second hand or re-leasing market for the rolling stock. Trends towards common multi-route designs of rolling stock should reduce the case for seeking protection or looking for a longer franchise to offset the perceived risk of being unable subsequently to realise the value of the new rolling stock. Some franchisees appear to be willing to take the risk of procuring new rolling stock without obtaining protection. The evidence currently available does not demonstrate that protection from new entry is essential to secure new rolling stock build.

Network benefits

The Franchising Director and the Rail Regulator have been anxious to ensure that the franchising process and the unbundling of the BR monopoly should not be accompanied by a loss of network benefits for passengers. The most important of these are:

- through ticketing;
- inter-availability of tickets;
- an integrated national timetable;
- impartial retailing and;
- national codes of conduct.

These policies sit relatively easily with the high degree of market protection enjoyed by franchisees until the year 2002. Are they appropriate or sustainable if franchisees perceive themselves in head on competition with one another? Will there not be a widening tension between policies which require franchisees to collaborate and the innovative competition in product offers expected in a liberalised passenger rail access market?

Integrated transport policy

The Government is now proposing that an integrated transport policy be developed. This is perfectly compatible with the franchising process. However, if Government wants to:

- synchronise bus and rail services to encourage use of public transport;
- encourage maximum use of rail on congested commuting and long distance corridors;
- retain maximum capability to cross subsidise in pursuit of decreasing car usage and financial objectives;

then it is not clear that liberalised access in the market is the best way forward. There will be pressures to continue with a high degree of protection to facilitate co-ordination.

Moderation of competition and the next track access charges review

In his December 1994 Policy Statement on Moderation of Competition, the Regulator rejected pricing instruments as the means for protecting the Franchising Director's budget from the impact of competition. It is reasonable to anticipate that the linkage between prices and competition will be revisited as part of the next Railtrack charges review.

The current administered charges regime for franchised services results in franchisees (indirectly the Franchise Director) meeting the great majority of the costs of the network via fixed 'capability' charges which do not vary with volume of activity. The Franchising Director may be expected to argue that such charges are inappropriate if:

- he is to face higher subsidy costs because of on track competition, or if;
- there is some diminution in the volume of train services which he underwrites relative to unsupported open access operators.

The possibility of access deficit charges has been canvassed, and there are a number of methods available. While these might succeed in insulating the Franchising Director's budget from the potentially adverse impact of liberalised access, they would be complex to operate.

A simple vehicle or train mileage related charge has previously been canvassed but unless the emerging marginal price equals marginal cost the outcome is likely to distort economic signals and restrict output.

It would also be possible for the Regulator to 'target' Railtrack with an open access income target and reduce the call on the Franchising Director's resources by a corresponding amount. At least in concept this is an extension of the current policy. However, a number of questions would arise. How is the target open access income set and what is the relationship to the Passenger Service Requirement set by the Franchising Director? Would Railtrack be truly free to negotiate rates for open access? What constraints might it face?

Conclusion

Other policy objectives may well determine what progress is made towards greater liberalisation of access to the network. The impact on the cost of franchise payments and on the achievement of an integrated transport policy may be key considerations. Liberalisation also appears to bring with it more complex regulation, rather than less, with commercial opportunities and economic outcomes dependent on the precise 'rules' the Regulator adopts.

Notes

1. I would like to acknowledge the assistance of Stephen Gibson and William Heath in the preparation of this paper.
2. Section 4(1) (d) Railways Act 1993.
3. Condition 6, Railtrack Network Licence 31, March 1994.
4. Directive 91/440/EEC.

5. Competition For Railway Passenger Services, A Consultation Document (July 1994).
6. Competition For Railway Passenger Services, A Policy Statement (December 1994).

CHAPTER 9

Privatisation and deregulation in railways: an assessment of the British approach

C.A. Nash

Introduction

Railways have long been seen as a major problem worldwide, combining declining market share with increasing financial difficulties. In the past few years, interest in privatisation as a solution to these problems has grown rapidly. Privatisation has been seen as promoting efficiency and innovation, by freeing the railway from government control and by removing the prop of subsidies. At the same time governments have been keen to relieve their budgets by transferring the financing of rail investment to the private sector.

The next section considers some of the alternative approaches to privatisation adopted around the world, before focusing on the British approach and explaining its rationale. We then consider some specific problems with that approach and the way in which they have been tackled. Finally we point the way to a consideration of the success of the approach, and identify the key difficulties involved.

Alternative models of privatisation

New Zealand led the way amongst recent rail privatisations, and certainly to an outsider the transformation of New Zealand Railways into a commercial organisation appears a remarkable success. However, New Zealand Railways, as a freight dominated organisation, was able to follow the North American model of a privately owned freight company owning its own infrastructure and disciplined by competition from road haulage, with a modest passenger service much of it run under contract to the government. For freight traffic, it may be considered that the strength of competition from road haulage removes any 'natural monopoly' problems, whilst in the case of passenger services, the minor role played by rail in the passenger market means that if the railway is not efficient, the government will simply cease to support it. Thus providing for competition within the rail sector has not been a major issue. A similar solution was found in Argentina, and later extended to other South American countries, although on the basis of offering substantial regional railway companies on long (30 year) franchises.

The situation in Japan is rather more like that in Europe, in that railways are dominated by heavy passenger traffic, although the traffic density and market share make this much more a commercial operation than many European rail passenger services. Japanese National Railways were divided into six regional companies integrating passenger services with infrastructure. Debt and excess staff were left in the hands of a separate 'resettlement corporation'. Whilst other companies still receive income from a government established fund designed to offset their losses, those companies operating in the more densely populated areas are now profitable. Freight operates as a separate company over the infrastructure of the passenger companies. Although it is often said that the railways were privatised in 1987, the process is better described as commercialisation. Sale of shares to the private sector has only taken place more recently and only for the more profitable companies.

Again in Japan, achieving competition within the rail sector was not seen as an issue, and Japanese railway management is adamantly opposed to any separation of infrastructure from operations. Presumably the financial arrangements and remaining regulation of fares and services are thought to guard adequately against any risk that the new railway companies might prove to be inefficient monopolies, and their performance to date has been impressive.

European railways offer bigger problems then any of these when it comes to privatisation. Not only is passenger traffic important, but also much of it is very unprofitable. Rightly or wrongly, governments are unwilling to see the sort of rationalisation that would be needed to make it profitable; on the contrary high quality urban and inter urban rail services are seen as an essential element in policy to cope with congestion and environmental impact on roads. This leads governments to wish to have considerable influence over the passenger services provided and to be reluctant to see rail passenger services left to private sector monopolies. Consequently a unique European model of rail privatisation has emerged. (CEC, 1996).

What is actually happening varies in detail from country to country, but it generally combines elements of:

a) Separation of infrastructure from operations: The intention here is to permit competing operators to run trains over the same infrastructure. It is argued that if one of them had control of the infrastructure this would put them in an unduly favourable position. In most cases, infrastructure has so far simply been placed in a separate division of the national rail company; only in Sweden and Britain is it with a totally separate organisation.

b) Franchising out of some or all passenger operations: If the government wishes to continue to control the level of passenger service operated and as a consequence to provide subsidy, then it is argued that competitive franchising offers the best way of using the forces of competition to ensure that operators are efficient. However, whilst franchising has been adopted for some local and regional services in Sweden and Germany, only Britain so far has adopted this for all passenger services, including main lines.

c) A degree of open access for other operators to operate competing services over the same infrastructure: Where possible it is thought desirable to encourage direct 'on the rails' competition. This is probably most readily achieved in the freight sector, where there are generally no explicit subsidies to particular rail services. There is open access for freight in a number of countries, including Britain, Germany and the Netherlands, and for certain international operations it is required by the EU, but entry so far has been very limited. In Britain, as in most countries, open access for passenger services is currently not permitted. (The only open access passenger operator of which I am aware is `Lovers Rail' in the Netherlands.)

However, Britain offers a number of features which are not currently possessed by any other European railway:

a) Privatisation of the infrastructure: Generally the approach of other European countries has been to retain the infrastructure in public hands, with investment planned on social cost-benefit grounds and pricing on short run marginal cost. Germany is an exception in having near-commercial charges with high kilometre based levels, but since the infrastructure is still owned and operated by a division of the national railway (DBAG) this merely servcs to deter entry. Separation of the infrastructure into a private company as in Britain may introduce private sector disciplines and innovation, as well as opening up access to the capital market and freeing infrastructure investment from the direct control of the public spending round. But it also leads to infrastructure being a private sector monopoly whose prices and arguably investment need tight regulation.

b) Extensive subcontracting: Although some subcontracting of activities such as rolling stock and infrastructure maintenance is common, and rolling stock leasing deals are not unknown in other countries, only Britain has adopted a comprehensive policy of subcontracting out the supply and maintenance of rolling stock and the renewal and maintenance of infrastructure to the private sector. This should increase competitive pressures but also adds to the complexity of the system. It certainly appears to be reducing costs in some areas, but there are also concerns that ownership of the current rolling stock, some of which is fairly specialised, by such a small number of leasing companies without any regulation may leave some monopoly power in their hands. There is currently no regulation of this sector.

c) An independent rail regulator: Although in some countries a track access authority with a degree of autonomy has been set up, no other country has an independent rail regulator as in Britain. The effect is for instance that in Germany DBAG basically sets the charges for competitors to use its infrastructure with no control other than through the political process.

Some difficulties in the implementation of the British system

The logic of the British approach is clear - introduce competition for the contract to provide services or inputs to the production process wherever possible and regulate the remaining natural monopoly element. i.e. the infrastructure. Nevertheless there are certainly some practical problems in its implementation. Perhaps the most serious are in determining a track access pricing regime, in ensuring a reasonable degree of integration of the 25 passenger franchisees without unduly constraining their commercial freedom, in ensuring that socially worthwhile investment takes place, and avoiding undue transactions costs. We shall discuss each of these issues in turn.

Infrastructure charges and quality of service

The recent EC Green Paper Towards Fair and Efficient Pricing (CEC, 1995) essentially identified four criteria which infrastructure charges should meet:

- they should be fair, and in particular non-discriminatory;
- they should promote efficiency both in the use of existing infrastructure and in promoting worthwhile investment;
- they should be transparent, so as to be seen to be fair;
- they should cover the total costs of the infrastructure in question.

Meeting all of these criteria is exceedingly difficult in practice. The system implemented in Britain for passenger franchises relies on a two part tariff, which involves a marginal cost based solely on wear and tear and where appropriate, electric traction costs, and a large fixed element based an avoidable costs and an allocation of joint costs (ORR, 1995). Given the arbitrariness of the allocation, the system does not necessarily provide good information on the relative profitability of different services. Moreover, the system has been criticised for the very low variable element in the charges. It is understood that the variable charges include no element either to allow for accelerated renewals as a result of additional use or for externalities such as air pollution. From the point of view of efficiency, the system may also reasonably be criticised as having no mechanism to ensure efficient use of scarce capacity, although any form of peak pricing is likely to be very difficult given the complicated way in which services interact and competition for space occurs at a multitude of points in a complicated network. Adjustments in capacity or quality may be made via negotiation between Railtrack and the operators; some of the difficulties that this involves will be outlined when we consider investment below.

The biggest problem with the track access pricing system is however in adjusting it to the possibility of open access competition. If open access competitors were allowed simply to pay what they could bear subject to the floor of the running cost element then they would have an advantage which may allow them to win traffic from existing operators even if they were less efficient. The theoretical solution is well known; they should pay an amount equal to the loss of surplus on the traffic they win from the existing operator (Baumol, 1983). How to implement this is, however, a big problem; obviously the incumbent will have an incentive to exaggerate the revenue at stake. A similar difficulty in ensuring fair and efficient charges exists in the case of freight operators, which are the subject of individual negotiations but again subject to the approval of the Regulator.

A major concern early in the process of privatisation was with the quality of service the monopoly infrastructure operator would provide. In practice, track access agreements include substantial penalty payments for failure to perform, and average train punctuality has actually improved substantially in the past year (OPRAF, 1997). However, train operators continue to complain that they do not find Railtrack as flexible and responsive as they would wish; concerns on the issue of investment are discussed below.

Integration versus competition

There is a clear trade-off here between having an integrated network with comprehensive information and tickets available on all services and allowing individual operators commercial freedom to decide what to provide. In Britain, the position is that the Franchise Director has laid down minimum requirements for levels of service, which are typically some 70-90 per cent of what previously existed in terms of frequency, and which cover other issues such as journey times and times of first and last trains; many franchises have committed themselves to more than this. Open access passenger operations will be strictly limited, at least for the first few years, and will not be able to attack the Franchisee's core business as they will only be allowed to enter on routes not currently served by through trains or which only account for a small proportion of the franchisee's revenue. Selected fares, including both season ticket and standard `leisure travel' fares are controlled. As part of their licence conditions the Regulator requires all passenger operators to provide through and interchangeable tickets and comprehensive information. All this may sound like a very heavily controlled situation, and certainly some operators have expressed concern at the lack of commercial freedom it offers.

Nevertheless, consumer organisations do see cause for concern. The more decentralised planning of the timetable may lead to a further deterioration of connections between services and a poorer spacing of trains on routes served by more than one operator. Already there has been a reduction in the willingness of one operator to hold connections with the trains of other operators and there have been criticisms of the accuracy and comprehensiveness of information. The actual operation of stations is generally leased to the major operator at that station which therefore becomes responsible for information and bookings on all services including those of its rivals. To some extent these developments are a continuation of trends that had already emerged with increased decentralisation of decision taking to sectors and profit centres under the previous British Rail organisation, but there is a fear that there may be some repeat of experience in the British bus industry, where deregulation and privatisation certainly achieved a dramatic reduction of costs and an increase in bus miles run but failed to prevent a further substantial loss of passengers. A deterioration in the integration of services and in information on them is widely blamed for this loss of traffic (Mackie, Preston and Nash, 1995).

Investment planning

In the new structure, it is expected that most investment will be carried out either by Railtrack (in the case of infrastructure) or, in the case of passenger rolling stock, by leasing companies (ROSCOs) - which might include manufacturers. Train operating companies were obliged to sign contracts to lease their existing rolling stock, mostly for five to seven years, but it appeared that either much longer contracts or guaranteed continued use of the rolling stock will be needed to encourage investment. In the short run, rolling stock investment virtually ceased. The key to overcoming this problem was the willingness on the part of OPRAF to grant longer franchises of up to 15 years where rolling stock investment was promised as part of the franchise agreement, and indeed in some cases to require this. The result is that OPRAF has a major role in decisions on rolling stock investment, although there are some cases now of ROSCOs or operators being willing to take the risk of ordering new rolling stock without any guarantee of their deployment beyond the end of the current franchise.

Regarding infrastructure renewals, the incentive for Railtrack to do this comes mainly via the performance regime, in which Railtrack will have to pay penalties for deteriorating performance if it allows the infrastructure to deteriorate. Whether this long term effect will provide an adequate incentive to renew assets may be of some doubt. The Regulator is now seeking other ways of influencing Railtrack's investment programme in order to ensure that renewals take place (for instance via his approval of the network management statement being extended to incorporate a commitment actually to undertake the investment), (ORR, 1997a).

In the case of infrastructure enhancement, one difficulty is the number of organisations that need to be involved in any decision. Some of the problems to be faced are illustrated by the West Coast Main Line. This route required heavy investment just to renew life-expired assets; in practice the project involves upgrading for higher speeds and cab signalling. This will obviously affect the many operating companies who use parts of this route not just in terms of their revenues and costs, but also by requiring investment in rolling stock on their part. The existing approach whereby the net benefit of any improvement is to be shared between the parties involved on a negotiated basis sounds reasonable but has some perverse incentives. For instance Railtrack, which is in the best position to know the costs of any improvement has an incentive to exaggerate them in order to raise its share of the revenue, whilst the train operators, who should know the final market best, have an incentive to understate the benefits in order to minimise their charges (especially where a number of operators are involved so that free-riding may be an attractive strategy). For the Regulator to determine the reasonableness of charges in this situation is extremely difficult.

Investment to achieve higher standards of service would alter access charges not just for the duration of the current franchise but possibly for up to thirty years into the future and may thus require the agreement of the Franchise Director to underwrite them beyond the life of the initial franchises, and of all the operators concerned. Again it is clear that OPRAF has a key role to play. Perhaps the biggest difficulty may actually arise with smaller schemes coming forward during the franchise period. It may well be that none of the operator, Railtrack or OPRAF will be willing to pay for them. There is also an issue concerning the interests of freight operators and customers in such negotiations. Will Railtrack, or the major freight operator, bear the risk of providing for enhanced freight capacity over such a long period?

Transaction costs

The degree of fragmentation of the former British Rail network is unparalleled anywhere in the world. Some 25 passenger train operating companies have been franchised separately. Each has contracts with the Franchise Director, Railtrack and fellow train operating companies (for instance regarding through ticketing and facilities provided at stations leased by one train operating company but served by another). They lease rolling stock from (initially) three rolling stock leasing companies, who secure their maintenance from specialist maintenance companies. Railtrack subcontracts maintenance to infrastructure renewal and maintenance companies. The logic of this approach is that competition will be introduced in all those functions such as rolling stock provision and maintenance and infrastructure renewals and maintenance in which it is feasible, even though it is accepted that infrastructure and (initially at least) provision of a particular passenger service must remain a monopoly. But it does require an intricate web of contractual arrangements in which specifying and monitoring responsibility for quality of service will not be easy. In other Western European countries such as Germany and Sweden, it is proposed to maintain a single major operator of main line passenger services, integrated except for provision of infrastructure, although local authorities will be free to contract out local services to other operators. Of course it is likely that a degree of reintegration will occur in Britain - it appears likely that companies will seek to hold a portfolio of franchises covering a variety of sets of services and of varied franchise length (some already do). Partly this would be to reduce the risk that a single failure at refranchising could put them out of business as well as to exploit potential economies of scale. Moreover, the purchase of one of the rolling stock leasing companies by Stagecoach represents a first move back towards vertical integration. Operation of bus and local rail services in an area by a single operator also offers a new form of integration which might offer considerable benefits in terms of integration of bus and rail services. Although it also raises concerns for competition policy, as long as OPRAF retains reasonable control over rail services and fares, and the bus and coach market is reasonably contestable, these concerns may not be too great. (Monopolies and Mergers Commission, 1996).

The main way of seeking to minimise transaction costs that has been implemented appears to be by incorporating user committees and appeals procedures into controversial decisions (such as allocation of slots). The costs of recourse to legal action are certainly to be avoided if possible.

Judging the success of British privatisation

In any privatisation there are likely to be three parties affected - the taxpayer, the consumer and the producers (including the workforce). To judge the overall success of the privatisation, we need to consider its effects on all three.

The franchising process started with Great Western and South West Trains, which started operation in February 1996 and was completed early in 1997. It is thought that the franchise was almost always awarded to the lowest credible bid; the successful franchisees and their bids for subsidy in 2002/3 are shown in Table 9.1. Whilst the first two bids promised relatively low rates of reduction of subsidy, bids have become progressively more optimistic, culminating in the biggest and most complex franchise, Inter City West Coast, which was won by Virgin, promising to turn a £77m subsidy in the first year into a £220m premium

Table 9.1 Rail franchises

Franchise	Owner	Length of Franchise (yrs)	Subsidy (£m Feb 1997 prices)	
			1996/7	2002/3
Great Western	MBO/Firstbus	10	61.9	36.9
South West Trains	Stagecoach	7	63.3	35.7
Great North Eastern	Sea Containers	7	67.3	.1
Midland Main Line	National Express Group	10	17.6	-4.4
Gatwick Express	National Express Group	15	-4.1	-12.0
LTS Rail	Prism	15	31.1	19.3
Connex South Central	Connex	7	92.8	35.9
Chiltern Railways	MBO/Laing	7	17.4	3.3
Connex South Eastern	Connex	15	136.1	32.6
South Wales & West	Prism	7	84.6	44.0
Cardiff Railways	Prism	7	22.5	14.3
Thames Trains	MBO/Go Ahead	7	43.7	3.8
Island Line	Stagecoach	5	2.3	1.0*
North Western	Great Western Holdings	10	192.9	129.7
Regional Railways North East	MTL Trust	7	231.1	150.6

North London Railways	National Express Group	7	55.0	20.0
Thameslink	Go Ahead/Via	7 yrs 1 mth	18.5	-27.0
West Coast Trains	Virgin	15	94.4	-3.9
Scotrail	National Express Group	7	297.1	209.3
Central Trains	National Express Group	7	204.4	136.6
Cross Country	Virgin	15	130.0	50.5
Anglia	GB Railways	7 yrs 3 mths	41.0	6.3
Great Eastern	First Bus	7 yrs 3 mths	29.0	-9.5
West Anglia Great Northern	Prism	7 yrs 3 mths	72.6	-14.6
Merseyrail Electrics	MTL Trust	7	87.6	61.8

Source: OPRAF Annual Report 1996-7.

Note:
Negative Subsidies indicate payment of a premium.
MBO stands for Management Buy Out.
* assumes constant subsidy after year 5.

payment to OPRAF in the last year of a 15 year franchise. Whilst the more ambitious bids clearly rely heavily on generating substantial increases in passenger revenue (in this case as a result of introducing faster services operated by a new fleet of tilting trains), it appears that also substantial cost reductions are anticipated and a start has already been made. For instance, several operators report a reduction in the number of drivers of the order of 30 per cent, resulting from measures such as greater flexibility in shift length and an ending of the requirement that trains travelling at more than 110 m.p.h. have two drivers in the cab.

Although a relatively small number of organisations were involved in bidding, the bidding appears to have been very competitive, with several serious bids for all but some of the earliest franchises. The British bus industry dominates the scene (being involved in 16 of the 25 successful bids) with a small number of successful management buyouts, a French conglomerate, Virgin, Sea Containers and a consultancy led company the other players. The dominance of the bus industry has raised concerns about lack of competition where the franchisee is also the major bus operator in the district. One case, the takeover of the Midland Main Line by National Express, which also operates almost all the express coach services from the area in question, has been referred to the Monopolies and Mergers Commission, but National Express was permitted to retain both sets of services on giving undertakings that trends in future price and frequency of express coach services on the routes in question will be no less favourable than on its network as a whole (MMC, 1996).

Table 9.2 Estimated subsidy (£m)

	1996/7	**2002/3**	**7 yr subsidy reduction**
Total (25 franchisees)	2090.1	920.3	1169.8

Source: Table 1.

Assuming the proposed rates of decline of subsidy are achieved, over the first seven years the annual demands on the Exchequer should be reduced by some £1,000m (Table 9.2). However, it is worth remembering that the new basis of charging for the use of infrastructure and rolling stock described above led to the subsidy bill rising from £1.1bn in 1993/4 to £2bn in 1994/5 (Table 9.3). It will thus be several years before subsidies return to the level they were at before the process started in 1993/4. In addition, there have been major transition costs, and the operating costs of OPRAF and ORR must be taken into account. On the other hand, the taxpayer has benefited from the proceeds of the sale of Railtrack, the ROSCOs and the other constituent parts of BR (maybe some £4.3bn, but the costs of the privatisation process of at least £0.25bn must be deducted from that - see Modern Railways Informed Sources, January 1997). The payments should provide for a higher level of investment than has been the case in the past. Under the old regime, support payments only paid for depreciation at historic cost, plus a special replacement allowance which failed to bridge the difference between this and replacement cost. The new subsidy payments include replacement cost depreciation and any enhancements that were committed to under the franchise agreement). It thus appears that, unless a high rate of subsidy reduction could have been achieved by British Rail without privatisation, the net outcome should not be the sort of big increase in costs that was initially feared, and will ultimately be beneficial for the taxpayer, although not nearly as much so as implied by a simple examination of the trends in support of the franchise agreements. This also presupposes that these reductions in support are actually achieved. Some commentators (e.g. Harris and Godward, 1997) include writing off of debt as part of the cost of privatisation, but we are only interested in the net effect on future cash flows, and it is unlikely that interest on debt would have exceeded future borrowings.

Table 9.3 Grants to British Railways Board (£m)

	1993/4	**1994/5**
Public Source Obligation	930	1645
PTE's	105	259
Level crossings	32	35
Pensions	54	45
Total	1121	1984

Source: Transport Statistics Great Britain 1996. Table 1.17.

Whilst there was a virtual halt to new projects, particularly rolling stock replacement, whilst the privatisation process was underway, many of the franchise agreements do provide for substantial investment and the new regime may make this less susceptible to short term Treasury cuts. These investments include substantial amounts of new rolling stock on the London Tilbury and Southend Line, South East Trains and Cross Country, and tilting trains for the West Coast Main Line, in conjunction with substantial renewal and upgrading by Railtrack. Elsewhere, innovative service patterns and higher frequencies have been offered, including the provision of a semi-fast service on the Midland Main line which will virtually double the number of train miles run. It should be remembered that British Rail itself had a record of introducing innovatory new and improved services particularly in the late 1980s so it should not be assumed that none of these innovations would have happened without privatisation. Moreover there are some developments which disadvantage passengers, such as more restrictions on the availability of fares by alternative routes, fewer cases of holding of connections (of course this actually benefits some passengers) and problems with the provision of passenger information. Fares may also become more differentiated with rises in some basic fares countered by more advanced purchase or other bargains. Overall, however, it seems unlikely that passengers will be disadvantaged by the changes if franchisees fulfil the conditions of their franchise agreements.

In addition to the three main business sectors described above, the privatisation of many other companies formed from parts of British Rail has taken place. Foremost amongst these are the infrastructure maintenance and renewal companies and the rolling stock heavy maintenance companies. These were sold to a mixture of existing engineering firms and management buyouts. Amongst the other companies privatised are included BR Business Systems, (responsible for computer and ticket issuing systems), BR Research, Rail Operational Research, engineering design offices, marketing organisations and many others.

It appears from the above description that the privatisation process has been completed remarkably smoothly, in an extraordinarily short period of time. In part this has been the result of a pragmatic approach to actual implementation which has seen many departures from original intentions - for instance, OPRAF has been willing to award longer franchises in return for promises of investment and open access has been limited, at least until 2001. There are certainly areas which remain of concern. For instance, surveys undertaken by the Consumers' Association have found that the quality and impartiality of information on fares and services provided by one operator about another has been poor. Whilst many of their examples are extreme cases where cheaper fares available on very limited and unattractive services have not been mentioned, some are not, and the Regulator clearly perceives there to be a problem (ORR, 1997b). Another concern surrounds the inability of some of the franchisees - and in particular South West Trains - to fulfil its Passenger Service Requirement regarding levels of service following a too rapid reduction in the number of drivers. An emergency timetable, cancelling many services, was introduced. Of course, Stagecoach will pay penalties to OPRAF for failing to fulfil the terms of its franchising agreement, but this early example of a new operator appearing to place cost cutting above its duty to provide services has renewed fears that service levels may suffer as a result of privatisation.

Looking ahead, there remain potential problems. One surrounds the intention of the Regulator progressively to move towards open access for passenger operators (with the exception of Inter City West Coast services, where in return for the high level of investment required, protection from competition will continue throughout the 15 year franchise). Other work we have undertaken suggests that, whilst head-on competition will tend to be unprofitable for the entrant, cream skimming entry with a few key trains may be profitable, and reduce the profits of the incumbent even if they are successful in retaliation. Scope for this may be limited by lack of track capacity unless incumbents are forced to surrender paths, however, as obviously cream skimmers would be looking for peak hour paths into the main termini (this is a likely reason for the intensification of services promised by many existing franchisees). The most likely entrants of this type would be neighbouring franchisees.

Table 9.4 BR performance 1979 - 1992/3 (1991/92 prices)

	1979	1983	1989 /90	1991 /92	1992 /93
Total Grant (£m)	1237	1430	705	1035	1243
Passenger routemiles	8955	8932	8897	8880	8896
Passenger miles (m)	19000	18350	20908	19920	19709
Fare per passenger miles (p)	9.14	9.69	10.81	10.51	10.43
Passenger stations	2365	2363	2483	2473	2482
Passenger train miles (m)	196	203	225	231	228
Train miles per member of staff	1421	1686	2043	1996	1975

Source: British Railways Board, Annual Reports and Accounts

Note: Total grant includes all payments from central and local government to BR, including capital renewal provision, transport police, level crossings and exceptional items. Number of passenger stations affected by transfers to Tyne and Wear Metro (25) and Manchester Metrolink (16).

Several franchisees are committed to a seven year subsidy reduction which is more than 50 per cent of current turnover. For some, this will rest heavily on cost reductions but others are projecting big increases in revenue. This raises the more general issue of what will happen if the franchisees are unable to secure the ambitious targets in terms of revenue increases on the basis of which many of them have made their bids. Apart from increased competition, the most likely cause of this is a downturn in the economy. In this situation, they have the right to reopen negotiations with OPRAF on the terms of their franchise, and if the problem is genuinely due to circumstances outside their own control, it appears likely that OPRAF will agree to some combination of cuts in services and increased subsidy. Should a franchisee become insolvent, then OPRAF would also obviously have to secure a replacement operator, and again the cost of this might be increased subsidies, poorer services or both. The cost and difficulty of this might well incline OPRAF to renegotiate subsidies and service levels rather than face this situation. Many sources in the industry believe that bidders assumed this to be the case when they made such favourable bids. If they are only able to secure this performance in the face of favourable economic circumstances, then the achievement of British Rail in the favourable economic climate of the second half of the 1980s, where it halved subsidies whilst expanding traffic, should also not be forgotten (Table 9.4). It is quite possible that a major reduction in subsidy with improved services could have been achieved without privatisation, given the sort of combination of a favourable economy, clear objectives and management restructuring which existed in that period.

Conclusions

Privatisation of the British rail network has taken place at a speed which seemed inconceivable only a couple of years ago, and the new system is working more smoothly than many commentators feared. In part this is because the degree of control by the twin public sector authorities of the Regulator and the Franchise Director greatly exceeds what was originally anticipated. At the start of the process, what was envisaged was a lot closer to what had already been undertaken in the bus sector. There would be open access for commercial operators to provide what services they wished, and to charge what prices they wished. Franchising would be designed to provide services which were seen as socially necessary. Cooperation between operators on matters such as through or interchangeable ticketing and the provision of information would be voluntary. ORR has played a major role in integrating services by requiring all operators to join through ticketing and information arrangements as part of their licence conditions, whilst OPRAF for instance is taking a major role in investment decisions. Whether this is the best solution to the planning and integration of the rail network is however doubtful. For instance, while OPRAF is indirectly responsible for financing rail infrastructure, it is the Regulator who has the power to seek to force Railtrack to spend the money it receives for investment in an appropriate fashion. Long term investment in enhancing the system generally involves both bodies; OPRAF underwriting access charges and the Regulator approving them. It is certainly arguable that it would be more effective for OPRAF (or a successor National Rail Authority) to contract directly with Railtrack for the level and standard of infrastructure it requires than to do this indirectly via the franchisee; if it were to take on such a role it would seem important for the National Rail Authority to have freight as well as passenger interests. OPRAF is developing appraisal criteria based on cost benefit analysis to assist in considering both appropriate service levels and investment (OPRAF, 1996).

Regarding the degree of success of the privatisation, not only is it too soon to judge, but it will always be very difficult to do so. In part this is because of the problem of forecasting the counter-factual: in the absence of privatisation would BR have been able to sustain its achievements of the 80s in improving services and cutting subsidies, or would the situation actually have deteriorated without privatisation (as it did in the early 1990s)? If franchisees achieve what they are committed to, then it is likely that in the long run both the taxpayer and the passenger will benefit, but many expect a number of franchisees either to have to renegotiate their franchises or to go out of business. Finally for those in the industry it has often been a painful experience, but so far there is little sign of the cutting of wages that for instance followed bus deregulation; cost cutting seems to be being pursued much more by changing conditions and working practices to improve productivity.

Overall then it is to soon to attempt to draw any overall conclusions on the success of the process, but events so far have largely failed to confirm the fears of many commentators on the process. This is largely because much firmer public sector control of the rail system has been maintained then was originally envisaged, and it appears likely that the trend will be to intensify that control rather than towards liberalisation. What is unclear, however, is what will happen at the next round of franchising. If franchisees achieve their aims, when the franchises come up for renewal all the inter city companies and many of the London commuter ones will be profitable. There may therefore be arguments for the implementation of open access, and for the removal of much of the existing regulation, if the companies concerned are no longer in receipt of public funds. My own view is that this would be a mistake. It is inevitable that most rail services will be a virtual monopoly with competition the exception rather than the rule, and tight regulation in the public interest will continue to be needed.

References

W. J. Baumol (1983) 'Some Subtle Issues in Railroad Deregulation,' *International Journal of Transport Economics* 10, Nos. 1-2.

Commission of the European Communities (1996) White Paper, *A Strategy for Revitalising the Community's Railway,*. COM (96)421 Final, Brussels.

Commission of the European Communities (1995) *Towards Fair and Efficient Pricing in Transport*, Brussels.

N. Harris and E. Godward (1997) *The Privatisation of British Rail,* Railway Consultancy Press.

P. J. Mackie, J. M. Preston, and C. A. Nash (1995) 'Bus Deregulation: Ten Years On', *Transport Reviews* 15, No. 3.

Monopolies and Mergers Commission (1996) National Express Group PLC/Midland Main Line Ltd. Report on the Merger Situation, HMSO, London.

Office of Passenger Rail Franchising (1996) Appraisal of Support for Passenger Rail Services. A Consultation Paper, London.

Office of Passenger Rail Franchising (1997) *Annual Report 1996-7,* London.

Office of the Rail Regulator (1995) Railtrack's Charges for Passenger Rail Services: The Future Level of Charges. A Policy Statement, London.

Office of the Rail Regulator (1997a) Railtrack's Investment Programme. Statement by the Rail Regulator, London.

Office of the Rail Regulator (1997b) 'Accurate and Impartial Retailing. Meeting the Needs of Passengers', London.

CHAPTER 10

Regulation and deregulation in a subsidised industry: Passenger rail franchising

Bob Stannard[1]

Introduction

This paper reviews some of the key issues which arise in designing a Passenger Rail Franchise and describes how these issues have been addressed in Great Britain. The paper covers the design of franchises and some of the trade offs which need to be made between the objectives of franchising and other aims, for example, promoting competition in the provision of passenger rail services. It also describes some of the incentives incorporated into franchises to address significant market failures. It does not deal with enforcement, which is an integral part of successful delivery of rail services under a franchised railway.

The key elements of a franchise are:

(a) service specification;
(b) risk allocation;
(c) price controls and competition;
(d) franchise term;
(e) bidding for subsidy.

The paper addresses each of these in turn.

Service specification

Under the OPRAF franchise agreement, a franchisee undertakes to secure the provision of specified passenger rail services. These must include at least those services in the Passenger Service Requirement (PSR). The specification includes, for some franchises or service groups, load factor standards or minimum capacity requirements. These may require services above the PSR to be run. The PSR is not a set timetable and leaves flexibility for franchise operators to decide how the services should be delivered. It therefore permits fine tuning in response to cost and revenue considerations.

The service specification forms the basis of the franchise competition and a benchmark for appraising bids. From the perspective of bidders, there is a premium on reducing uncertainty. An overly rigid service specification may increase this by restricting their opportunity to respond to changing commercial circumstances. Passengers, of course, wish to see their service protected. But public officials are not well placed to assess consumers' requirements, and therefore at deriving a satisfactory specification. OPRAF has approached this issue by:

a) setting a PSR which leaves discretion with the operator to vary the current pattern of services in response to commercial requirements;
b) constraining train operators' discretion to reduce services, particularly in cases where the fare box provides a poor incentive to retain them to meet customer requirements.

The cost structure of TOCs is important in influencing this approach. First, TOCs are constrained by charges fixed in lease agreements for rolling stock and for access to infrastructure, over a period of several years. This means that there are strong incentives to run services additional to the PSR which cover variable costs and contribute to these fixed costs. Second, where TOCs have monopoly power and peaks of commuter demand, load factor regulation requires them to respond to changes in this demand by altering capacity - although the nature of the response is left open to the TOC. Some TOCs earn little revenue from passengers on socially necessary services. The PSR is likely to cover a greater proportion of the services in the present timetable in these circumstances, compared with more commercial routes. OPRAF has also consulted on proposals to undertake appraisals of proposed changes in service specifications. These provide a framework for allocating subsidy between loss-making services, taking account of a wide range of potential outputs.

The incentives to maintain standards which are efficient, in terms of cost, are encouraged through performance linked changes to the level of support a TOC receives from OPRAF. This mirrors the Railtrack Performance Regime in those markets where passenger revenue provides an inadequate incentive to providers, either because of monopoly power or because of low fare box income. It penalises or rewards TOCs for changes in performance - punctuality, capacity shortfall and cancellations - against pre-set benchmarks. The financial effect of this approach is reflected in fares for those London commuter markets where the tariff basket method of fares regulation is applied. The benefit of improved performance is reflected in higher fares and a shortfall in performance will result in lower fares than if performance were unchanged.

The difficulty of defining a complete contract for services is a further reason for leaving some discretion with the TOC. It also means it is desirable to permit changes to the PSR during the franchise term, which could be up to 15 years. The information to inform such changes will generally be held by the TOC and will be expensive for OPRAF to collect. OPRAF therefore needs to provide incentives for TOCs to bring forward proposed changes which are value for money, by allowing them to share in any financial benefit. However, it would not be equitable to share the benefits with the operator in cases where OPRAF itself proposes a change. This mechanism preserves some incentive to innovate, even for services covered by the PSR and supplements the commercial incentives which will affect the planning of other services.

Risk allocation

Risk allocation will influence the cost franchisees perceive and be factored into their franchise bids. Some commercial risks are associated with the new industry structure. TOCs lack a record of performing within it. OPRAF, therefore, made large volumes of business information available to bidders and provided analysis of the operation of the new contractual arrangements. This helped to reduce uncertainty.

The capacity of a bidder to absorb risks was a factor at the bid pre-qualification stage. In practice, the uncertainty implicit in the franchise proposition led to a number of risks being retained by OPRAF. These were cases where TOCs had no control over outcomes and very little opportunity to substitute inputs in response to changes in regulated costs; or to reflect, for example, future changes in the regulatory regime affecting rail competition.

OPRAF has retained general inflation risk and provided for some 'cost pass through' arrangements in the Franchise Agreement.

Price controls and competition

The case for regulation of fares depends on the absence of a threat of competition in some markets. There are markets where rail has some monopoly power, at least in the short run. In other markets, substitution by consumers is a real possibility and it would be onerous and inefficient to impose regulation in these circumstances. In addition, the structure of fares inherited by the franchisees contained anomalies. The tariff basket approach to fares regulation in the major regulated markets avoids ossifying for the life of a franchise what might be an obsolete and inefficient fares structure. It also reassures passengers about the magnitude of changes they might face by capping the annual increase in individual fares within the tariff basket. Elsewhere, regulation was extended to cover important commuting and leisure fares, partly to reassure passengers about the consequences of the new regime.

Fares regulation sets the maximum increase in the average level of fares for fares baskets and the maximum annual increase for some other fares, notably 'savers' and weekly season tickets. The regulation of fare levels represents a break with the recent past. The reductions in cost implied by the regulatory regime for Railtrack and cost changes in ROSCO leasing charges are reflected in the future level of regulated fares which will decline in real terms. This raises an interesting issue for public policy. If marginal costs decline, should prices decline too? This would seem efficient but many passenger services are heavily supported by the taxpayer. This is costly in financing terms and may result in higher marginal tax rates than otherwise required. Fares set below long run marginal costs also pose a dilemma when new capacity, or replacement, is required. An increase in capacity will increase support. At the end of the day, judgements about fare structures are not wholly economic in an industry which has been heavily regulated for many years and where passengers have made important decisions based on present practice.

Franchising is about competition for the market. Competition on the tracks is the preserve of the Office of the Rail Regulator (ORR). It is important for OPRAF that the extent of on rail competition is known with some certainty. The centrally driven objectives of the British Railways Board meant that little guidance on the potential impact of competition on rail was available for franchisees. It was, however, evident that railways are prone to potential 'cream skimming' and that competitors could 'free ride' on the marketing efforts of other rail businesses. The Regulator has therefore applied an incremental approach to the development of competition, taking into account the effect of his policies on the budget of the Franchising Director. The interests of ORR and OPRAF will remain close. ORR has the ability to increase or decrease burdens on OPRAF's budget. OPRAF, however, would face a conflict of interest if it was responsible for regulating Railtrack, whose charges account for about 40 per cent of TOCs operating costs. The regulatory risks for franchisees are mitigated by the offer of a franchise review at the time a general change in the competition regime occurs. But the piece-meal extension of open access competition within a given competition regime is treated as an ordinary commercial risk.

However, there are issues surrounding the efficiency of competitive market driven incentives. For example, are the structures which would support intensive 'on-rail' competition the same as those necessary to underwrite worthwhile investment on fixed assets? Is regulation or moderation of on-rail competition a price worth paying to secure investment? How intrusive should OPRAF and ORR be in deals struck between franchisees and, for example, Railtrack? Some of these issues are addressed in the next section.

Franchise term

The design of the initial passenger franchises addressed the following questions:

a) what length should the first franchise term be?
b) would the term be fixed?

The franchises have to be saleable, provide continuity for customers, facilitate investment and be enforceable. It is also necessary to make provision for reletting the franchises. Generally, long term contracts also reduce the likelihood that a fixed price contract will be delivered without periodic renegotiation, or provision of 'cost pass through' mechanisms to reduce risk for the supplier. These considerations were taken into account in designing the franchise proposition.

The development of a lease market for rolling stock and the access charge regime for infrastructure makes possible a franchise term which is independent of the lives of major assets. However, industry parties, including OPRAF, still have to manage residual value risk. Without this, innovation could be stifled and under investment occur.

The new railway industry is still developing. ROSCOs, for example, are now prepared to take residual value risk for some types of new rolling stock which have alternative uses around the network. Railtrack, too, is capable of taking a portfolio approach to new investment although, like other suppliers, it is a commercial entity seeking a commercial return. Experience of working within the new structure will take time to develop and the contracts which underlie it will change over time.

OPRAF's role reflects its function as a purchaser, not a supplier. It is prepared to develop a partnership approach but also relies on initiatives coming through franchisees. However, the Franchising Director does have a role and powers under the Railways Act 1993, notably S.54 which enables him to procure future franchisees to use new facilities or other railway assets, thereby underwriting their financial return. This undertaking has only been given once, for Thameslink 2000, the infrastructure element of which costs over £500m. It will be completed beyond the life of present franchises. In this case, the Franchising Director is taking a risk on the value of the new capacity to franchisees in exchange for various undertakings from Railtrack, the infrastructure supplier.

Other instruments[2] available to the Franchising Director include a long initial franchise term. This may be conditional on taking forward investment, and may revert to a shorter period if this fails to occur. The MML franchise is of this form. The primary objective is to give the franchisee a period of enjoyment of leased assets, in return for managing the procurement process. This provides an incentive for the franchisee to achieve value for money.

At present, the Franchising Director is not permitted to extend a franchise without open competition. Clearly, this is a possible area of change. In other franchise situations, it has been possible to lengthen an initial franchise in return for investment, cost reductions or for service enhancement. An argument against this approach is that it excludes new entrants and may reduce value for money in the long term. This may be particularly true where capital markets, through threat of take-over, are not an effective discipline on incumbent operators. Mergers and take-over activity could be constrained by the competition authorities and by ORR who may also seek to enhance on-rail competition if the structure of the industry changes.

Bidding for subsidy

Many of the issues which needed to be addressed during the initial franchising round have already been discussed. However, the successful completion of the sales process was, in practice, a highly labour intensive activity. It required innovation and a wide range of skills. It also required bidders to climb a steep learning curve and to react quickly to a demanding timetable. This paper does not attempt to discuss theoretical issues surrounding the organisation of the bidding. In practice, the system seems to have worked well. A total of 13 bidders have acquired 25 franchises, more franchisees than many would have anticipated. The bids have been attractive in financial terms and have promised improvements to services. The chart shows the decline in contractual payments committed by franchisees.

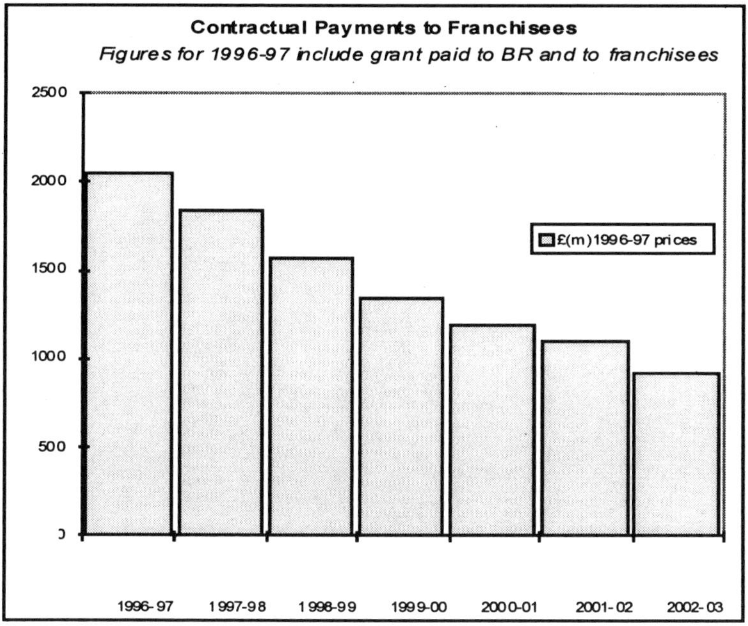

Figure 10.1 Contractual payments to franchisees

It is of key importance that franchise compliance is monitored and that OPRAF's franchise managers are effective in delivering the contracts on behalf of passengers and taxpayers. This will not be straightforward. However, the structure of the Franchise Agreement, combining commercial incentives, incremental exposure to external competition and fixed contractual elements, provides a framework in which this can be achieved. Clarity of responsibilities will be an important discipline on suppliers and government, as will the range of market based incentives retained or established by recent reforms. OPRAF, however, will continue to review the effectiveness of its enforcement powers and adapt to changing demands as they emerge. The franchising process was only a means to an end - better outcomes for passengers and taxpayers. Final judgement will have to wait. The task of managing the franchises is only just beginning and this experience will provide the basis of another, more interesting paper, in about seven years time!

References

OPRAF (1996) Appraisal of Support for Passenger Rail Service - A Consultation Paper, London.

Notes

1. This paper represents the views of the author and does not necessarily reflect Government policy.
2. See also OPRAF (1996).

CHAPTER 11

Competitive behaviour in Britain's privatised rail industry

Michael Schabas[1]

Abstract

Although there remain many parts of the passenger railway industry which are natural monopolies, the Government's chosen fragmentation and franchising structure has created competition on many levels. This paper first surveys the success at creating and maintaining competition in the bidding for passenger franchises and for the acquisition of other businesses formed from British Rail. Second, we look at apparent moves by industry players to gain market control, in ways which might be considered anti-competitive. Third, we look at the regulatory responses to date.

Competition for the supply of railway services

In devising the restructuring of the British railway industry, Government sought to create effective competition wherever possible, both in the supply of railway services to passengers and freight shippers, and in the inputs to the operators of the railway services. The unitary British Railways Board was broken into approximately 100 companies. These were then vested and sold, usually by competitive tender.

Vesting and shadow running

During the vesting process, Government attempted to balance two fundamentally conflicting objectives. On the one hand, proceeds to the

Government could usually be maximised if companies were sold as protected monopolies. On the other hand, experience had shown that the benefits of privatisation would be greatest if the new companies were exposed to competitive as well as (or even instead of) regulatory discipline.

The pragmatic response was to vest the ex-BR companies with contracts in place, ranging up to ten years in length although usually only lasting two to four years. Often the extent or scope of the vested contracts also declines within the overall contract term. As there are in most cases lengthy lead times for competitors to acquire the necessary equipment and know-how, the contractual and regulatory restrictions that are in place may be more in the nature of comfort to purchasers than materially delaying emergence of real competition.

A distinction is required between the 25 franchised passenger Train Operating Companies (TOCs), in which Government has a continuing financial interest, Railtrack which is a regulated monopoly, and the 'non-core' service and supply companies which are not regulated but which are nevertheless major suppliers to the TOCs and Railtrack.

If the TOCs and Railtrack were locked into monopolistic suppliers, then this would limit the potential to reduce subsidies to TOCs. Thus the Infrastructure Maintenance Units were vested with contracts protecting most of their initial workload for five to seven years, but with a declining schedule of prices to reflect expected efficiency gains. The Track Renewal Units have their protected workload decline more rapidly, with none contracted at vesting beyond the end of the fourth year. The majority of rolling stock leases extend until 31 March 2004, however some expire as early as 1998 and others can be terminated without penalty by 2000.

All companies were vested over a two year period 1994-1996. Management teams drawn primarily from within British Rail were assigned to each company, to negotiate internal trading contracts with other ex-BR companies. Only a few non-core companies were sold in 1994-1995, with BR insisting that a complete contract matrix (or nearly so) be in place before sales of the TOCs could proceed in earnest. At the time, this insistence on a 'shadow running' period was criticised as causing delay, however it appears to have proven successful as few problems have subsequently emerged and prices obtained for the businesses generally exceeded expectations.

Management buy-out bids

Management teams were invited to form companies to bid for their own newly-formed businesses, and British Rail was authorised to provide up to

£100,000 towards each Management Buy Out (MBO) team to cover out-of-pocket expenses such as legal and accounting fees. This served three purposes. First, it satisfied a political desire to empower railway managers, so that privatisation could be described as a devolution of responsibility. Second, it increased the competition for the businesses, as there would then usually be at least one serious bidder. Third, and perhaps most important, it gave management a strong incentive to ensure that a robust matrix of contracts was indeed in place. This gave comfort to outside bidders that the businesses were, indeed, sound enterprises.

Sale of the franchised TOCs

The actual sale of the TOCs commenced in 1995 and continued through into early 1997. The sale process was managed for the Franchising Director by the Office of Passenger Rail Franchising (OPRAF), of which he is the head. OPRAF was supported in the sale process by the Corporate Finance divisions of various merchant banks and accounting firms.

The franchising process was structured as a series of competitive tenders, but was in fact a combined sale of the TOC business units together with award of a Franchise Agreement which is really a form of service contract. After an initial screening to establish a level of technical and financial competence, pre-qualified bidders were provided with a 'Long Form' report, disclosing detailed financial information on the performance of the TOC over the period of shadow running and before, if available. There was also a Data Room containing TOC contracts and other materials, and an opportunity to meet with TOC management under OPRAF supervision.

Bidders were invited to submit an Indicative Bid setting out their approach to the franchise, and presenting detailed financial projections including the level of subsidy that would be required. A shortlist of three or four bidders were then invited to complete final 'due diligence' and submit final, binding bids. The successful bidder purchased the TOC, together with all contracts, employees and assets which were vested in it, for a nominal £1 (after "zeroisation" of the TOC balance sheet, so that there are no net assets). At the same time, the bidder entered into a Franchise Agreement with OPRAF normally for a period of seven to fifteen years.

Although the Franchising Director has considerable discretion in award of franchises, in practice it appears that he followed quite strict and straightforward criteria. In all cases it would appear that the franchise was awarded on a financial basis. All bidders were required to commit to deliver at least the full Passenger Service Requirement,[2] and to maintain or improve existing performance standards. While many bidders volunteered additional service enhancements, in most cases these seem to have been financially self-supporting. The author is not aware of any case where a franchise was awarded other than to the bidder asking for the lowest subsidy or offering the highest level of premiums, when discounted by OPRAF to a Net Present Value at 6 per cent real. Many successful bidders have made binding commitments also to invest in improved services, but it appears that these commitments are (or were when they were made) believed to be commercially worthwhile or, at worst, financially neutral.

During the passage of the Railways Act, there was much scepticism that Government would find companies willing to take on the TOC franchises. Clearly, these were new businesses without an obvious model, and with a somewhat unusual mix of risks and opportunities. In practice, Government always received at least four initial bids for each franchise, although for a few of the early franchises only two or three reasonably compliant bids were actually received. Early in the process, declared interest was limited to a few existing transport companies. This included bus companies which had emerged from the earlier privatisation of that industry, together with Sea Containers, a shipping company which had played a role in privatisation of British Rail's cross-channel ferries in the 1980s, and Virgin Group which is probably best known for its airline business.

Later, others joined in the bidding although many did not publicise their interest until well into the process.

Table 11.1 presents information on the 25 franchises and their new operators. While there is a heavy presence of bus companies, who are involved directly or indirectly in 18 of the 25 franchises, some of the largest franchises are owned by non-bus companies. The largest player in the London suburban market is Connex, a subsidiary of the French utility Groupe Générale des Eaux. More that half of the long distance market is controlled by Virgin Group, followed by Sea Containers. National Express has a mix of local and long distance routes, to complement their local and long distance bus operations and regional airport portfolio. Besides companies with existing transport interests, two completely new companies were launched, both now listed on the Alternative Investment Market (AIM), the 'junior market' of the London Stock Exchange.

Table 11.2 presents subsidies for the 25 franchised TOCs, for selected years. Note that the actual subsidy that is paid will vary as follows:

- annual subsidy payments will vary by a few percent depending upon actual train performance;

- subsidy payments may be adjusted after 2002 to reflect possible adjustments in Railtrack charges and change in on-track competition policy. In each case the Regulator has not yet set his policy;
- actual subsidy payments are subject to inflation.

It is possible that some franchises will be terminated early, either because the operator is in breach or, perhaps more likely, because he is in danger of financial insolvency. Before awarding each franchise, OPRAF has reviewed the financial projections to ensure that bidder assumptions are reasonable. However, the Franchising Director has at least four mechanisms to protect against financial insolvency:[3]

1. First, before a franchise is awarded the Franchise Operator (FO) must provide capital equivalent to about 15 per cent of passenger income. Half of this must be in the form of a bond, while half may be used as working capital. The requirement to put up millions or tens of millions of pounds must guard against cavalier bidding. In the event that an operator does choose to 'walk away' from a money-losing franchise, OPRAF can use this capital to re-let the franchise. The capital also provides a financial cushion should an otherwise sound franchise encounter short-term problems;
2. Second, an operator with more than one franchise may be required to cross-collateralise, so that it may not withdraw from one franchise without forfeiting all of its franchises;
3. Third, a franchise operator with ambitious investment commitments, such as rolling stock replacement, will likely need to enter into commitments to pay lease charges at least for the full duration of the franchise. Thus the financial loss to a company which 'walks away' from a money losing franchise can be substantially greater than the franchise capital;
4. Fourth, all franchises are subject to regular audit by OPRAF which can, therefore, take remedial action if financial problems are evident.

Note that the 25 franchises are now controlled by 11 to 13 distinct groups, depending upon how one treats interlocking interests.

Table 11.1 The franchised train operators

Franchise	Operator	Passenger Revenue 1997 £m
South Central	Connex (CGEA, a division of Group Generale des Eaux, a diversified French utility	177
South Eastern		248
Great Eastern	First Bus Plc (urban bus operator)	130
Great Western	Gt Western Holdings (First Bus with Management Buyout)	183
NW Regional Railways	Gt Western Holdings	50
West Coast	Virgin Group (Venture/Private Capital)	241
Cross Country		116
Scotrail	National Express Group Plc (diversified transport business - coach/bus/airports)	110
Central Trains		72
Midland Main Line		71
North London		56
Gatwick Express		32

South West Trains	Stagecoach Holdings Plc (coach/bus/rolling stock leasing)	285
Island Line		1
Great North Eastern	Sea Containers Ltd (shipping/hotels/luxury trains)	257
LTS Rail	Prism Rail Plc, new company floated on Alternative Investment Market	55
W Anglia Gt Northern		130
South Wales & West		48
Cardiff Railway		7
Thames	Victory (Go Ahead Group Plc wth MBO)	61
Thameslink	GOVIA (Go Ahead Group with Via GTI, a French bus and rail operator)	93
Regional Railways NE	MTL Trust Holdings, an urban bus operator	75
Merseyrail		20
Anglia	GB Railways Group Plc, a new company floated on Alternative Investment Market	41
Chiltern	MBO with John Laing, a construction contractor/infrastructure developer	29

Table 11.2 Subsidies for the train franchises

Franchise	Length (years)[a]	Subsidy (Premium) 1997 £millions[b]			Other key franchise commitments
		1997	2003	2011	
South Central	7	85	35	(35)	
South Eastern	15	125	27	(3)	Replace slam door stock by 2006
Great Eastern	7	41	(5)	(24)	
Great Western	10	60	35	(10)	Half hour frequency to Bristol
NW Regional Rlwys	7	192	130	108	Replace old trains
West Coast	15	93	(4)	(202)	New tilting train fleet by 2002
Cross Country	15	127	49	(5)	Total fleet replacement by 2004
Scotrail	7	292	208	170	Replace old trains
Central Trains	7	198	137	118	
Midland Main Line	10	17	(4)	(20)	New fleet for local stops
North London	7	55	20	(20)	Replace old trains
Gatwick Express	15	(5)	(12)	(23)	New fleet by 1999
South West Trains	7	60	40	(25)	
Island Line	5	2	2	2	
Great North Eastern	7	65	0	(19)	

LTS Rail	15	30	19	11	New fleet by 1999
W Anglia Gt Northern	7	71	(14)	(32)	
South Wales & West	7	77	43	29	
Cardiff Railway	7	22	14	6	
Thames	7	44	4	(11)	
Thameslink	7	16	(27)	(30)	
Regional Rlwys NE	7	223	151	124	Replace old trains
Merseyrail	7	91	62	48	
Anglia	7	41	9	(10)	New fleet by 2000; half-hourly to Norwich
Chiltern	7	17	3	(11)	New trains; hourly to Birmingham
PTE Revenue		(86)	(97)	(114)	
Net Total Subsidy		1,952	820	23	

Notes to Table11.2

a) Extension of the LTS, Midland, and Great Western Franchises beyond seven years are conditional upon investment in improved rolling stock.

b) Where a seven year franchise has been awarded, subsidy for 2011 is the author's projection, shown in italics.

c) For the North London, Central, North East, North West and Scotrail franchises,some or all of the revenue risk was retained by local Public Transport Executives. Subsidy for the franchises is presented before PTE revenue, which must be subtracted to give the net subsidy from Government.

Sale of other businesses

In parallel with the sale of the franchised TOCs, British Rail sold off the 60-odd subsidiary companies which had been established. These supply everything from trains to sandwiches to the franchised TOCs. It also sold off the companies which maintain tracks and trains, and which operate freight services. Altogether, the companies on sale had turnovers including trading with each other worth in excess of £4 billion. Given the complexity of internal trading, the process could be compared with privatisation of the economy of a middle-size eastern European state.

Where practical, two or more companies were created for each business activity, often following pre-existing management divisions. This would provide an immediate level of competition, if sold to separate buyers, and would also make the companies more 'bite size' and thus suitable for a range of buyers with different financial and market appetites.

In practice, a degree of reconsolidation has been tolerated and was probably always expected. This will be considered later.

Privatisation balance sheet

The privatisation of British Rail is remarkable in that a money-losing, subsidised industry has been sold in a manner which is cash-positive to Government and which will remain so compared with other alternatives.

Since nationalisation, British Rail has required increasing levels of subsidy, at least in nominal terms. Although there has been some significant investment and services have, slowly, improved, this was never at a pace that was considered acceptable to the public or indeed to Government. Despite continuing rationalisation and productivity improvement, it was difficult to reduce the operating subsidy below £1bn per year. Although BR managers had much to be proud of, they had reached the limits of what could be achieved with a centrally-managed, government owned monopoly. Allowing for 'steady state' renewal of infrastructure and other capital assets such as trains, without restructuring and privatisation Government could expect a liability of about £1.5 bn per year indefinitely, with only slow service improvement and no prospect of satisfying public or political expectations.

Restructuring is understood to have cost about £400 million, mostly in fees to lawyers, accountants, and other advisers. Sale of assets, including the ROSCOs, Railtrack, and various non-core businesses, has generated about £4 billion in cash. Annual subsidies have increased, from the (inadequate) £1 billion which government was paying to BR in the early 1990s, to almost £2 billion in 1996-7.

Some rail privatisation critics have pointed to this increase in 'headline' subsidy as evidence that privatisation is costing taxpayers money. However, net contracted subsidies will decline rapidly, to below £1 bn in 2002 and, potentially, to nothing by 2012. In that year Britain's railways will, taken as a whole, require no government subsidy whatever. In fact, since most of the railway companies will be paying tax, the industry will be cash-positive to the Exchequer from an earlier date. Treating the £4bn sales proceeds as a "sinking fund" to pay the higher subsidies in the early years, the privatisation will always be cash positive and will bring fiscal benefits worth several billion pounds in later years.

Privatisation should also bring real benefits to rail passengers and the country as a whole. Most franchise operators have made commitments to deliver service improvements far in excess of what could have been reasonably expected if BR remained in public ownership. Britain already has one of the most modern train fleets in Europe, with high frequency services on most routes. Franchise operators are committed to refurbishing or replacing most of the remaining aged rolling stock, and to increase service frequencies on many routes. Just as privatisation has brought Britain the best telecoms and air transport services in Europe, with profits for shareholders and tax receipts for Government rather than subsidies, so rail privatisation will do the same.

Figure 11.1 Contracted and projected subsidy for the franchises

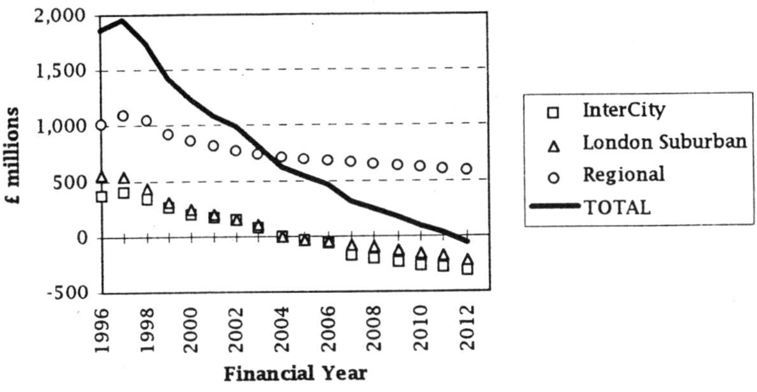

Notes
1996 to 2003 - based on contracted subsidies
2003 to 2012 - combination of contracted subsidies and author's projections

Anti-competitive behaviour and regulatory protection

On offer at the beginning of 1995 were:

- twenty five passenger franchises, based on 'bundles' of routes;
- three rolling stock leasing companies, newly created;
- one network operator (Railtrack);
- six heavy maintenance depots;
- six infrastructure maintenance units, based on geographic territories;
- five track renewal units, based on geographic territories;
- three trainload (bulk) freight companies, based on geographic territories;
- one domestic container freight company;
- one parcels business;
- one international (channel tunnel) freight company;
- one express freight (mail) train operator;
- two signal engineering companies, based on pre-existing separate design offices;
- five civil engineering design companies, and three rolling stock engineering companies, based on separate design offices operated by BR's InterCity, Network Southeast, and Regional Railways business sectors;
- one computer systems provider.

There has subsequently been a degree of consolidation although only a few cases have attracted attention or concern.

Vertical 'integration' of ROSCOs and TOCs

The three rolling stock leasing companies were sold separately and Government insisted from the start that no purchaser would be allowed to acquire more than one. However one ROSCO, Porterbrook, was subsequently acquired by Stagecoach which also operates the South West Trains franchise and was, at the time, bidding for several other franchises as well. Concern was expressed at that time that Stagecoach would use its market power in an anti-competitive manner. However it remains unclear how it could do so. The Franchising Director already has safeguards in place to ensure incumbent operators must offer rolling stock to successor franchise operators on reasonable terms. A 'Direct Agreement' is required between OPRAF and any major supplier to a franchised TOC, to facilitate re-letting of the franchise on a competitive basis.

While it would be difficult to ensure that Porterbrook did not offer its parent more favourable terms than to competitors, if this is a concern it is likely that the Franchising Director would simply refuse to invite Stagecoach to submit a bid for the franchise renewal.

As an apparent 'sweetener' to encourage approval of the Porterbrook acquisition, Stagecoach entered into a commitment to acquire some new rolling stock for the SWT franchise. This contract was in fact awarded to Porterbrook although Stagecoach did invite competitive bids.

Around the same time as Stagecoach was acquiring Porterbrook, Connex (a division of Compagnie Generale des Eaux) was acquiring the Southeastern Franchise. This franchise is for 15 years and includes a commitment to replace about half of the rolling stock, with a notional value of £250 million. One reason Connex may have won this franchise is its ability to acquire new rolling stock on its own balance sheet, taking residual risk itself. Connex can take a positive view of residual value because it expects to remain in the business of operating trains beyond the life of the initial franchise.

Connex has not, yet, acquired the new trains and it remains to be seen whether they will acquire them on its own book. Whatever it decides, there is no obvious reason why it should be more confident of winning the franchise extension than another bidder. It must enter into the same direct agreement, making the trains available in fifteen years to any subsequent franchise operator.

Connex has taken two risks, one to operate a franchise and one to own and lease trains, albeit to itself for the initial period. If ownership of the trains does offer a real competitive advantage when the franchise is up for renewal in 2011, then OPRAF may need to exclude Connex from the bidders as otherwise it will be unable to create a genuine competition.

Although Stagecoach and potentially Connex have taken ownership of two levels in the industry, they have not yet succeeded in vertical integration in a way that should seriously limit participation at either level of non-integrated players.

Consolidation of freight

Five of the six rail freight businesses have been acquired by one group, English Welsh & Scottish Railway which is led by Wisconsin Central. EW&S initially acquired Rail Express Systems (Res) and then subsequently put in the highest bid for all three trainload freight companies. It later purchased Railfreight Distribution, operating intermodal services through the Channel Tunnel (although this remains subject to EU approval). Of the ex-BR freight companies, only Freightliner escaped the EW&S net, going instead to an MBO team.

The public debate on this focused on two issues:

- First, why had Government spent money splitting trainload freight if reconsolidation was then to be allowed?
- Second, was competition between the freight companies always misguided if most competition was actually with road haulage?

Both debates obscured more meaningful issues. The costs incurred in splitting Trainload Freight into three businesses were tiny, compared to the sale proceeds which exceeded £250m. By splitting Trainload Freight into three companies, Government made it into pieces that were digestible by more bidders, thus creating keener competition. All three Trainload Freight companies had in place management teams which each could and did launch bids for their own companies and for each other. Had Trainload Freight been kept as a single company, Government would probably have received much less from the sale as there would have been few interested bidders.

The road versus rail debate was also confused. It was never suggested to split Res, Freightliner, or RfD. All three operate as integrated national networks serving many small flows. These businesses do indeed operate in direct competition with road for most of their traffic, and splitting them would reduce their ability to offer a complete product range to shippers. Indeed, these businesses will probably now expand both vertically, into road haulage, and horizontally into international routes.

It seems clear that splitting Trainload Freight into three companies helped Government achieve a higher sale price. At the same time, reconsolidation has not posed a problem as there is a demonstrated ability of new operators to enter the bulk rail freight market.

Multiple franchises and on-track competition

The 25 passenger rail franchises have now been acquired by 11 groups or companies. It was always expected that companies would seek to acquire franchise 'portfolios'. Apart from sharing head office overheads, there are probably few economies in owning more than one franchise. Considerable care has gone into protecting individual operators from predatory or exclusionary behaviour, and there is indeed a specific prohibition on exclusionary behaviour in the template Franchise Agreement.

However there are advantages to investors in owning portfolios with different risk profiles. For example, traffic on London commuter franchises is affected directly by changes in central London employment. In 1990-1994, traffic on some routes declined by as much as 35 per cent. In contrast, traffic on InterCity and Regional (non-London) routes declined much less.

While 'Head Office' costs as such are fairly small, there appear to be real advantages in sharing head office skills and strategic resources. Just as a small number of large property developers are active in each major city, so a small strategic team can apply its skills to acquiring and guiding a portfolio of franchises.

When shortlisting bidders for each franchise, OPRAF was concerned primarily with getting the best value for the taxpayer, that is minimising subsidy or maximising premiums. However the Regulator made it clear that he wanted to encourage the growth of on-track competition.

While the 25 TOCs are often characterised as protected monopolies, there are a significant number of flows where they already compete with each other.[4] For example, commuters often have a choice between using suburban and long-distance trains. Since management performance within BR was monitored in part by revenue growth, there has always been a degree of covert 'raiding', but competition with another sector of BR was not in the corporate interest and was frowned upon, even if it might benefit passengers. Once the TOCs were split and sold to separate companies it was inevitable real competition would begin to emerge.

The Regulator has taken specific steps to encourage and protect on-track competition. For each major flow, the dominant TOC is designated as the 'lead' operator, which must offer and accept only inter-available fares. Other operators can accept these interavailable fares, receiving a share of revenues proportional to the passengers they carry. Competing operators can also offer specific non-interavailable fares. Thus a secondary operator can capture a share of the business without the requirement to operate a full, regular service.

At the same time, the Regulator has applied rules to 'moderate' competition, at least until 2002. Without some protection against unlimited new entry, it was thought that it would be difficult to sell the franchises. Although the mechanics are somewhat complex, new entry is generally restricted up to a cap of 20 per cent of flows, weighted by value, for each franchise.[5]

Quite a few franchises already face competition on flows reflecting more than 20 per cent of their revenue, and on these routes there is only limited protection against further competition. Gatwick Express, Anglia, Thameslink, Cross Country, and Thames are some of the operators facing red-blooded, on-track competition, on a significant share of flows. Others such as LTS Rail and Great Eastern compete on significant flows from nearby stations, for example at Southend, although they do not share the same tracks. In the sale of the franchises the Regulator made it clear he would not look favourably on substantial reduction in on-track competition, through common ownership of franchises. This had the effect of narrowing the eligible bidders for each franchise.

For example, once LTS Rail was awarded to Prism, this company could no longer be shortlisted for Great Eastern. Award of Great Western to an MBO backed by First Bus precluded that company from bidding for Thames, although it was shortlisted for the adjacent Chiltern franchise. Award of both Cross Country and West Coast to Virgin Group has reduced on-track competition on some routes in the Midlands and Northwest, however it appears that the Regulator was willing to accept this as a reasonable price to pay for the massive investment Virgin has committed to make on both franchises.[6]

It surprised some that the Regulator had no difficulty with Connex acquiring both South Central and then the adjacent South Eastern. In fact, these two franchises compete on very few flows. Whether the dominant presence of Connex in Kent and Sussex will make it difficult for any new entrant remains to be seen. Thameslink and Gatwick Express both operate in the area, and with completion of the Thameslink 2000 and Channel Tunnel Rail Link schemes there will be additional paths into the area which will be sold separately. Connex may find that it has not bought quite such a protected monopoly as it hoped. Given that the traffic risks faced by South Eastern and South Central must be very similar, one must also ask whether Connex might not have done better with a more diverse portfolio of franchises.

There has been much talk of re-consolidation, with all franchises eventually owned by two or three companies. This seems unlikely. While it is possible that there will be some consolidation of the existing franchises, it is difficult to see how the 25 franchises could be owned by less than six or seven groups without reducing on-track competition to an extent that the Regulator would not accept. The Regulator has recently stated that he will not support further integration, vertical or horizontal, unless he can see 'substantial public interest benefits'.[7]

Interlocking interests

Several of the franchises are owned by consortia with interlocking interests. For example, Eurostar is owned jointly by Virgin and National Express Group (NEG), among others. Virgin in turn owns the West Coast and Cross Country franchises, both of which compete with NEG's Midland, North London and Central franchises as well as with NEG's intercity coach operations. It appears that the Regulator is satisfied that these overlapping interests will not result in reduced competition, although this remains to be seen.

Many franchises are owned by bus companies, which sometimes have strong local presence in the same area. First Bus dominates local services in Essex and Suffolk, and also owns the Great Eastern franchise which serves the same area. Stagecoach operates many bus routes in the same area as its South West Trains franchise. MTL operates the Merseyrail franchise and is also the dominant bus operator in the Merseyside area.

There is much talk among politicians of the benefits of 'integrated' transport. The nationalised BR was reluctant to enter into through ticketing arrangements, or to introduce any bus feeder services which might be perceived as competing either with its own railways or with wholly private services. These constraints no longer apply, and several rail franchise operators have introduced bus services to feed passengers onto their trains. Anglia has introduced a coach link from Colchester to Stansted airport that is, at least in theory, in competition with rail. Great North Eastern has introduced a feeder bus from Lincoln to Newark, which, depending upon one's viewpoint, either complements or competes with Central's rail service between these two points.

It may be that ownership of a rail franchise is seen as strengthening the defensive arsenal of a bus operator against the threat of competition. National Express has come under scrutiny for its ownership both of several rail franchises serving long distance passengers (in particular Midland, Scotrail, and Central) which serve similar flows to its coach businesses.

Barriers to entry: EWS's new Track Access Agreement

Government designed the franchising process so as to retain a long term 'carried interest' in the passenger businesses. Thus the long term requirement to subsidise unprofitable routes will be offset by premiums earned on more lucrative services. Similarly, the Regulatory regime for rail freight was designed so Government could capture any large margins from the use of what remains a natural monopoly.

Essentially, the freight operators are meant to offer prices as high as the market can bear. In most cases, this means marginally less than road transport, taking into account adjustments for aspects of service as perceived by the shipper. Railtrack, in turn, negotiates track access on the same basis, earning large margins on some traffic, particularly long hauls and heavy bulk loads where road is not an attractive option. Railtrack's margins contribute to its bottom line profits, which are then captured in the Regulator's 'single till'. Every five years, the Regulator reviews

Railtrack's passenger access charges and has the opportunity (which he will presumably take) to reduce them so that Railtrack's long run return is as low as possible, consistent with attracting capital for investment. OPRAF in turn has a right to review Franchise payments, to adjust for changes in access charges. So the freight operators are, in effect, collecting money for the Government. Negotiated margins from shippers is an ideal, non-distorting substitute for taxes.

The British approach to freight regulation is unusual. In other countries with privatised rail systems, the operators keep any windfalls to themselves, although perhaps paying some to Government in taxes. EW&S, who purchased most of the BR freight companies, would very much like to change the rules so that they could keep more of their money. They would also, at the same time, like to strengthen their defences against competitors. EW&S have expressed their preference for vertically integrated freight railways, conveniently ignoring the need to share track with passenger trains.

Of course EW&S could not ask directly to be given, in effect, a lucrative monopoly position. Instead, EW&S have argued that the need to negotiate freight rates with Railtrack for each flow is cumbersome, and limits its ability to compete with road transport. Instead of negotiated contracts for each flow, EW&S has offered to pay a standard tariff for each train, plus a fixed charge which would be negotiated on a periodic basis. EW&S could then sell new capacity without needing to negotiate rates with Railtrack.

EW&S's proposals are seductive, not least because very few people even within Railtrack seem to understand the regulatory framework within which the industry operates. EW&S's argument is, however, a bit misleading. First, although EW&S will not need to negotiate charges for new services with Railtrack, they will still need to negotiate access within the timetable. It is not at all clear that Railtrack has been slow to quote charges; what is clear is that there will always be a lead time of weeks if not months in timetabling trains. So the claim that this will make it easier for EW&S to respond to customers seems questionable.

Second, EW&S are not offering to quote a standard tariff to shippers: EW&S will still charge what it thinks the customer can bear, but it will not pass on any large margins to Railtrack at least between periodic reviews. There is no clear mechanism by which Railtrack will adjust the 'fixed' charge, short of auditing EW&S's books and establishing in effect a regulated rate of return. Of course, any margin captured by Railtrack is, in turn, captured by the Regulator and passed to Government. While the mechanism as it stands seems perfect, Railtrack only have an incentive to act as 'tax collector' because it keeps any excess margins between its own quincennial reviews. If its contract with EW&S means there are no margins between these reviews, then Railtrack has no interest in the deal which is done.

Third, the proposed Railtrack-EW&S access deal is likely to be a material barrier to entry. Only EW&S has a portfolio of freight flows with which to do this sort of deal. It will allow EW&S, on the margin, to win business at a lower price than a competitor. And once EW&S wins a flow, it will be hard for competitors to win it away.

Conclusions

The new rail industry structure is three years old, although most companies have only been in private ownership for less than one year. There has been some vertical 'integration', with two companies attempting to own both rolling stock and franchises. However the other two ROSCOs are firmly owned by groups without interests in franchising. Railtrack is widely held on the Stock Exchange.

There has been some re-consolidation of the operating companies, EW&S dominates rail freight but not yet to an extent that threatens the viability of other players. There was lively competition for passenger franchises, with eleven distinct groups active in the industry where a few years ago there were none. Although there have been rumours of takeovers or mergers between the franchise operators, there is no evidence that any consolidation is actually underway.

With healthy, dynamic competition emerging the new rail industry is off to a good start.

Notes

1. Michael Schabas is a founder Director (non-executive) of GB Railways Group Plc, a newly formed company which was recently awarded the Anglia Railways franchise. Mr Schabas holds a Master's degree in City & Regional Planning from Harvard University's JF Kennedy School of Government. He came to London in 1988 for the Canary Wharf project. As an independent consultant, he has advised a wide range of public and private clients on matters relating to passenger railways, mostly in Southern England.

2. The PSR specifies, *inter alia,* minimum service frequencies for each route, maximum load factors, first and last trains, and maximum journey times. See paper by Bob Stannard in Chapter 10.
3. All Franchise Agreements are public documents, available for inspection at the Office of Passenger Rail Franchising. Financial terms and certain commercially sensitive clauses may not be available in the public versions.
4. The franchised train operators also, of course, face significant competition from other modes of travel. Long distance operators compete mostly with car, air and coach, while urban commutre operators face more limited competition from commuter coaches, private car and sometimes, London Underground.
5. For a fuller explanation of the mechanics of moderation of competition, see paper by Philip O'Donnell in Chapter 8.
6. Note that it is the Regulator who is concerned with maintaining and extending competition. OPRAF is concerned primarily with selling franchises so as to procure the PSR at minimum cost to the taxpayer and, as such, may be much less concerned with consolidation of franchise ownership.
7. See Regulatory Objectives for Passenger Train and Station Operators, paragraph 26, Office of the Rail Regulator, June 1997.

CHAPTER 12

Japan railways since privatisation

Fumitoshi Mizutani and Kiyoshi Nakamura

Introduction

Ten years have passed since 1 April 1997, when the Japan National Railway (JNR) was privatised and broken up into six regional passenger railway companies and one freight company, called the Japan Railway (JR). Although Japanese researchers have paid little attention to the JNR privatisation, it would be to our benefit to look closely at how privatisation occurred and what its results have been so far. Because the Japanese privatisation process seems to differ from that in other countries, it might be of special interest to the international community to see what has happened in Japan.

The main purpose of our paper is to summarise the effects of the JNR privatisation and to consider some implications for policy. We have been involved in the economic analysis of privatisation for the past several years, and although this paper is based on much of the work we have previously done, especially Mizutani and Nakamura (1997), what we would like to add here is an analysis of the performance results of individual JRs and what lessons can be garnered for other countries involved in the privatisation of their railways.

Besides the introduction, this paper consists of four parts. Firstly, we will offer a brief explanation of the privatisation of the JNR, an explanation which itself is divided into four parts: the reasons for privatisation; the process of privatisation; the main characteristics of Japanese style railway reform; and regulatory and organisational differences between JNR and JR. Of these four parts, the latter two are especially important in characterising the Japanese privatisation and revealing the differences between before and after privatisation.

Secondly, we will summarise the performance changes that have occurred in each JR in the post-privatisation period. JNR was divided into six regional passenger JRs and one nationwide freight JR, and we will look at how each company has improved its performance during this period. We will focus on performance differences between Honshu JRs, the three largest JRs operating on the main island (i.e. JR East, JR Central, JR West), and Three-Islands JRs, operating on the smaller three Japanese islands (i.e. JR Hokkaido, JR Shikoku, JR Kyushu).

Thirdly, we will discuss lessons learned from the ten years' experience of the Japanese rail privatisation. This chapter will be the most important in our paper. We will explore five vital issues. The first is non-open access and competition, related to the yardstick competition scheme and its effectiveness. The second comprises horizontal separation and optimal size. The third is an inquiry into issues related to vertical integration, discussing the advantages and disadvantages of operation and infrastructure integration. The fourth is an investigation of how investment in railway facilities has changed since privatisation. Lastly, we will address the pros and cons of the lump-sum subsidy scheme. The final chapter will cover remaining issues from the Japanese point of view.

Before we begin our discusion, it is necessary to clarify our use of the word *privatisation.* Our use of the word would differ from that of a researcher analysing, for example, the privatisation of British Railways. The Japan Railway cannot be said to have been fully privatised, as much of the stock of the privatised organisations known as JRs have not been completely turned over to the private sector. However, *privatisation* can encompass many meanings, for example the sale of existing state-owned enterprises, the use of private financing or management, or the contracting out of certain tasks to the private sector (Gomez-Ibanez and Meyer, 1993). Although the privatisation of the JNR cannot yet be called complete, we would like to use the words *privatisation* and *privatised* in reference to the JRs, as these organisations are on the way to being fully owned and managed by the private sector.

Privatisation of the Japan National Railway

Causes of privatisation

JNR, originally operated by the Ministry of Railways, began as a public corporation in 1949 after World War II (Ministry of Transport, 1992).

Although JNR expanded along with the economic growth following World War II, financial problems began to occur in the mid 1960s. In 1964, JNR began incurring operating deficits, which grew worse into the 1980s. In 1980, the annual financial deficit was about 1,008.4 billion yen (about 30 per cent of revenue), and long-term debts amounted to about 25 trillion yen (Mizutani and Nakamura, 1997). It was mainly these continuing financial difficulties which led to calls for the privatisation of JNR in the 1980s.

Researchers have offered many hypotheses as for the causes of JNR's financial problems (see Kusano, 1989; Fukui, 1992; Ministry of Transport, 1992; Fukui et al, 1994). Of the causes for JNR's decline, we consider the following important. First was the rise in the use of other transportation modes, most notably the private car. The construction of an extensive modern highway system made car use more convenient, diverting share from the rail sector. Air transportation also began to rival rail travel for longer trips.

These outside factors were beyond the control of JNR, but internal problems existed as well. First, there was the public corporation itself, with its implicit government guarantee, leading to a lack of localised responsibility, and low priority for the promotion of efficiency, expressed in the well-known phrase *Oyakata Hinomaru* or 'the Japanese rising sun [the government] stands behind us'. What this meant was that the government would support the company, no matter how inefficiently it was run. Consequently, there was almost no concern at JNR with the possibility of bankruptcy, in contrast to the situation at private railway companies, where the consequence of inefficiency was lost business and possible bankruptcy. This lack of a reward and punishment system eventually damaged work incentives, encouraged shirking and eventually led to other problems. Lack of cost consciousness, for example, was evident in the number of JNR employees, which peaked at a staggering 400,000.

Another problem was the strained relationship between management and labour. To stem its spiralling financial problems, JNR repeatedly imposed large scale organisational reforms, which instead of producing the intended productivity improvements, did nothing but aggravate tensions. When JNR tentatively began reducing the number of its employees, who were accustomed to stable working conditions and secure long-term employment, they joined massive strikes led by radical labour unions, although strikes by public corporations were illegal according to Japanese law.

It could be argued that the government and rail users themselves were also at fault for the chronic problems at JNR. The government bureaucracy and the Diet (Japanese Parliament) controlled rail fares, investment plans, and personnel matters such as the appointment of directors and other rail officials. A possible increase in fares, for example, became used for political deals in the Diet or between elected officials and voters, without consideration for economic reality. Furthermore, because JNR was a government owned organisation, tax paying citizens assumed that the company was there to serve their every need, and thus they pressured Diet members to provide rail services. The Diet complied, often constructing new rail lines in areas where there was no possible justification in economic terms for such projects.

All these problems came to a head with the appointment under the Cabinet's jurisdiction of the Provisional Committee on Administrative Reform (the Provisional Committee) in March of 1981. The Provisional Committee discussed how to achieve an effective, flexible administrative system, and in July of 1982 proposed privatisation with sub-division of the JNR (Fukui et al, 1994; Kusano, 1989).

Debate about whether to privatise JNR revealed many fears and concerns (Kamakura, 1986; Tateyama, 1989). The most frequently voiced were related to safety issues or the loss of local service. Opponents of privatisation were concerned that safety would suffer because the profit-oriented private sector would spend less money on safety precautions, and would reduce the number of employees to dangerously low levels in the name of efficiency. Others were concerned that the newly privatised rail companies would ruthlessly eliminate any unprofitable lines, leaving the transportation poor such as children, the elderly, and the handicapped immobilised. Some also feared that the regional sub-division would destroy the nationwide rail network. Rail employees were concerned that working hours and responsibilities would increase and wages decrease. However, in spite of all the anxious debate in the committee and the Diet, the privatisation went forward, and on 1 April 1987, JNR became six passenger JRs and one nationwide freight JR.

The process of privatisation

The crucial problem at JNR, according to policy makers, was the public corporation itself and the nationwide control of a gigantic railway network under a single management system (Ministry of Transport, 1992). Over objections both within and outside JNR, it was decided to divide the mammoth corporation into smaller, more manageable parts, the six regional passenger companies, each a separate JR (Japan Railway) and the freight company called JR Freight. Two non-operator organisations were established at the same time, the JNR Settlement Corporation and the Shinkansen Holding Corporation.

The Japanese railway privatisation perhaps differs from that in other countries in that JNR was not wholly converted to the private sector from the beginning. The government was concerned that JNR's image had been so tarnished by years of deficits and inefficient operations that a stock offering would not interest enough investors, a debacle that would further add to JNR's public relations woes. Therefore, the government created a temporary holding company called the JNR Settlement Corporation, where JNR stocks could be held until the newly privatised companies could prove their worthiness to the public by increasing efficiency and profits and thereby attracting a larger number of investors when stock was eventually offered for sale.

In 1993, six years after privatisation, 62.5 per cent of JR East's stock was sold to the private sector. The subsequent recession delayed the issue of the JR companies' remaining stock, but with the recovery of the stock market, JR West and JR Central shares went on the market in 1996 and 1997 respectively. As of August 1997, 62.5 per cent of JR East's and 68.3 per cent of JR West's shares were held by the private sector. Furthermore, about 67.0 per cent of JR Central's shares will go to the private sector in October, 1997. On the other hand, all shares of JR Hokkaido, Shikoku, Kyushu, and Freight are still held by the government.

Main characteristics of Japanese style railway reforms

The Japan National Railway reform has six distinguishing features: horizontal separation (or regional subdivision); passenger-freight separation; vertical integration (or operation and infrastructure integration); lump-sum subsidy for low density JRs; establishment of an Intermediary Institution; and allowance of non-rail business (Mizutani and Nakamura, 1997).

Horizontal separation refers to the separation of JNR into six passenger railway companies and one nationwide freight railway company (Table 12.1 gives a profile of these JRs). The basic rationale for horizontal separation is to accurately address regional needs and to introduce competition in management among JRs and with private railways. Regional sub-division was based on geographical distribution of demand.[1]

Table 12.1 Profile of JR companies at privatisation (1987)

	Passenger						Freight
	Honshu island JR			Three-Islands JR			-
	JR East	JR Central	JR West	JR Hokkaido	JR Shikoku	JR Kyushu	JR Freight
route-km (km)	7,657	2,003	5,323	3,176	880	2,406	10,010
passenger-km or ton-km (million)	104,491	41,148	45,782	3,920	1,673	7,664	20,026
train-km (million)	227	78	151	34	16	46	75
Shinkansen operation	yes	yes	yes	no	no	no	no
employees	82,469	21,410	51,538	12,719	4,455	14,589	12,005
total assets (billion yen)	3,884.5	553.0	1,316.3	976.2	323.9	738.1	163.8
% of stability fund to total assets	-	-	-	69.9%	64.3%	52.5%	-
transferred long-term debt (billion yen)	3,298.7	319.1	1,015.8	0	0	0	94.3
capital (billion yen)	200	112	100	9	3.5	16	19

Source: Mizutani and Nakamura (1997)

However, in order to balance the financial benefits and burdens among the JRs, the most profitable markets absorbed unprofitable lines. As a result, newly established JRs vary by size and number of operators less than is the case in the BR's sub-divisions.

Passenger and freight services were separated for several reasons. Firstly, it was desired that management responsibility for the freight division should be made clear. Under the old system, although it was clear that freight was unprofitable, because it was buried under passenger services the extent of the problem was hard to determine. Secondly, separating freight from passenger services would allow the fledgling private passenger companies to start unfettered by the burden of unprofitable freight service. The separation might be further supported by the argument of economies of scope, several researchers having noted that there are diseconomies of scope with passenger and freight operations in the railway industry (Kim, 1987; Preston, 1996).

Although European style privatisation separated railway service operation from the ownership of rail infrastructure, in Japan there was almost no question that companies would own both rail service and the infrastructure on which it operated (e.g. Suga, 1997). Control of rail tracks and signal systems as well as rail operation by the same organisation is considered vital to effective service scheduling and makes it easier for a railway company to expand with increased traffic volume. In the Japanese market especially, with its relatively high traffic volume, such integration might be necessary. Only JR Freight operates mostly without its own tracks, paying fees to the passenger JRs whose facilities it uses.

The fourth characteristic of Japanese railway reform is the lump-sum subsidy scheme that came with the creation of the Management Stability Fund. It was clear from the outset that the Three-Islands JRs, JR Hokkaido, JR Shikoku, and JR Kyushu, would be handicapped by geographical locations with smaller, lower density populations than the Honshu JRs, so interest revenues from the fund were to be used to provide support for these operators. Interest gains from the fund would be limited and were to be used only to compensate for operating deficits, so that railways would have incentives to keep costs down. The fifth characteristic of the privatisation was the establishment of an intermediary institution called the JNR Settlement Corporation, which was set up to supervise the handling of liabilities and redundant employees (Goda, 1992), and to serve as a sort of buffer so that the new JRs could avoid the burden of direct deficits.

The final characteristic of railway reform was that the newly established JRs were allowed to operate non-rail businesses,[2] such as real estate development and tourism. Although it is necessary to investigate how, several researchers have stated that these side businesses have helped rail operation (Saito, 1993; Killeen and Shoji, 1997). Japanese private railways have long been engaged in such outside activities, with the assumption that real estate development along railway lines, for example, would lead to increased ridership.

Regulatory and organisational differences between JNR and JR

Railway related laws were largely revised with privatisation in 1987. Before privatisation, the two main railway laws were the Japan National Railway Law (Nihon Kokuyu Tetsudo-ho) for the JNR and the Local Railway Law (Chiho Tetsudo-ho) for private railways and public subway systems. Since 1 April 1987, a new law called the Railway Enterprise Law (Tetsudo Jigyo-ho) has been in effect to regulate all railway companies, regardless of ownership.

In describing the competition policy inherent in the regulatory reform of 1 April 1987, we will focus on the regulatory and organisational differences in the JNR before and after privatisation. A summary of the differences can be seen in Table 12.2. Liberalisation seems to have been accomplished in many respects.

Firstly, approval by the Diet is no longer necessary for changes in fare, diminishing the use of such changes for political purposes. Secondly, there was no competition at JNR as it was a publicly owned monopoly organisation. With the division into regional companies, the government hoped to encourage competition and increase managerial efficiency. Several years after privatisation, there is not much information about competition policy in the JRs, but in January 1997 the government introduced an evaluation scheme called yardstick competition, which we will discuss below. Further, it is no longer necessary for the Diet to approve the investment plans of rail networks, nor is there interference in the appointment of managing directors.

There have been many changes in organisation. First, the organisational form has been converted from a public corporation, which was one governmental body, to a special corporation, a commercialised body but still regulated by special laws. Second, service areas were changed from a single nationwide passenger service to several regionally based passenger services. Passenger and freight services were separated. Newly established JRs were allowed to operate non-rail services. The most important changes have occurred in management. In the JNR era, decision making was so complex that in order to approve a single initiative, it was necessary to circulate documents among twenty to thirty people, a time-consuming process. The new JRs introduced private company-type performance rating systems. While wages at JNR were generally based on a seniority system applied to all government employees, wages at JR have to some degree begun to be determined by employees' ability and performance.

Performance changes at individual Jrs

The previous study by Mizutani and Nakamura (1997) reported on the differences in overall performance of JRs as a group between the before- and after-privatisation periods. Here, we will complement that study by providing results on the performance of individual JRs since privatisation. Our main interest is in how performance changes vary among the JRs. We selected these performance measures: financial situation; fare and service quality; demand change; accident rate; wage and labour productivity; and average operating cost. We selected three time periods for the analysis, 1987 (the first year), 1991 (the fifth year), and 1995 (the ninth or most recent year). Details of performance change at each JR are summarised in Appendix-1.

Overall performance changes as a result of privatisation

Based on our previous study (Mizutani and Nakamura, 1997), we can say that the most remarkable results of privatisation occurred in productive efficiency gains, improving the financial situation of JRs. While the operating revenue-cost ratio in the before-privatisation period was 0.771, in the after-privatisation period, it reached 1.167, a 51 per cent improvement. Such financial gains occurred on both the revenue and cost sides.

As for fares, the most important fact is that while fares had been increasing yearly before privatisation, the only fare revision in nine years after privatisation occurred in April 1989, when it was necessary to accommodate the consumption tax introduced in Japan. However, it is worth noting that the average fare after privatisation was higher than that before privatisation, perhaps partially accounting for the improved financial circumstances at the JRs.

Table 12.2 Regulatory and organisational differences: before and after privatisation

items	before privatisation (JNR)	after privatisation (JR)
regulatory aspects		
(1) fare system	approval by the Diet.	approval by Transport Minister.
(2) competition scheme	lack of competition scheme.	conceptual installation of yardstick competition scheme effective after January 1997.
(3) investment plan	government investment in capital subject to approval by the Diet.	freedom by JR to invest without the Diet approval but Transport Ministry assesses.
(4) appointment of directors	chairman of JNR was appointed by the cabinet directors were appointed by the chairman subject to approval of the Transport Minister.	no regulations except that it is required that representative directors be approved by the Transport Minister.
organisational aspects		
(1) organisational form	public corporation (national government).	special corporation.

(2) service areas	one corporation for all Japan.	six regional passenger corporation and one freight corporation for all Japan.
(3) rail service	mixed passenger and freight service.	separated passenger and freight services.
(4) scope of business	rail-related service only non-rail business was very limited.	non-rail business (e.g. real estate development, tourism) allowed.
(5) management of organisation	complex decision-making system essentially government-type seniority system.	simplified organisation introduction of private company-type performance rating system.

Note:

This table was made by the authors based on several sources: Mizutani and Nakamura (1996), Mizutani and Nakamura (1997), and Mizutani (1997b).

In terms of service quality, frequency was much improved among JRs. On the other hand, car availability measured by the number of cars per passenger or freight did not improve. JRs took the strategy of using cars more efficiently by increasing load factors. A more micro-based analysis by Sone (1992), who did a service quality evaluation on the five years after privatisation, showed that service quality in major rail markets has improved in terms of frequency, travel time and speed, departure time of the first and last train, and so on. However, he noted that service problems remained, such as inadequate face-to-face customer assistance and queuing congestion at ticket booths. In addition to these problems, we would like to note the particularly heavy overcrowding on trains in Tokyo and Osaka.

Fourth, wages increased by 11-12 per cent after privatisation, possibly to give incentives to improve efficiency, to compensate workers for their new, multi-faceted positions, or to recruit more capable workers who might otherwise be drawn to private railways.

Finally, productivity has improved dramatically. According to Mizutani and Nakamura (1996), labour productivity increased remarkably after privatisation improving by about 29 per cent over ten years, eliminating factors such as network and output differences by using regression models. Labour productivity improvement represents the biggest factor in efficiency gains. As a result the cost reduction effect of privatisation was about 3.2 per cent to 26.5 per cent.

Individual JR's performance change in the post-privatisation period

Revenue side

The important question here is whether or not there are differences among the JRs in performance change in the post-privatisation period. First, rail fare as measured in 1995 value has continued to decrease at each JR. There was no fare level difference between the Honshu JRs and the Three-Islands JRs in 1995: the average fare level was 16.41 yen per passenger kilometre for Honshu JRs and 15.86 yen per passenger kilometre for Three-Islands JRs.[3] However, it should be mentioned that because of the decrease in interest revenue from the Management Stability Fund, the Three-Islands JRs, JR Hokkaido, JR Shikoku, and JR Kyushu increased fares by about 6.4 per cent to 7.3 per cent (Syukuri, 1996), perhaps indicating that the fare level difference among JRs, especially between Honshu JRs and Three-Islands JRs, might be greater in the future.

As for service quality in the post-privatisation period, it can be clearly shown that each JR has been increasing frequency. The train density of all JRs, measured by dividing train kilometre by route kilometre, has been increasing constantly in the post-privatisation period. There are still big differences in the measure of train density between Honshu JRs and Three-Islands JRs: the train density in 1995 was 40,391 for Honshu JRs but 22,917 for Three-Islands JRs, reflecting the different demand conditions between them. However, an interesting result shows that the growth rate of the Three-Islands JRs in train density surpassed that of the Honshu JRs: the growth rate between 1987 and 1995 was 1.232 for Honshu JRs but 1.420 for Three-Islands JRs. This result corresponds with the idea that typical options taken by JRs to improve service quality have included increasing frequency (Choami, 1992; Ishi et. al. 1994). Among JR lines competing with private railways near large cities, a remarkable increase in frequency has occurred (Nakamura and Mizutani, 1995). However, car availability as measured by the number of cars per passenger did not increase much in the post-privatisation period, although the Three-Islands JRs seem to have had a slight increase.

Both passenger service and freight have certainly shown an increase in demand since the beginning of privatisation in 1987. However, some JRs seem to have experienced less demand quite recently, in 1995, than at their peak (e.g. JR Central, JR Shikoku, and JR Freight). JR Freight and JR Shikoku, operating on the smallest of the four major Japanese islands, are concerned about loss in demand to their rivals, the private auto and truck.

Finally, the accident rate, as measured by dividing the number of all accidents by train kilometres, has been declining in the post-privatisation period. As of 1995, the Three-Islands JRs had a higher rate than the Honshu JRs. However, the improvement rate between 1987 and 1995 is high for both groups.

Cost Side

Wages as measured by standard monthly salary per employee have grown continuously. In 1995, the wage level of each JR was 3 to 32 per cent higher than in 1987. This rate varies among individual JRs, however, Honshu JRs having a higher growth rate than Three-Islands JRs or JR Freight. Although in order to analyse wage differences it is necessary to analyse the details of corporate structure, such as age distribution and organisational structure, JRs in a better financial situation seem to have the potential for further wage growth.

Table 12.3 Summary of performance improvements since privatisation: comparison between 1987 and 1995

classification	Honshu JR			Three-Islands JR			
JR operator	JR East	JR Central	JR West	JR Hokkaido	JR Shikoku	JR Kyushu	JR Freight
operating revenue cost ratio	0	++	+	++	++	+	-
average fare improvement	++	+	++	++	+	++	++
service quality (1) improvement	-	+	-	+	0	++	- -
service quality (2) improvement	++	++	++	++	++	++	++
demand improvement	++	++	++	++	++	++	++
labour productivity improvement	++	++	++	++	++	++	++
wage change	++	++	++	++	++	+	+
ave. operating cost (1) improvement	+	++	++	++	++	++	++
ave. operating cost (2) improvement	+	++	++	++	++	++	0
accident rate change	++	++	++	++	++	++	++

(Note):

(1) Symbols are as follows: ++ (more than 10% improvement), + (between 10 and 2% improvement), 0 (between 2 and - 2% improvement), - (between -2 and -10% improvement), - - (less than -10% improvement)

(2) These results are based upon the numbers between FY 1987 and FY 1995.

(3) The definition of these measures is as follows:

(a) operating revenue-cost ratio = total operating revenues / total operating costs in rail service,
(b) average fare = fare revenues / passenger-km (or ton-km),
(c) service quality (1) = car availability (car-km / passenger-km),
(d) service quality (2) = train density (train-km / route-km),
(e) demand = passenger-km (or ton-km),
(f) labour productivity = passenger-km (or ton-km) / employees
(g) wage = standard monthly salary per employee
(h) ave. operating cost (1) = operating costs (excluding depreciation) / passenger-km (or ton-km)
(i) ave. operating cost (2) = operating costs (excluding depreciation) / car-km
(j) accident rate = number of all accidents / train-km

Labour productivity has grown since privatisation. Although strictly speaking it is necessary to adjust productivity measures by considering network factors and output condition differences (Mizutani and Nakamura, 1996), the productivity growth rate of the Three-Islands JRs and JR Freight in the post-privatisation period was generally higher than that of Honshu JRs. The Ministry of Transport uses labour productivity as an evaluation measure in its yardstick competition scheme, prompting individual JRs to improve their productivity. Three-Islands JRs and Freight have had higher growth rates perhaps because there had been more organisational slack in these areas before privatisation than in the Honshu area. Less efficient companies have a tendency to improve their performance when yardstick competition is applied. However, it is worth noting that some proportion of productivity growth in maintenance activity was possibly caused by the transfer of input use from labour to capital (e.g. more use of machines in track maintenance and increased use of contracting out).

Finally, the average operating cost has also decreased in the post-privatisation period. The improvement rate in operating costs is similar to that in labour productivity, showing greater improvement among the Honshu JRs than among Three-Islands JRs. Even if all JRs achieve the same level of cost-saving in average operating costs, the average operating cost of the Honshu JRs is smaller than that of the Three-Islands JRs, giving Honshu JRs a higher cost reduction rate.

Financial situation regarding rail services

As a result of the changes in costs as mentioned above, each JR has seen improvements financially, as measured by the operating revenue-cost ratio. JR East and JR Freight, however, may have reached their limits in financial improvements, with JR Freight in the uneasy position of having recorded operating deficits in 1995. As for the rate of improvement, the Three-Islands JRs have seen a slightly higher growth rate than the Honshu JRs.

Several lessons from Japanese railway reform

Non-open access and competition

Competition factors

First, we must determine whether or not competition would occur as a result of privatisation. Some might argue that competition cannot occur with the Japanese privatisation because the privatised railways still have the characteristics of publicly held institutions, being highly regulated and with a licensing system protecting incumbent operators from potential entrants, lessening the threat of market loss (Mizutani, 1997b). Most stock is still held by the government indirectly, which would lower the incentive to improve performance, according to property rights theory. These arguments are partially valid. However, we believe that although a perfectly competitive situation does not exist among JRs and between JRs and other private railways, the 1987 Japanese railway reform has succeeded to some degree in creating competition in privatised JRs.

Increased incentives are most likely related to the causes mentioned above, such as the release of management from government control, thereby reducing the possibility of directors and managers being appointed for political reasons. Employees have begun to be evaluated by a private company-type system and are given higher wages. Competition has occurred among JRs, much of whose stock is held by the government. Full privatisation, that is the sale of all remaining shares of the JRs, depends upon the efficiency improvements brought about by competition. Furthermore, because the Japanese fare system is based on the full cost principle, subsidies are virtually unavailable for private railways (Mizutani, 1994), so that the rail companies are eager to find ways to increase ridership for increased revenue. Even if there is no competition from new entrants into the Japanese rail market, there is room for competition to occur to some degree among existing railways.

Empirical evidence

Yardstick competition among JRs

The government has not made public its assessment of the degree of competition among JRs since privatisation, so we will use statistical data to evaluate whether or not yardstick competition is working. Our method is the same as in a previous study (Nakamura and Mizutani, 1995), which is as follows. If yardstick competition works, then the average performance of the group increases while the differences among operators decrease. Therefore, conventionally, we can assess whether or not the sample mean of the operator group improves and the coefficient of variation of the operator group becomes smaller. We take three performance measures: operation revenue-cost ratio in railway service; labour productivity; and average operating cost. We evaluate whether or not these measures improved between 1987 and 1995. Table 12.4 shows the results. The average performance of JRs increases over time and their

Table 12.4 Summary of results of yardstick competition

Measure	Hypothesis		Actual Results Honshu JRs		Actual Results Three-Islands JRs	
	mean	difference	mean	difference	mean	difference
operating revenue-cost ratio	increase	smaller	increase	**larger**	increase	smaller
labour productivity	increase	smaller	increase	smaller	increase	smaller
average operating cost	decrease	smaller	decrease	smaller	decrease	smaller

Note:
(1) The definition of these measures is as follows: operating revenue-cost ratio (operating revenues / operating costs), labour productivity (passenger-km / number of employees), average operating cost (operating costs excluding depreciation / passenger-km).
(2) The 'mean' is the sample mean and 'difference' is coefficient of variation.
(3) These results are based on the numbers in 1987 and 1995.
(4) Honshu JRs include JR East, JR Central and JR West. Three-Islands JRs include JR Hokkaido, JR Shikoku and JR Kyushu.

performance difference becomes smaller in most cases. Although our method is simple, we believe that it indicates the existence of competition to some extent. However, it is worth noting that some other measures related to service quality do not always support our hypothesis. Measures related to productive efficiency show reasonable results.

Competition between rail lines

The competition effect was not simple when considering direct competition among rail lines (Nakamura, 1995; Nakamura and Mizutani, 1995). This kind of direct competition would likely occur in parallel lines where JR operates alongside private railways, usually typical commuter railways connecting central cities with neighbouring 'suburban' cities. To compete with private railways, new JRs typically took the strategy of improving service, notably of increasing frequency and speed. However, fare competition was limited by continuing regulation by the Ministry of Transport, which most likely feared damaging the financial situation at the JRs by reducing fare revenue.

Severe competition has occurred on some parallel lines where JRs' passenger share had been low, and which had been dominated by private railways. Several reasons have been offered as to why these lines had been so thoroughly dominated by private railways. As a single, gigantic, nationwide organisation of urban as well as intercity, passenger as well as freight, JNR had not been able to devote sufficient attention to the details of local markets. Meanwhile, private railways had been developing ridership from early on by diversifying business, branching out into housing development and retailing. Some private railways even developed their own bus lines, so that the market was secured simply by the convenient accessibility of bus transportation, even though the market was relatively close to the JNR station (Yoshikawa et. al., 1983).

Compared to the competition of JRs with parallel private lines, the situation where JNR had an adequate share does not show as marked a change. Even though performance results of both JRs and private railways have not changed much since privatisation, this does not mean that there is no competition in this market. Improvements such as increases in frequency and speed have already reached their limit, showing a balance of attractiveness between the JRs and private railways. Very recently, aggressive behaviour has been reported to have occurred at JR West, which has been working to better its services in the Kyoto-Osaka-Kobe area by increasing frequency, increasing speed, and introducing direct-train services, among other changes. This behaviour has slightly affected the share balance in the Kyoto-Osaka-Kobe area, which could perhaps provoke surrounding private railways to reorganise themselves into a unified railway network in competition against JR West.

Intermodal competition

An example of intermodal competition would be that between air transportation and the Shinkansen network system operated by the JRs. Our earlier study (Nakamura and Mizutani, 1995) analysed JR-air competition in three typical intercity markets for the years 1975 and 1992. There was no clear statistical evidence that the JRs' share had improved much in the interval. Only the Osaka-Fukuoka market changed significantly. However, we believe that lack of statistical evidence does not necessarily indicate a lack of competition. We can see evidence of aggressive behaviour by JR in some markets, particularly in the case of JR East, which has introduced a direct Shinkansen service between Tokyo and Yamagata, Northeast of Tokyo (a journey of about 3 hours). This service was created by using both the Tohoku Shinkansen and a conventional rail line. A regular super-express train, called the Mini-Shinkansen, is coupled to a regular Shinkansen and travels along the special Shinkansen track, then is decoupled in order to travel along conventional tracks for the remainder of the journey. This scheme obviates the need for new construction of Shinkansen tracks. While conventional tracks must be modified for Mini-Shinkansen use, the cost of such modification is minimal relative to that of new construction. Another advantage is that regular tracks can be used by both regular trains and the mini-Shinkansen, while regular Shinkansen tracks cannot. The Mini-Shinkansen's running with the regular Shinkansen has reduced travel time considerably, perhaps taking away market share from air transport. JR East has also introduced such Mini-Shinkansen service between Tokyo and Akita, Northeast of Tokyo (a journey of about 5 hours), providing further evidence that competition may exist between rail and air.

Horizontal separation and optimal size

Regional subdivision of the passenger rail service seems to be functioning well so far. First, yardstick competition has improved the overall performance of the JRs. Second, more regional needs have been met, particularly with improvements in frequency. Third, as for the integration of railway services into different regional organisations, not many problems have been reported, although the number of inter-regional rail services has decreased.

One problem, however, is related to the size of the different organisations created from the former JNR. The six regional passenger rail companies differ greatly in size, population and density. The Honshu JRs (JR East, JR Central and JR West) are in a highly advantageous position compared to that of the Three-Islands JRs (JR Hokkaido, JR Shikoku and JR Kyushu). The government has taken note of the differences in demand conditions, and has installed a lump-sum subsidy scheme under the auspices of the Management Stability Fund, interest revenues from which supply subsidies to the less fortunate Three-Islands JRs. This dependence on interest revenue, however, makes the Three-Islands JRs susceptible to dips in the stock market. and these JRs have certainly been adversely affected by the recent low interest rates in Japan. The tenuous situation of the Three-Islands JRs has made some critics advocate their absorption by the more stable Honshu JRs.

In fact, as Preston (1996) noted, the optimum railway size for minimising operating costs might have a network of around 4,000 km and run 120 million train-kms per annum. His results would imply that the Three-Islands JRs are at a disadvantage (Mizutani and Nakamura, 1997). These railways, especially JR Shikoku, may simply be too small to be fully privatised independent rail operators.

Vertical integration: operation and infrastructure integration

The relationship between operation and railway infrastructure can take many forms (Brooks and Button, 1995). In the case of the British Railways privatisation, rail operation was separated from infrastructure. One important characteristic of the newly established JRs is that such separation does not take place; the same organisation both owns the tracks and operates trains on them. Of the new companies, only JR Freight does not own most of its infrastructure but borrows it from the other JRs. Cases involving only operation or only ownership of infrastructure are very limited in Japan, so we will discuss vertical integration or separation issues based upon the relationship of passenger and freight JRs.

The most striking problem with operation and infrastructure separation is scheduling difficulty. JR Freight has complained that its schedule is of lower priority than the passenger services of the JR companies which own the tracks, showing perhaps that track holders in profitable, high volume markets can afford to give track renters less than optimal use of facilities.

A second problem with separation is the conflict between operation and track maintenance activities. An operator wants a track holder to choose a time most advantageous to itself for rail operation activities, which in Japan's highly dense railway industry would be the time of least service activity - from midnight to early morning. However, in order to attract quality maintenance employees in Japan's expensive labour market, the track holder might need to offer daytime work instead. A compromise would have to be reached between the track holder and the operations company, but transaction costs would be huge, and train scheduling would be very difficult were the operations company to insist upon interrupting daytime scheduling with daytime maintenance activity.

Finally, companies which own their own infrastructure as well as operations would likely have a long-term commitment to the full development of the communities served. Private companies in Japan have long shown this commitment by developing real estate and shopping establishments along rail lines. A company involved only in operations might not share this long-term commitment, in the uncertainty that it may lose its operating licence at some point. It is conceivable that an operations-only company, with its attention focused only on fulfilling the minimal promises of a written contract, might lose the entrepreneurial behaviour characteristics of a private company. Service quality might therefore suffer. In the long-run, the location of households would change. On the other hand, it is not easy to predict a track holding company's behaviour. If the track holding company were to receive revenues from the rail operations company as a track user charge, the track holding company would probably be less interested in real estate development than a company which owns tracks as well as operations. If the track holding company were to receive revenues directly from rail users, it would have an incentive equivalent to a company which owns tracks as well as operations. Conversely, an operations-only company would lose the incentive to increase market share in the long run. Thus, it would seem that the separation of operations from infrastructure might damage long-term commitment to the community. Railways which have been private from the beginning have shown this commitment to the development of areas along their lines and are often considered preferable in terms of both housing environment and rail service quality.

Investment in rail facilities

Have the privatised JRs invested more or less in their railways since privatisation? The former JNR's investment practices were often the result of political deals, so that it engaged in economically unjustifiable projects such as building rural lines for which there was clearly insufficient demand. This political meddling was the cause of JNR's huge deficits. Thus, it would seem that the new JRs would spend less money than the old JNR, at least for political reasons. However, the Averch-Johnson effect (1962) might apply here, that a private company under rate-base regulation tends to overcapitalise in order to make more profits. The three Honshu JRs, to which rate-base regulation is applied, as well as large private railways, have incentive to overcapitalise.

Cost function is necessary for the assessment of optimal capital size and capital adjustment but here we simply show the annual investment, our main concern being the magnitude of investment changes since privatisation. To clarify this behaviour we divided the time into three periods: before (1980-83), during (1984-86) and after (1987-95) privatisation. We adhere to the calculation method which the Japanese Economic Agency used to evaluate Japanese railway infrastructure.[4] As there had been massive investment in Shinkansen construction (Tohoku and Joetsu Shinkansen) during the 1970s and the beginning of the 1980s, we subtracted the amount of each year's Shinkansen investment. However, because the years 1982, 1987 and 1991 had abnormal circumstances, such as the completion of the Shinkansen networks from the Shinkansen Holding Corporation, these years are eliminated from the evaluation.[5]

Based on the calculation of annual investment in railway infrastructure and facilities, we got the result that investment after privatisation clearly became smaller than before privatisation. The annual investment of JRs was on average about 482 billion yen after privatisation (1988-95) in 1990 value, greatly contrasting with the investment 906 billion yen during the JNR era (1980-83). This means that annual investment after privatisation decreased by about 47 per cent. The large investment during the JNR era can also be seen in another measure. Annual investment per annual fare revenue was 0.307 before privatisation, but decreased to only 0.127 after privatisation. Other sources with similar results support our findings. For example according to the Ministry of Transport (1996), annual investment was about 751 billion yen before privatisation but 494 billion yen after privatisation, both in 1990 value. Although more detailed study is necessary in order to show to what extent the reduction can be attributed to over-investment in the JNR era, these results show that JRs have spent less money since privatisation.

The management stability fund as lump-sum subsidies

There are two issues to be discussed concerning the Management Stability Funds, given to the Three-Islands JRs (JR Hokkaido, JR Shikoku, and JR Kyushu). First, while traditional operating subsidies tend to cause wasteful cost escalation (Pucher and Markstedt, 1983; Pickrell, 1985), the limited revenues from the Fund seemed to work effectively to improve performance (Mizutani and Nakamura, 1997). Although it is necessary to do a more detailed analysis to determine whether our hypothesis is correct, our first estimation using statistical data supports our idea that the subsidies have helped to improve performance. When we applied regression analysis,[6] we got these results: the average operating cost in the year following receipt of the subsidy has a positive relationship with the average subsidy, but with the lump-sum subsidy dummy variable, the relationship is negative. We believe this shows that the lump-sum subsidy scheme, which provides limited funds regardless of performance, might suppress the effect of increases in operating costs.

An aforementioned disadvantage of the Fund is that the revenues are subject to variations in the external market. If interest rates are stable, JR companies can expect stable revenues and increases in their profits, as operating deficits would presumably be covered by Fund revenues. This mechanism worked during the several years after privatisation, but the recent depression in the Japanese market has demonstrated the vulnerability of the Three-Islands JRs, which were forced to increase fares in January 1996. An amended or alternate subsidy plan is desirable, but so far none has been proposed.

Concluding remarks: current issues in Japan

The privatisation of the Japan National Railway has brought about improved efficiency. Productivity increased dramatically after privatisation because redundant employees were moved to the JNR Settlement Corporation and partly because JR management and employees became cost-conscious, making a commitment to improving service quality. As a result, operating costs decreased (Mizutani and Nakamura, 1997), greatly improving the JRs' financial situation. The three larger Honshu JRs (JR East, JR Central, JR West), have maintained good performance levels throughout the privatisation period and seem to be on the road to success. The smaller Three-Islands JRs (JR Hokkaido, JR Shikoku, JR Kyushu), and JR Freight have been improved in many ways, but face problems which make their future dimmer than that of their Honshu counterparts. The Management Stability Fund in particular causes financial concern, as it is susceptible to market fluctuation. The way to full privatisation is unclear in the case of the Three-Islands JRs.

The biggest challenge in the privatisation process is the liquidation of the long-term debts of JNR Settlement Corporation. As of 1996, remaining long-term debt had swollen to 27.6 trillion yen, up from the 25.5 trillion yen debt in 1987 at the time of privatisation. The debt grew for several reasons. First, most of the debt consists of loans from government investments and loan programmes with high interest rates, making interest payments formidable. Second, the planned sale of JNR land was postponed during the economic boom of 1980s, when fears were expressed that such sales would accelerate the inflation of land prices. This delay has now reduced the value of land held by the JNR Settlement Corporation. It is inevitable that taxpayers will be called upon to share the burden of the debt. Finally, the Japanese government must bear much of the blame for the staggering debt the JNR Settlement Corporation faces. Even after reluctantly acknowledging that taxpayers' money must be used to help repay the debt, because of fear of the political consequences, the government has allowed debt to grow by procrastinating in setting up a detailed repayment plan. Several proposals for dealing with the debt problem have been offered, but the solution remains elusive.

References

H. Averch and L. Johnson (1962) 'Behaviour of the Firm Under Regulatory Constraint', *American Economic Review,* Vol. 52, pp. 1053-1069.

M. Brooks and K. Button (1995) 'Separating Transport Track From Operations: A Typology of International Experiences', *International Journal of Transport Economics,* Vol. 22, No. 3, pp. 235-260.

Y. Choami (1992) 'Five Year's Progress of JR and Its Future Tasks: JR Kyushu' (JR 5 Nen no Kiseki to Kadai: Kyushu Ryokyaku Tetsudou Kabushiki Kaisya), *Transportation and Economy (Unyu to Keizai),* Vol. 52, No. 12, pp. 53-61 (in Japanese).

K. Fukui (1992) *Japanese National Railways Privatisation Study: Japanese Experience and Lessons for Developing Countries*, World Bank

Discussion Paper No. 172, The World Bank, Washington D.C., U.S.A..

K. Fukui, K. Nakamura, T. Ozaki, H. Sakamaki, and F. Mizutani (1994) *Japanese National Railways Privatisation Study II: Institutionalising Major Policy Change and Examining Economic Implication,* CFS Discussion Paper No.107, The World Bank, Washington D.C., U.S.A..

N. Goda (1992) 'Five Year's Progress of JR and its Future Tasks: JNR Settlement Corporation' (JR 5 Nen no Kiseki to Kadai: Nihon Kokuyu Tetsudo Seisan Jigyodan), *Transportation and Economy (Unyu to Keizai)* Vol. 52, No. 12, pp. 77-81, (in Japanese).

J. Gomez-Ibanez and J.R. Meyer (1993) *Going Private: The International Experience with Transport Privatisation,* The Brookings Institution, Washington D.C., U.S.A..

Y. Ishii, K. Okada and T. Yada (1994) 'Railway Management in the Age of Changes: Kyushu Railway Company' (Henkakuki no Tetsudo Keiei: Kyushu Ryokyaku Tetsudou Kabushiki Kaisya), *Transportation and Economy (Unyu to Keizai),* Vol. 54, No. 4, pp. 5-40 (in Japanese).

T. Kamakura (1986) *Objection to 'Reform of the Japan National Railways' ('Kokutetsu Kaikaku' wo Utsu),* Ryokufu Syuppan, Tokyo, Japan (in Japanese).

B.J. Kileen and K. Shoji (1997) 'Diversification Strategy and Urban Transportation: The Case of Japan', paper presented at the *5th International Conference on Competition and Ownership in Land Passenger Transport,* Leeds, U.K., 27-30 May.

H.Y. Kim (1987) 'Economies of Scale and Scope in Multiproduct Firms: Evidence From US Railroads', *Applied Economics,* Vol. 19, pp. 733-741.

A. Kusano (1989) *Reform of the Japan National Railways (Kokutetsu Kaikaku),* Chuo Koronsya, Tokyo, Japan (in Japanese).

Ministry of Transport, Railway Bureau (ed.) (1992) *Issues and Achievements in the Five Years Since the Japan National Railways Reform,* Japan Transport Economics Research Centre, Tokyo, Japan.

Ministry of Transport, Railway Bureau (ed.) (1996) *Compendium of Railway Law (Chukai Tetsudo Roppou),* Daiichi Hoki, Tokyo, Japan (in Japanese).

Ministry of Transport, Railway Bureau (ed.) (1996) *Rail Fact Book '96 (Suji de Mirn Tetsudo '96),* Japan Transport Economic Research Centre, Tohyo, Japan.

F. Mizutani (1994) *Japanese Urban Railways: A Private-Public Comparison,* Avebury, Aldershot, U.K./ Brookfield, U.S.A..

F. Mizutani (1997a) 'Empirical Analysis of Yardstick Competition in the Japanese Railway Industry', *International Journal of Transport Economics,* Vol. 24, No. 3, pp. 367-392.

F. Mizutani, (1997b) *Research on Railway Competition: Phase 1 (Regulation and Organisation of Japanese Rail Industry),* paper prepared for the Dutch Ministry of Transport, and Dutch Railway , September 1, mimeo.

F. Mizutani and K. Nakamura (1996) 'Effects of Japan National Railways' Privatisation on Labour Productivity', *Papers in Regional Science,* Vol. 75, No. 2, pp. 177-199.

F. Mizutani and K. Nakamura (1997) 'Privatisation of the Japan National Railway: Overview of Performance Changes,' *International Journal of Transport Economics,* Vol. 24, No. 1, pp. 75-99.

K. Nakamura (1995) 'An Analysis of Benchmark Competition: The Case Study of Privatised Japanese Railways,' *Waseda Commercial Review,* No. 363, Waseda University, Tokyo, Japan.

K. Nakamura and F. Mizutani (1995) 'The Effects of Railway Privatisation on Competitive Performance: A Case Study of Japanese Railways,' *Journal of the Eastern Asia Society for Transportation Studies*, Vol. 1, No. 1, pp. 85-102.

D.H. Pickrell (1985) 'Rising Deficits and the Uses of Transit Subsidies in the United States', *Journal of Transport Economics and Policy,* Vol. 19, pp. 281-298.

J. Preston (1996) 'The Economics of British Rail Privatisation: An Assessment,' *Transport Reviews,* Vol. 16, No. 1, pp. 1-21.

J. Pucher and A. Markstedt (1983) 'Consequences of Public Ownership and Subsidies for Mass Transit: Evidence from Case Studies and Regression Analysis,' *Transportation,* Vol. 11, pp. 323-345.

M. Tateyama, (1989) *JR's Bright Side and Dark Side (JR no Hikari to Kage),* Iwanami Syoten, Tokyo, Japan (in Japanese).

M. Tateyama (1992) *Five-Years Assessment of Privatisation of the Japan National Railways ('Kokutetsu Minei Bunkatsu' Gonenme no Kessan),* Kenyukan, Tokyo, Japan (in Japanese).

T. Saito (1993) *Private Rail Industry: Development of Japanese Style Railway Management (Shitetsu Sangyo: Nihongata Tetsudo Keiei no Tenkai),* Koyo Syobo, Kyoto, Japan (in Japanese).

S. Sone (1992) 'JR's Diagram, Rolling Stock and Operations in the Past' (Daiya, Syaryo, Eigyo ni Miru JR no 5 nenkan), *Transportation and Economy (Unyu to Keiai),* Vol. 52, No. 8, pp. 39-51 (in Japanese).

T. Suga (1997) 'The Separation of Operations from Infrastructure in the Provision of Railway Services: Examples in Japan', In the European Conference of Minister of Transport (ECMT)(ed.), *The Separation of Operations from Infrastructure in the Provision of Railway Services,* OECD, Paris, France, pp. 155-176.

M. Syukuri (1996) 'Fare Revision of JR Railway Companies: JR Hokkaido, JR Shikoku, and JR Kyushu (Kokutetsu Kaikakugo Hajimete no Unchin Kaitei: JR Hokkaido, JR Shikoku, and JR Kyushu),' *Transportation and Economy (Unyu to Keiai),* Vol. 56, No. 4, pp. 5-17 (in Japanese).

K. Yoshikawa, M. Haruna, F. Mizutani, and T. Mori (1983) 'Analysis of Impacts of Railway Facility Improvement for Railway Network Planning: Focusing on Neighbours of Station' (Tetsudo Shisetsu Seibi Keikaku no tame no Eikyo Kouka no Bunseki: Ekiseiken wo Chuushin to shite), *Proceedings of 37th Annual Conference of Japan Society of Civil Engineers, Kansai Division,* Section-IV, pp.137-138, Japan Society of Civil Engineers.

Notes

1. Ninety-five per cent of all trips would be completed within the borders of these JR passenger operators' regions.
2. It is important that Japanese accounting rules strictly separate these activities from rail services, but these businesses work to enhance the main rail business.
3. Among the six passenger JRs, only JR Central seems to have a higher fare level (i.e. 21.10 yen per passenger kilometres), due to the fact that most of the rail services by JR Central are Shinkansen services with normally higher operating costs, so that the average fare tends to be higher than regular train services.
4. Gross investment is calculated as follows:

   $I_t = (K_t - K_{t-1}) + (T_t - T_{t-1}) + r(K_{t-1} + T_{t-1}) + D_t$

   where: I = gross investment

   K = tangible fixed asset (property, plant and equipment)

   T = construction in progress

   D = depreciation

   r = rate of consumption in fixed assets

   t = year.
5. In 1982 the Tohoku and Joetse Shinkansen were opened. Although the Shinkansen networks had been leasing to JR companies since

privatisation, in 1991, the Shinkansen networks were sold to the threc Honshu JRs (JR East, JR Central and JR West). With the privatisation of the JNR in 1987, all assets held by the JNR were divided among several new organisations such as the JR companies and the JNR Settlement Corporation. These incidents show some statistical abnormalities in JR (JNR capital stock).

6. The regression formula is as follows:
   $AVOC_{t+1} = a + (b + g\ LUMP)\ AVSUB_t + d\ T$
   where:
   $AVOC_{t+1}$= operating cost excluding depreciation per car-km in year t+1
   $AVSUB_t$ = subsidies per car-km in year t
   LUMP = dummy variable (if the lump-sum subsidy = 1, otherwise = 0)
   T = trend variable (T = year - 1974)
   The sample before privatisation is from 1975 to 1986 for all JNR data set and the sample after privatisation is from 1988 to 1995 for Three-Islands JRs. Our hypothesis is that the lump-sum subsidy dummy has the negative effect on the average operating cost.

Appendix -1 Performance Results of Each JR After Privatisation

	Fiscal Year	JR East	JR Central	JR West	JR Hokkaido	JR Shikoku	JR Kyushu	JR Freight
operating revenue cost ratio	1987 (1st year)	1.233	1.089	1.104	0.575	0.684	0.819	1.069
	1991 (5th year)	1.288	1.342	1.156	0.652	0.896	0.869	1.032
	1995 (9th year)	1.255	1.515	1.165	0.692	0.817	0.888	0.985
	1995/1987	0.994	1.391	1.055	1.203	1.194	1.084	0.921
average fare (yen / passenger-km or yen / ton-km) (1995 value)	1987 (1st year)	15.20	22.74	16.83	18.02	17.93	15.80	8.88
	1991 (5th year)	14.08	21.27	15.73	16.24	18.00	14.26	7.52
	1995 (9th year)	13.30	21.10	14.84	16.09	17.35	14.14	6.93
	1995/1987	0.875	0.928	0.882	0.893	0.968	0.895	0.780
service quality (1) car availability (passenger(freight) car per 1000 passenger (ton))	1987 (1st year)	17.58	19.40	22.96	29.62	31.54	27.32	66.40
	1991 (5th year)	17.07	19.12	22.02	30.98	28.78	31.45	54.83
	1995 (9th year)	16.80	19.90	22.39	31.80	31.84	31.26	57.99
	1995/1987	0.956	1.026	0.975	1.074	1.010	1.144	0.873
service quality (2) train density (passenger(freight) train per route)	1987 (1st year)	30,038	39,406	28,910	10,584	18,217	19,619	7,434
	1991 (5th year)	36,298	48,964	36,505	14,114	22,523	27,951	9,034
	1995 (9th year)	34,908	48,764	37,500	14,968	24,240	29,543	8,773
	1995/1987	1.162	1.237	1.297	1.414	1.331	1.506	1.180
demand (passenger-km or ton-km) (million)	1987 (1st year)	104,491	41,148	45,782	3,920	1,673	7,664	20,026
	1991 (5th year)	125,974	52,110	53,690	4,800	2,123	8,333	26,698
	1995 (9th year)	128,599	49,508	55,484	4,787	1,986	8,633	24,702
	1995/1987	1.231	1.203	1.212	1.221	1.187	1.126	1.234
labour productivity (passenger-km or ton-km per employee)	1987 (1st year)	1,293	2,034	918	324	444	562	1,762
	1991 (5th year)	1,868	2,545	1,162	442	679	796	2,594
	1995 (9th year)	1,969	2,300	1,213	435	661	869	2,659

(thousand)	1995/1987	1.523	1.131	1.321	1.343	1.489	1.546	1.509
wage	1987 (1st year)	257,653	265,336	257,381	265,874	234,881	253,650	282,291
(yen per employee	1991 (5th year)	298,825	270,408	292,409	298,343	215,485	258,094	330,665
per month)	1995 (9th year)	340,338	308,066	323,115	309,589	291,351	261,153	291,153
(1995 value)	1995/1987	1.321	1.161	1.255	1.164	1.240	1.030	1.031
operating cost (1)	1987 (1st year)	10.32	19.52	14.21	31.00	23.87	18.73	8.18
(yen per passenger-km	1991 (5th year)	10.01	13.33	12.49	26.14	18.25	16.52	7.55
or ton-km)	1995 (9th year)	9.36	10.43	12.04	24.79	20.43	16.54	7.30
(1995 value)	1995/1987	0.907	0.534	0.847	0.800	0.856	0.883	0.892
operating cost (2)	1987 (1st year)	587	1,006	619	1,047	757	686	123
(yen per car-km)	1991 (5th year)	586	697	567	844	634	525	132
(1995 value)	1995 (9th year)	557	524	538	780	642	529	126
	1995/1987	0.948	0.521	0.869	0.745	0.848	0.771	1.022
accident rate	1987 (1st year)	1.639	0.640	1.421	1.273	3.489	2.190	0.980
(all accidents per	1991 (5th year)	n.a.	n.a.	n.a.	n.a.	n.a.	n.a.	n.a.
million train-km)	1995 (9th year)	0.630	0.321	0.805	0.695	2.844	1.321	0.806
	1995/1987	0.382	0.502	0.567	0.546	0.815	0.603	0.822
rail specialisation	1987 (1st year)	0.981	0.993	0.979	0.790	0.869	0.975	1.000
ratio	1991 (5th year)	0.962	0.996	0.983	0.842	0.843	0.851	1.000
(rail revenues per	1995 (9th year)	0.958	0.994	0.980	0.889	0.835	0.824	1.000
total revenues)	1995/1987	0.977	1.001	1.001	1.125	0.961	0.845	1.000

CHAPTER 13

The impact of deregulation of the planning system on the ridership potential of local rail networks: 1979-95[1]

Russ Haywood

Introduction

Except for the niche market of leisure travel, transport is not something that people engage in for its own sake but in order to gain access to perceived benefits such as work, retail goods and personal services. The locations of these perceived benefits are influenced by the spatial impacts of transport networks, and the locations, in turn, affect the demand for transport; the interrelationship is two way. This paper is concerned with these locations, the decision-making processes by which they are fixed, and the outcome in terms of transport influences, by reference to the product of research in Manchester and Sheffield.

The management of the locational decision-making process is the prime function of the modern land use planning system, introduced in 1947 by the Town and Country Planning Act. This was part of a raft of statutes introduced by Attlee's Labour Government which marked a deepening of state intervention, characterised by the setting up of the Welfare State. After a long debate about how state control over land development processes should be increased (Ashworth,1954) (Cherry,1974), the Planning Act left land in private hands whilst nationalising the right to develop it. Since that time, although the detail of the planning system has changed considerably, this underlying legal structure is still, with one significant exception, the basis of the system fifty years on; 'the exception being the repeal of (and failure to replace) the 'financial provisions' which sought to solve the compensation and betterment problem' (Cullingworth, 1994, p. 277). That problem arose, and indeed still arises, as the value of land is affected by its development potential which is a product of planning policy.

The driving force behind political support for regulation of land development was concern over issues arising directly out of the use of land, rather than the indirect impacts of land use patterns on transport demand. These issues ranged from the need to protect urban housing areas from noxious industry, to the need to protect farmland from suburban encroachment. Land use planning was also seen as a tool for implementation of the Welfare State; an adequate supply of suitable land was needed for public sector housing and for the development of the various welfare and educational facilities. With regard to the transport aspects of land use planning, planning ideology was dominated by the garden city model of urban form with its notions of decentralisation, self-containment and zoning (Barlow, 1940). Although road building was a central element of contemporary plans (Forshaw and Abercrombie, 1943; Abercrombie, 1944), the view was that self-containment would keep the lid on transport demand. British planning was not influenced by transit oriented ideology, even in London (Haywood, 1997).

The Conservative Government elected in 1951 emasculated the 1947 Town and Country Planning Act by the repeal of its clauses dealing with the taxation of betterment. However, in line with the 'post-war consensus' (Smith, 1990), the rest of the Act remained intact although the Conservatives encouraged the initiative for development to pass to the private sector, this being reinforced by growing national prosperity. By the early 1960s, the vision of public sector led planned dispersal had been replaced by the reality of private sector led suburban growth, characterised by growing dependency on the private car (Hall, et al, 1973). With further projected growth in incomes car ownership and road traffic were expected to rise substantially and in light of this the Ministry of Transport published 'Traffic in Towns' (Buchanan, 1963). This contained mixed messages; one about the potential for urban road building to accommodate more traffic, and the other about the environmental limits to such traffic growth. Pointedly it was not a part of the brief of the report to make recommendations about public transport and its major policy impact was to facilitate traffic growth. There was awareness amongst planners of the potential impacts of the effects of car usage with regard to accelerating the process of urban decentralisation (Clark, 1958), but the dominant policy concern was loss of farmland and country landscapes. Green belt policies were introduced (Ministry of Housing and Local Government, 1955) to limit urban sprawl but in practice, over the longer term, development leapfrogged over them and green belts have become barriers through which car drivers commute, thereby increasing average journey lengths (Department of the Environment (DOE), 1993a and b).

During the 1960s public transport entered a period of crisis, typified by the severe rationalisation of the railway network (British Railways Board, 1963) and growing financial problems for bus operators (Ministry of Transport, 1966). The debate about public transport was moving towards the development of the social and environmental case for public subsidy, rather than the reliance on absolute commercial profitability (Dickinson, 1964).

The political opportunity to put a base under public transport services was seized by one of the few women to be associated with transport policy in Britain, 'the tough and pugnacious Barbara Castle' (Garbutt, 1985, 9), who was Transport Minister in the mid-1960s Labour Government. Her 1968 Transport Act led to the setting up of Passenger Transport Authorities (PTAs) and Passenger Transport Executives (PTEs)[2] in the seven main conurbations outside London, and which came to include Manchester and Sheffield. The PTEs took over the municipally owned local bus services and were given powers to purchase local rail services from British Rail and to invest in local rail infrastructure. This marked a significant shift in power from the national to the sub-regional level.

At the same time concern about the way in which the post-war planning system had lost sight of strategic issues, owing to the pressure of coping with detailed matters, led to passage of the 1968 Town and Country Planning Act which introduced the separation of plan making into structure (ie. strategic) planning and local planning. This was followed in 1972 by the reorganisation of local government which created a two tier structure; 'structure plans' were to be produced by county councils with local plans being produced by district councils, within the policy context fixed by the structure plan. In the major conurbations new 'metropolitan' county councils were created, whereas the 'shire' county councils continued to operate outside the conurbations, with some changes to their areas of jurisdiction (Hall, 1992). The metropolitan county councils also became the Passenger Transport Authority/Executive as well as the highway authority. This reorganisation was completed in 1974, and it then seemed that the right sort of structure was in place for effective land use-transport planning integration at conurbation level.

This potential was reinforced by the fact that the early 1970s marked a shift in urban transport planning as proposals for urban motorway construction, typified by those initially included in the draft Greater

London Plan, began to run into intense public opposition once some were built and their environmental impacts became clear (Hillman, 1971). The structure plans developed for the metropolitan counties in the late 1970s took pro-public transport stances with policies which sought to steer major developments, particularly offices and retail developments, to city centres and town centres so that they would be accessible to public transport users. In turn the PTEs began to develop proposals for major investments in public transport; for example Tyne and Wear embarked on the construction of Britain's biggest rail scheme in a provincial city for years (the Metro) and Merseyside began work on the underground lines under central Liverpool to connect up its existing suburban rail networks. As local bus services in the metropolitan counties were also owned and regulated by the PTEs, they were operated as feeders to these rail services. In 1975 the South Yorkshire PTE embarked on its cheap fares policy based on the injection of public subsidy into bus services, and Greater Manchester PTE (GMPTE) developed plans for an underground railway link similar to the Liverpool scheme, but was in the investment queue behind Newcastle and Liverpool (Williams, 1985). These positive trends in urban transport and land use planning were brought to an end for all practical purposes by the crisis which engulfed the British economy in 1976; this took major state investment in public transport off the agenda and led to a downturn in the property market which meant that the strategic land use policies had little to bite on.

Deregulation of the planning system in the 1980s

The election in 1979 of the Conservative Government led by Margaret Thatcher marked a watershed in British post-war history as this government was committed to breaking the consensus, rolling back the frontiers of the state, and giving the market its head. The impact of 'Thatcherism' on the planning system has been well documented (Ambrose, 1986) (Thornley, 1992) and was typified by the view that the planning system was a brake on enterprise and economic growth which 'imposes costs on the economy and constraints on enterprise that are not always justified by any real public benefit in the individual case,' (DOE, 1985, p. 10). The attack on planning started early; for example Circular 22/80 (DOE, 1980) reduced the scope of planning control and the Local Government Planning and Land Act 1980 introduced the concept of Urban Development Areas, where planning would be taken out of the hands of elected local councils altogether and replaced by market led property development, overseen by Urban Development

Corporations (Imrie and Thomas, 1993). Enterprise Zones were also introduced in 1980 as specially designated areas where the need for developers to obtain formal planning permission was relaxed. In theory the market centred ideology of Thatcherism should have led to the privatisation of the right to develop land, but in practice this was never likely. The reasons for this were political, illustrated by the backlash when the prospect of a relaxation of green belt policy was floated, from the Government's own supporters who resided in the rural shires protected from development by such regulatory policy. Thereafter deregulation of planning was geographically selective, focused on urban areas, particularly in the inner cities.

A significant impact of the steer against the sort of strategic intervention represented by structure plans came with the abolition of the metropolitan counties in 1986; this abolition did not include the PTA/PTE structure which remained, with the PTA membership coming from the remaining lower tier of district councils. However, the contemporaneous deregulation of stage bus services, under the 1985 Transport Act, meant that local bus services would, in future, compete with local rail services rather than act as feeders to them. By 1986 an Enterprise and Deregulation Unit was operating within the Department of the Environment, which was looking at further reducing the 'unnecessary burdens' of planning control and developing a more receptive approach to development, 'recognising that there is always a presumption in its favour, unless that development would cause demonstrable harm to interests of acknowledged importance' (DOE, 1986, 21). As part of this process an important statutory instrument, the Use Classes Order (DOE, 1972), came under the scrutiny of the Planning Advisory Group of the DOE which 'had a membership heavily weighted towards property interests rather than planning' (Home, 1992, 192), and was given a brief to produce recommendations for simplification with fewer use classes. The Government followed this up by issuing a new Use Classes Order in 1987 (DOE, 1987) which formally reduced the effectiveness of planning control.

In the property boom of the late 1980s, this deregulation of planning control led to structural shifts in patterns of land use, typified by the growth of out-of-centre retail and office developments which, because of their orientation towards the road networks, helped fuel the growth in road traffic and undermine the locational basis of public transport in general and local rail networks in particular. At the time the Government was relaxed about these impacts and, in light of forecasts of large increases in traffic (Department of Transport (DoT), 1989a) declared its intention to accelerate the road building programme to accommodate them (DoT, 1989b).

However, there was growing public concern at this time over the environmental impacts of road traffic through exhaust pollution and loss of countryside from road building and associated development. These concerns led to the development of what has been termed 'the new realism' (Goodwin et al., 1991) with regard to the transportation issue, whereby the view that ever rising volumes of traffic can be accommodated through road building and traffic management is being fundamentally challenged. The resultant political pressures from these transport related issues, and other environmental concerns focused around atmospheric pollution and climate change, led to significant changes in Government attitudes towards the environment and the planning system, which was marked by publication of the consensual White Paper 'This Common Inheritance' (DOE,1990). This was a crucial part of the move away from the deregulatory stance, towards seeing the planning system as having a positive role to play as part of what became the Government's 'Strategy for Sustainable Development' (DOE, 1994a). With regard to the relationsip between land-use planning and transport, initial research (DOE, 1993a) was followed up by more prescriptive policy guidance typified by Planning Policy Guidance Note 13 (PPG13) which states that:

> ...local authorities should adopt planning and land use policies to:
> - promote development within urban areas, at locations highly accessible by means other than the private car;
> - locate major generators of travel demand in existing centres which are highly accessible by means other than the private car. (DOE, 1994b, 3)

This paper will now go on to demonstrate that the impacts of deregulation have undermined the utility of existing fixed track public transport, and made it difficult to maximise the potential locational advantages of new systems, by reference to case study research in Manchester (Haywood, 1996) and Sheffield.

Office decentralisation in Manchester after 1979

Office location is important because of its impact on the nature of the journey to work; historically offices have been concentrated in city centres and have been strongly associated with rail based commuting. The Conservatives' deregulatory stance towards planning came to have specific implications for office location policy. Decentralisation of offices was a new feature of the early post-war property market, but it had been regulated through the planning process to create new surburban locations accessible by public transport, typified by the growth of Croydon as a satellite of the City of London (Marriott, 1967). Although Greater London dominated the office market, significant office development took place in provincial cities too, especially Manchester (Daniels, 1977) (Damesick, 1979). In Greater London this strategic approach to decentralisation was ended by the abolition of the GLC in 1986 and in the ensuing cyclical upturn in the commercial property market, fuelled by deregulation of the Stock Market in October 1986, Docklands was developed as an alternative focus to the City without any strategic overview as to the desirability of this or the way in which it should be done. This abandonment of strategic planning was 'most dramatically illustrated by the inability to link transport and land use and the impact this has had on the area' (Brownill, 1990, p. 177).

The development of the office market in the South East in the early 1980s was associated with the development of high technology research and business activities in locations such as Cambridge, Milton Keynes and Bracknell, and was seen as part of a wider process of deep-seated change taking place within world capitalism,

> a historically articulated complex of transformations which concerns, simultaneously, capitalism as a social system, informationalism as a mode of development, and information technology as a powerful working instrument. It is this complex socio-economic-technical matrix that is transforming societies, and thus cities and regions (Castells, 1989, 3).

The experiences in the South East suggested that the markets for industrial, warehousing and office buildings were undergoing rapid change, with evidence of convergence which led to a debate about the way in which the 1972 Use Classes Order (DOE, 1972) classified these land uses: offices and light industry were in separate classes with planning permission needed to move between them (Henneberry, 1988). The Government responded by creating a new use class in the amended Use Classes Order:

> Class B1. Business
> Use for all or any of the following purposes:
> (a) as an office other than a use within class A2 (financial and professional services)

(b) for research and development of products or processes, or
(c) for any industrial process,
being a use which can be carried out in any residential area without detriment to the amenity of that area by reason of noise, vibration, smell, fumes, smoke, soot, ash, dust or grit (DOE, 1987, 3).

As a result, large scale commercial office developments, not just high-technology manufacturing and research and development activities, were formally freed from restrictive locational policies. In the buoyant conditions for office development in the late 1980s the creation of the B1 class meant that the tremendous pressures from developers for out-of-centre developments, already encouraged by the presumption in favour of development, were irresistible. The locational advantage to developers was the potential to use cheaper land for high value development, with easy access to the major road network and abundant on-site parking as the marketing attractions.

The background in the city of Manchester to the introduction of B1

Manchester was the archetypal nineteenth century industrial city. By 1900 it had also developed into a business centre with a well-defined central business district (CBD) which served the whole of the North West of England. In the modern era the city experienced growth in office development in the 1960s and this accelerated in the 1970s. Decentralisation from the city centre was already occurring at that time (Damesick, 1979) and the process continued as, between 1974-86, Manchester city centre's share of total office floorspace in Greater Manchester fell from 60.6 per cent to 52.8 per cent. The biggest concentrations in other districts in 1986 being Trafford (9.6 per cent), Stockport (8.9 per cent), Macclesfield (Wilmslow) (7.7 per cent) and Salford (adjacent to central Manchester) (6.0 per cent) (Law, 1991). However 'In the early 1980s ... office location policies favoured the clustering of development in existing centres, which had good services including public transport, (Law and Dundon-Smith, 1994, 1).

In the 1960s and 1970s the only location specifically designated for business use outside Manchester city centre was Wythenshawe Civic Centre (City of Manchester, 1961, 10), but little development actually took place. In the 1980s strategic policy for Greater Manchester was laid down in the Structure Plan, first adopted in 1981, but with a later and final version containing only minor amendments five major themes were identified underpinning the land use strategy for the metropolitan county:

- to secure urban concentration and regeneration,
- to redirect development and investment towards the inner core of the conurbation,
- to maintain the regional centre, (i.e. Manchester city centre),
- to improve the environment and conserve amenity,
- to conserve resources and maintain and use existing infrastructure.

(Greater Manchester Council (GMC), 1986, 1).

Within this general strategy, office location policy was interpreted very precisely: 'Office developments will normally be expected to locate in or adjacent to town centres or in Trafford Bar office centre' (1986, 4). The City Centre Local Plan (Manchester City Council (MCC), 1984) welcomed the structure plan themes of concentration and support for the regional centre, and argued that:

> Office activity is a major and vital part of the regional Centre, providing substantial, wide-ranging employment opportunities and helping to sustain, both directly and indirectly, a wide variety of other uses and activities (MCC, 1984, 43).

Together the Structure Plan and the City Centre Local Plan comprised a robust regulatory policy on office location which recognised the need to bolster the city centre, that the location of office development in the centre was an important part of this process, and that the relative ease of access by public transport to the city centre was a feature to be nurtured. The structure plan sought to steer decentralised office development to the existing suburban office centres. Both these plans were formal, statutory plans (Kitchen, 1996) with their policy content endorsed by Central Government. But the Greater Manchester Council, like the GLC, was abolished in 1986 and strategic office location policy went with it. The brevity of the subsequent DOE strategic advice highlighted its absence: 'the Councils should identify a comprehensive range of development sites for manufacturing and service industries' (DOE, 1989a, para 5).

The location of office development in the city of Manchester 1983-95

Since 1985 Manchester City Planning Department has produced annual reports on major office developments (those over 1000 square metres) in the city and these provided the raw material for this research. Only since 1989 have the reports included developments outside the city centre because it is only since then that major office completions have occurred in significant numbers in suburbs within the city's administrative area. The data includes both new construction and substantial refurbishments of existing buildings; in all cases the result has been the release of modern office developments onto the market.

The methodology followed in the research was that for each development reported as completed in any one year, the relevant planning application was identified and the floor area and number of parking spaces associated with the development were quantified. The listed completions were verified and amended where necessary. In several cases the floor area was checked with managing agents or occupiers of buildings and in many cases the level of parking provision was verified by inspection on site.

The results are set out in Tables 13.1a and 13.1b and show that the cyclical upturn in the economy in the late 1980s led to a surge in completed office developments but, significantly, the total floorspace completed outside the city centre during the boom was considerably greater than that within it. Once the boom subsided the primacy of the city centre was re-asserted, although the rate of completions fell back considerably. Figure 13.1 shows that the locations outside the city centre were suburban, mainly in South Manchester and mainly not close to the local rail and Metrolink networks. (The latter is a light rail system which was opened in 1992; although it runs through Manchester city centre, for the most part its route lies outside the City's administrative boundary.) In fact, as shown in figure 13.2, most developments were located broadly in the Princess Parkway/M56 corridor, i.e. along Manchester's most important - and most heavily trafficked - primary radial road (MCC, 1993, Map 4a).

With regard to the permitted levels of car parking, for each year, and on average over the period as a whole, there were far more parking spaces provided per square metre of office floorspace outside the city centre than inside it; 3.43 spaces per 100 square metres as compared with 0.82 spaces per 100 square metres. This tendency for out-of-centre office developments to have more on-site parking than city centre locations is a trend which had already become apparent in the USA by the mid-1980s when it was stated that, 'currently, the average suburban office development provides 3.9 spaces per 1,000 square feet' (Cervero, 1986, 65).

Table 13.1a City of Manchester major office completions: city centre (1985-95)

	Total floor area square metres	*Total floor area square feet*	*Total parking spaces*	*Spaces per 100 square metres*	*Spaces per 1000 square feet*
1985	11 314	121 786	107	0.95	0.88
1986	12 913	138 998	48	0.37	0.34
1987	13 749	147 988	71	0.52	0.48
1988	6 540	70 398	65	0.99	0.92
1989	6 888	74 145	221	3.21 (1)	2.98
1990	33 930	365 231	248	0.73	0.68
1991	35 574	382 935	305	0.86	0.80
1992/93 (2)	65 685	707 054	624	0.95	0.88
1993/94	17 028	183 098	33	0.19	0.18
1994/95	11 194	120 366	43	0.38	0.36
TOTAL/ AVERAGE	214 815	2 311 999	1765	0.82	0.76

Notes

1. In 1989 the floorspace completed in the city centre comprised only two schemes; one included a large amount of parking for a pre-existing building and the newly completed extension to it, thereby producing a high rate of provision when related only to the part of the scheme completed in 1989.

2. In 1993 the City Planning Department moved from analysing data on a calendar year basis to a mid-yearly basis e.g. 1 July to 30 June, so data for the previous period ran for 18 months from 1 January, 1992, to 30 June, 1993.

Source: Manchester City Planning Department, *Office Development in Manchester*, Reports for each year 1985 - 95.

Table 13.1b City of Manchester major office completions: outside city centre (1989 - 95)

	Total floor area square metres	*Total floor area square feet*	*Total parking spaces*	*Spaces per 100 square metres*	*Spaces per 1000 square feet*
1989	18 516	199 311	680	3.67	3.41
1990	36 675	394 779	1737	4.74	4.40
1991	54 584	587 556	2007	3.68	3.42
1992/93	21 849	235 188	783	3.58	3.33
1993/94(1)	14 685	157 909	230	1.57	1.47
1994/95(2)	16 411	176 464	138	0.84	0.78
TOTAL/ AVERAGE	162 720	1 751 207	5575	3.43	3013

Notes:

1. Data for 1993-94 includes a large divisional police headquarters (3,170 sq. metres) and a large conversion on the edge of the city centre (2,663 sq. metres).
2. Data for 1994 - 95 includes a large conversion on the edge of the city centre (10,695 sq. metres). The downturn in commercial office completions outside the city centre was therefore larger than the data in the table suggests.

Source: Manchester Planning Department, Office Development in Manchester, Reports for each year 1989 - 95.

Figure 13.1 The location of recent major developments with regard to rail routes in Manchester

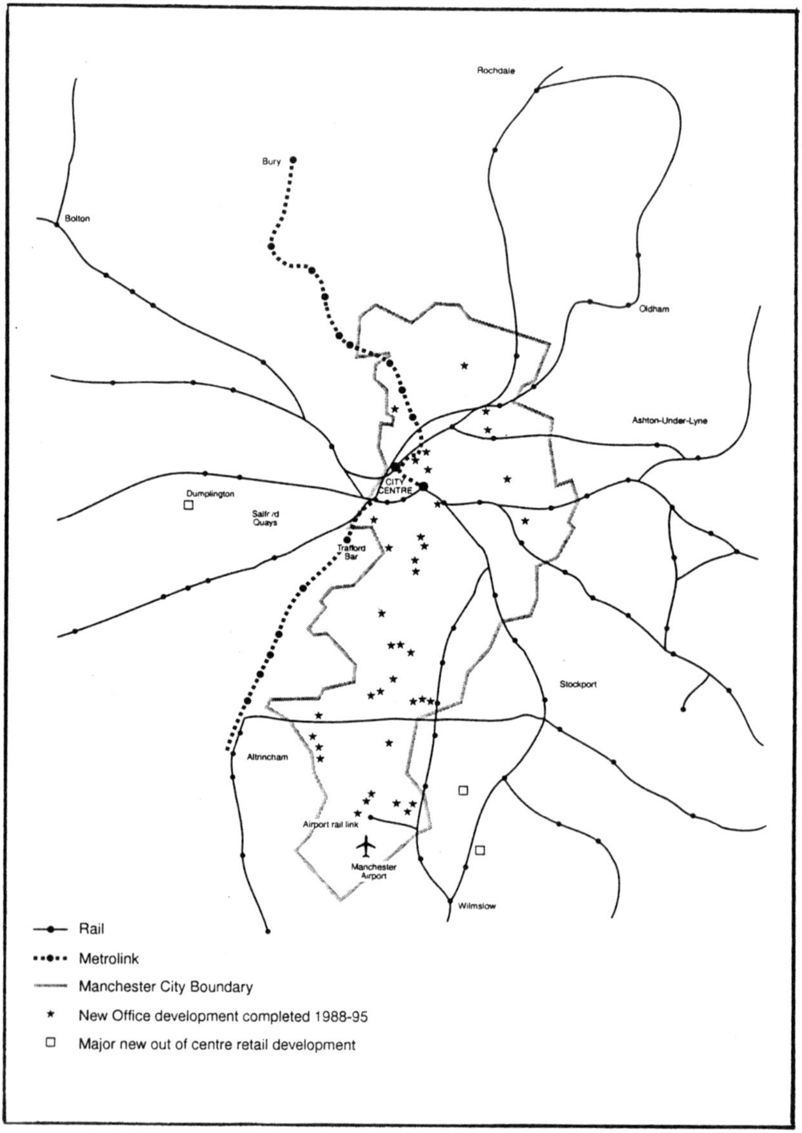

Figure 13.2 The location of recent major developments with regard to major roads in Manchester

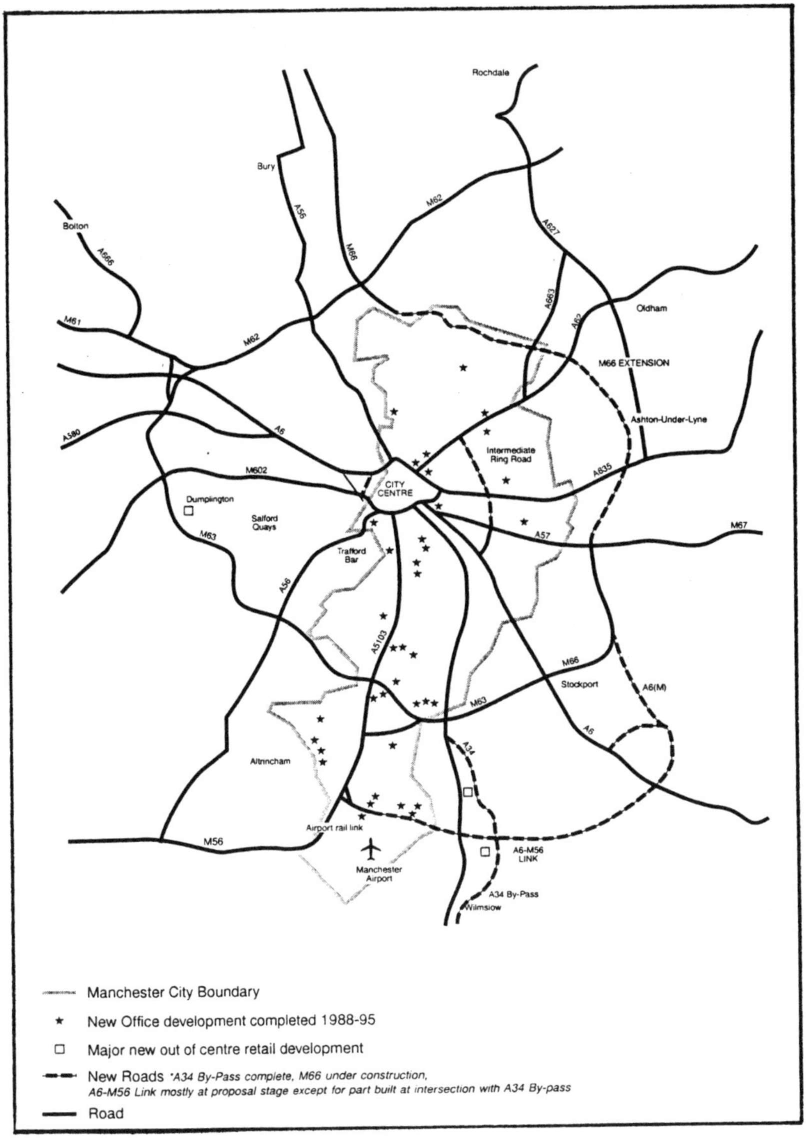

Office development trends in the Greater Manchester Conurbation: 1980-96

The research results show that the trends in the City of Manchester in the location of office developments and associated parking, support the view that the loosening of planning controls in the 1980s led to such developments occurring in locations which were qualitatively different to those which decentralised office developments occupied previously. These new locations are road oriented; this and the associated levels of on-site car parking provision will tend to increase the likelihood of employees and visitors travelling to them by car and decrease the likelihood of them using public transport. In general Manchester city centre is clearly the most accessible location for public transport in the conurbation being the hub for bus services, local rail services and Metrolink. The new suburban locations are not well located with regard to the local rail network and, even where they are reasonably close, the design and access details generally discourage rail access.

Manchester experienced strong competition in the office market from locations outside the city's administrative area at Salford Quays, in Salford District, and the adjacent area in Trafford District administered by Trafford Park Development Corporation (TPDC). These are known collectively as the Docks and are close to the long standing office centre at Trafford Bar. They contained an Enterprise Zone, designated in 1981 (Salford City Council, 1981). There was no prior passenger rail access to the Docks and none was developed in the 1980s despite the area being developed as a major office centre containing over 185,000 square metres of floorspace (Law and Dundon-Smith, 1994), in association with housing and leisure uses. These developments were the product of master planning involving the preparation of a strategy document (Salford City Council, 1985), but despite this, because of the lack of an overall interventionist strategy to co-ordinate land development with public transport infrastructure, the area was not connected to the Metrolink system and public transport access is poor (although construction of an extension to Salford Quays and Eccles is about to start at the time of writing (Modern Railways, 1997)). It has been a matter of policy at Salford Quays to achieve a minimum parking provision of 3 spaces per 100 square metres, which is comparable with the out-of-centre sites in the City of Manchester. Although the Docks is a much bigger concentration of office development than the suburban developments identified by the research in the City of Manchester and offers greater potential for access by public transport, its impact on car usage and traffic generation is likely to have been similar.

It is worth mentioning at this point that the Thatcher Government's deregulatory stance towards the planning system affected retail developments in the Manchester area as strongly as the office sector, but the impacts took much longer to come about. Figures 13.1 and 13.2 show the major out-of-centre retail developments completed in late 1995 in the south of the area in association with a new road scheme, the A34 by-pass (the developers contributed substantially towards the construction costs): these comprise 53,531 square metres of floorspace with 1909 parking spaces and although relatively close to the Manchester-Crewe railway they are not readily accessible from it. Similarly inaccesible to the local railway network is the major out-of-centre regional shopping centre (93,500 square metres of floorspace) currently under construction at Dumplington to the west of Manchester, though Metrolink may be extended out to this centre. Although these developments clearly contravene the planning policy shifts of the early 1990s they do postdate it, demonstrating the inertia in the policy implementation process.

Sheffield supertram and its regenerative impacts

The concept of light rapid transit in Sheffield originated with the Sheffield and Rotherham Land Use Transportation Study in 1976 (Martin and Voorhees et al, 1976) and was endorsed by the South Yorkshire PTE (SYPTE), then a part of South Yorkshire Metropolitan County Council, in 1978 (SYPTE, 1978). A Parliamentary Bill was promoted in 1985 to obtain powers to build the route, Line 1, from Hillsborough, through the city centre to Mosborough (see Figure 13.3); at this initial stage the rationale for the scheme was largely based on its transport benefits and its general relationship to patterns of urban form, particularly the link between Mosborough, Sheffield's 'new town' (Sheffield Corporation, 1969), and the city centre. With abolition of the South Yorkshire County Council in 1986, the support of Sheffield City Council for the scheme became crucial as the route was largely restricted to the city. Originally the city council objected but later gave their support as the proposal was expanded to include a new route, Line 2, from the city centre through the Lower Don Valley urban regeneration area to the new Meadowhall shopping mall (see Figure 13.3). Parliamentary approval for Line 1 was granted in 1988 and Line 2 was similarly approved in 1989. Line 2 figured prominently in the promotional material for the city council's bid to host the World Student Games in 1991, which highlighted the fact that regeneration objectives were becoming significant to the evolution of the project.

Figure 11.3 Sheffield: location of new transport infrastructure

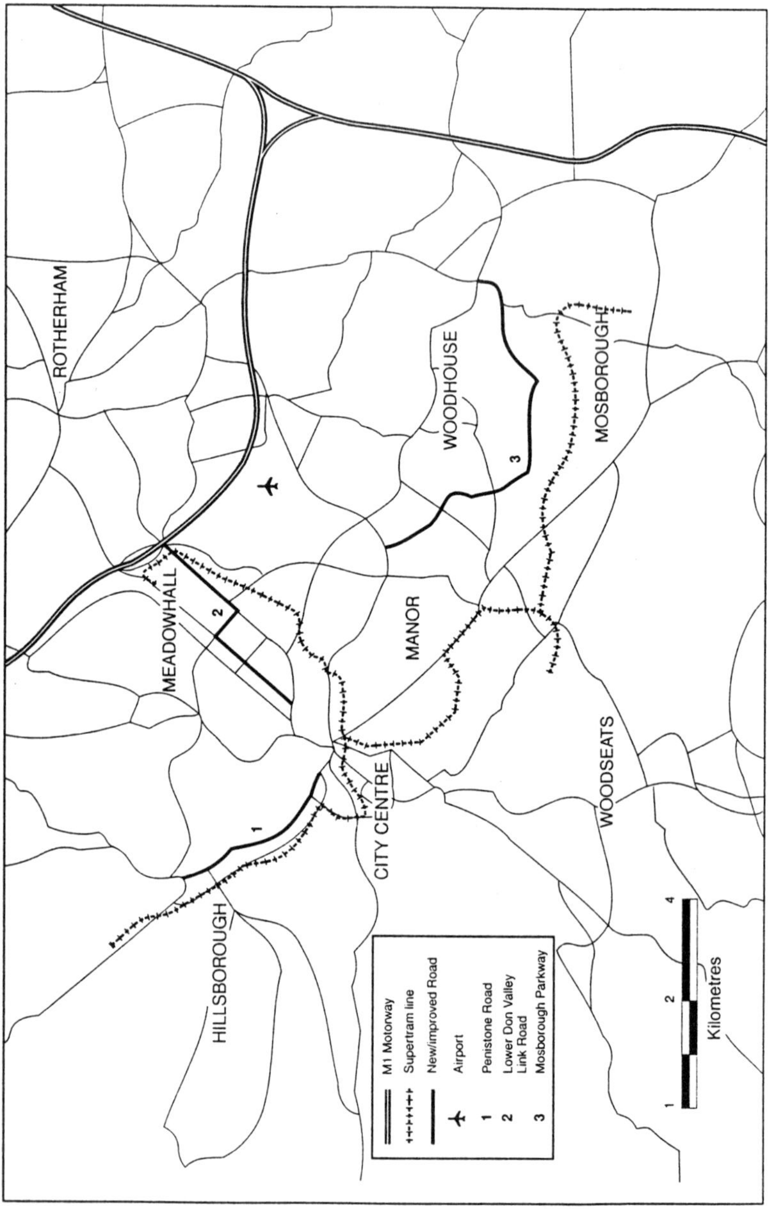

However, funding for the project was not forthcoming from the Department of Transport until 1990; although too late for the Games, the approval under Section 56 of the 1968 Transport Act was on the basis that, amongst other benefits, the scheme would provide benefits for non-users of the system owing to its positive impacts upon urban regeneration. The DoT's stance towards investment in public transport infrastructure had been changed in the late 1980s (DoT, 1989c). The users of new, and better, transport facilities would pay for their benefits through fares, but the benefits for the wider community, which could not be gathered through fares, could form the basis of the case for the public subsidy of the capital costs of constructing new infrastructure. Although it was expected that the major area of non-user benefits would be through reduced traffic congestion for those using the roads, nevertheless there was an expectation that there would be regenerative effects too and that the impact on the value and development of land would be one such benefit (Lawless and Dabinett, 1995). This hinges around the assumption that locations which experience improved accessibility resulting from investment in transport infrastructure, will be more attractive to business. However, the geographical impact of such increases in accessibility around the route will be variable, depending upon the nature of access to it. For example, passengers arriving at a station by bus or car and then inter-changing on to Supertram will create highly concentrated activity around the interchange point. On the other hand, where passengers walk to or from the station the effects will be more widely felt in an area surrounding the station. The effect will generally be weaker with increasing distance from the station - the 'distance'decay' effect· where stations are widely spaced then the effects will be concentrated around these points with little in between, but where stations are closely spaced then something approximating a corridor effect will result. These potential changes in activity patterns, and other influences, such as the improved image of locations adjacent to a modern light rail system, may influence the perceptions of decision-makers, which may in turn impact on property values, investment and, thereby, patterns of development. If such patterns of development around the route do emerge, then this is clearly a positive impact from the operator's point of view too, as it brings passengers to the system. It is clear therefore from the background to the development of the South Yorkshire Supertram that the SYPTE, the Sheffield City Council and the DoT all had an interest in ensuring that these potential impacts of the project on patterns of land development were realised. The expectation would therefore be that planning authorities would exercise their powers in ways to achieve this end. During the period of the research there were in fact two planning authorities in Sheffield (three if the Peak Park Planning Board is included, as part of the City of Sheffield is in the national park, although this part of the city is rural), these being the City Council and, in the Lower Don Valley Urban Development Area which was inaugurated in 1988, the Sheffield Development Corporation (SDC).

During the period of this research Sheffield City Council was progressing with its draft Unitary Development Plan (UDP) (Sheffield City Council, 1991) production of which had become a statutory requirement following abolition of the South Yorkshire Metropolitan County Council in 1986. As in Manchester, the context for preparation of the UDP was set out in a Regional Policy Guidance note (DOE, 1989b), although this too was very brief and generalised in line with the lack of Government commitment to prescriptive strategic planning and their relaxed stance towards the issue of road traffic congestion. Supertram received only a brief mention: 'Account should be taken of the plans for light rail transit in Sheffield' (DOE, 1989b, para.28). Sheffield's draft UDP contained a number of policies broadly supportive of Supertram, but although mention was made of the intention to develop park and ride sites in association with the lines there were no more prescriptive policies and proposals to bring about intensive development around them. Given the deregulatory stance towards planning of the Government of the day, such content would not have been acceptable in any case. On the other hand, the development proposals put forward at the same time by the SDC were much more specific. The SDC's area of jurisdiction, the Lower Don Valley, had experienced severe de-industrialisation in the 1980s and was characterised by large amounts of derelict and vacant land. The SDC was charged with the task of securing its development within a limited timescale and to achieve this was vested with strongly interventionist powers and given resources to reclaim land, assemble sites using compulsory purchase if necessary, and develop infrastructure. However, its stance was strongly market oriented and its proposals included extensive road building in parallel to the Supertram route. In a handbook on design principles for the redevelopment of the area Supertram was not mentioned, and parking standards were fixed at 4 spaces per 1000 square feet (approx 92 square metres) for office and business parks, a higher rate of provision than that at Salford Quays (SDC,1990). The SDC's development proposals were therefore generally oriented around the road axis rather than the Supertram axis (SDC, 1992). Thus it was clearly unlikely from the outset that there would be development in Sheffield of a transit-oriented suburban centre such as Scarborough in Toronto (Cervero, 1986) or of a business park development of the sort seen in Switzerland (Lehrer, 1994).

This paper will now go on to report on research in Sheffield using data on planning applications to investigate whether or not there is evidence of any strong association between the Supertram route and patterns of proposed development, and to draw out the effectiveness of planning control on the relationship.

Planning application research: methodology and results

Applications for planning permission are required for most categories of new building construction and for changes of use between categories of land use, as specified in the town and country planning legislation and associated statutory instruments such as the Use Classes Order. Not all planning applications are granted permission and not all permissions are implemented. Nevertheless the submission of a planning application can be seen as an indicator of an intention to invest in land and/or buildings and as such is an indicator of the general state of the property market. With regard to a consideration of the effects of Supertram on the property market, planning application submissions over a given period of time can be examined to see what geographical variations in patterns of submission emerge, if any, and the degree to which these changes may be associated with the opening of Supertram. This study is not concerned with whether these applications were approved or refused by the planning authority, it is the pattern of submissions which is being taken as a market indicator. This will be affected to some degree by the applicants' perceptions of the likelihood of their application being approved, and this in turn will be affected by the policy stance of the planning authorities towards development in the various localities of the city and their powers to influence this. A number of pieces of data about each application are recorded by planning authorities including: the DOE development type categorisation, the location, floor space area, site area and description of proposal. The DOE development types are shown in Table 13.2.

The categorisation embraces a wide range of development types and consideration had to be given as to what types of application would be appropriate to use for the research. As the development impact of Supertram could range from small scale property improvements to major development projects, it was decided that, in principal, most applications should go into the study. However, some planning applications reflect special planning policy considerations which have a particular spatial dimension and their inclusion might have led to distortions in the data; examples include applications to lop or fell trees and permission for advertisements. Similarly some 'other' types of applications, such as those for fences and satellite dishes, may be seen to reflect nuances of planning control rather than trends in the property market and so they were excluded from the study too. These omissions reduced the sample size by around a fifth, to approximately 4800 applications for 1992-93, and to approximately 4450 for 1994-95.

Table 13.2 DOE development type categorisation

Code	Category	Code	Category
AD	Advertisement	MO	Major Office
CA	Conservation Area	MR	Major Retail/Services
CU	Change of Use	MX	Major Other
HO	Household	ND	Minor Dwellings
LD	Listed Building Demolition	NI	Minor Industry/Warehousing
LE	Listed Building Ext/Alt	NO	Minor Office
MD	Major Dwellings[3]	NR	Minor Retail/Services
MI	Major Industry/Warehousing	NX	Minor Other
		X	Trees /Other

As evaluation of the spatial characteristics of the data was required it was necessary to have a postcode for each application; unfortunately few of the applications were initially coded beyond the three digit code and so it was necessary to manually assign a full six digit postcode to each application. The median value of users under any given six digit postcode is 12 (Raper et al, 1992, 37) allowing a relatively fine level of spatial analysis through the use of a GIS system (MapInfo) to allow the distribution with regard to the Supertram routes to be explored visually and quantitatively.

This quantitative analysis involved the creation of zones around the Supertram route at differing distances from it; these were 100m, 200m and 400m. By overlaying these zones on the spatial distribution of the planning applications, the applications could be sorted and allocated to zones using 'point-in-polygon' search functions. This allowed the spatial distribution of the planning applications to be considered with regard to the degree of clustering around the Supertram route. As the MapInfo package allows the area of defined polygons to be calculated, it was also possible to calculate the density of planning applications within the zones around the route. The expectation was that if Supertram was exercising a positive influence on the location of development, then the clustering of applications around the line of its route would increase over time. Similarly any negative influences would produce a reduction in clustering over time. This type of logitudinal study depends on the designation of a time period for examination of data 'before' construction of the transport infrastructure and designation of an 'after' period. Planning applications submitted during 1992-93 were taken as the 'before' situation and, given that the first tram route opened in 1994, data on applications for 1994-95 was taken as the 'after' situation. Examination of trends between the 'before' and 'after' periods was facilitated by producing an annual average from the two constituent years of each period.

The results

Table 13.3 provides a summary of the analysis of the change in spatial distribution of the planning applications around the Supertram route between the before and after periods. It shows that the proportion of all applications received which fell within the 400m buffer zone decreased from 17.7 per cent in 1992/3 to 17.3 per cent in 1994/5, and there was a general reduction in the density of applications within the buffer zones over the study period. However, whereas in the 1992/93 period the density of applications was lowest within the 100m zone, in 1994/95 the density was slightly higher in the 100m and 200m zones than the 400m zone, which is evidence of a slight intensification of applications around the route.

Table 13.4 summarises changes in the number of applications received over the study period broken down by DOE development type for the overall 400m buffer zone. Bearing in mind the overall reduction in the number of applications received, this shows that, nevertheless, there was a marked proportional increase in the major office and retail/services categories, although the actual numbers of applications concerned was very small. However this is suggestive of a positive impact of Supertram with planning policies securing an increase in significant development activity around the route. With a similar reservation with regard to the numbers of applications involved, the data shows a marked decline in the major industry/warehousing, minor dwelling and householder categories.

Table 13.3 Supertram corridor: changes in distribution of planning applications between the buffer zones: 1992/3 - 1994/5

Buffer zones	Area km2	Average applications received 1992/3 (% total)	Average applications 1992/93 per km2	Average applications received 1994/5 (% total)	Average applications 1994/5 per km2	% Change in no. of applications 1992/3 - 1994/5
100mbuffer zone	6.07	106 (4.4)	17.5	101 (4.5)	16.6	-4.7
200mbuffer zone	12.11	220 (9.2)	18.2	201 (9.0)	16.6	-8.6
400mbuffer zone	24.07	426 (17.7)	17.7	387 (17.3)	16.1	-9.1
Rest of city		1974		1843		-6.6
Total		2400		2230		-7.1

Table 13.4 Supertram corridor: change in the average number of planning applications by DOE development type in the 400m buffer zone: 1992/3 - 1994/5

DoE development type	Buffer zone 4 1992/3	Buffer zone 4 1994/5	%Change 1992/3 - 1994/5
Change of Use	91.5	97.5	6.6
Householder	124	100	-19.3
Major dwellings	8.5	8	-5.9
Major industry/ warehousing	5.5	3.5	-36.4
Major office	6.5	8.5	30.8
Major retail/services	5.5	12.5	127.3
Major other	11.5	10	-13.0
Minor dwellings	25.5	17	-33.3
Minor industry/ warehousing	18.5	16	-13.5
Minor office	17	13	-23.5
Minor retail/services	53.5	45.5	-14.9
Minor other	58.5	55	-6.0
Total	426	386.5	-39.5

Planning applications: road corridor as a counterfactual

Given that the transport modes most strongly associated with land development in Sheffield are road based, and almost exclusively so before construction of Supertram, then measures of the association of planning applications with the road network can be used as a counterfactual to put the research findings on the Supertram corridor in context. In fact there have been significant contemporary investments in the city's road network in corridors parallel to the Supertram routes (Player and Haywood, 1996), so an analysis of planning applications in these corridors, using the same methodology as that carried out for Supertram, will produce very relevant contextual data. The sections of new road are the Lower Don Valley Link Road (not completed at the time of the research), Penistone Road (Upper Don) (final stage opened 1995), and the Mosborough Parkway (final stage opened in 1994) (Figure 13.3). These sections of new road are connected by existing major roads and together this forms a network with three arms radiating from the city centre which parallel the Supertram network.

Table 13.5 shows that this analysis for the road corridor produced results which were very different to those for the Supertram corridor. The area of the overall 400m road corridor was 17.7 km^2 as compared to 24.07 km^2 for the Supertram corridor, that is the road corridor was 73.5 per cent of the area of the Supertram corridor. The total number of applications within the road corridor in the before and after periods was 54.5 per cent and 68.2 per cent respectively of those submitted in the Supertram corridor over the same periods, showing that the relative propensity for developers to submit planning applications in the road corridors increased. Despite the overall decline in the total number of applications submitted in the city as a whole, there was an increase in the number of applications in all the road corridor buffer zones. In addition the increase in the 100m and 200m zones, those nearest to the roads, was almost 20 per cent, whereas the increase in the overall 400m buffer zone was only 13.8 per cent. In both the before and after periods there was a steady fall off in the density of applications with increasing distance from the road lines, but the rate of fall off was greater in the after period than the before period. There is strong evidence here of a geographical association between these major road axes and the propensity of developers to submit planning applications. These patterns are in stark contrast to those shown for Supertram.

Table 13.6 summarises the changes in the number of applications received over the study period broken down by DOE development type for the overall 400m road corridor buffer zone. Despite the overall picture of an increase in the number of applications within the road corridor there was

Table 13.5 Road corridor: changes in distribution of planning applications between the buffer zones: 1992/3 - 1994/5.

Buffer zones	Area km2	Average applications received 1992/3 (% total)	Average applications 1992/3 per km2	Average applications received 1994/5 (% total)	Average applications 1994/5 per km2	% Change in no. of applications 1992/3 - 1994/5
100mbuffer zone	4.40	66 (2.7)	15.1	79 (3.5)	18.1	19.7
200mbuffer zone	8.87	123 (5.1)	13.9	147 (6.6)	16.6	19.5
400mbuffer zone	17.7	232 (9.7)	13.1	264 (11.8)	14.9	13.8
rest of city		2167 (90.3)		1965 (88.2)		-9.3
Total		2400		2230		-7.1

Table 13.6 Road corridor: changes in distribution of average number of planning applications by DOE development type in the 400m buffer zone: 1992/3 - 1994/5

DoE development type	Buffer zone 4 1992/3	Buffer zone 41994/5	%Change 1992/3 - 1994/5
Change of Use	68.5	72	5.1
Householder	18.5	21.5	16.2
Major dwellings	5.5	5	-9.1
Major industry/warehousing	5	7	40
Major office	7	12	71.4
Major retail/services	4	5.5	37.5
Major other	9	12	33.3
Minor dwellings	7	4.5	-35.7
Minor industry/warehousing	22	29	31.8
Minor office	7.5	14.5	93.3
Minor retail/services	41.5	31	-25.3
Minor other	36.5	50	37.0
Total	232	264	13.8

a decrease in three categories; major dwellings, minor dwellings and minor retail/ services. Overall this suggests a looser association with residential development in the road corridor as compared with the Supertram corridor.

The road corridor as compared to the Supertram corridor shows an increased association over the study period with the following categories of applications; major industry/warehousing, major office, major retail, minor industry/warehousing and minor office. Notwithstanding the fact that the Supertram corridor performed more strongly than the road corridor with regard to major retail developments, this evidence is suggestive of a closer association between the road corridor and industrial/commercial development than the Supertram corridor.

Conclusions

The research in Manchester shows that, with regard to the very specific category of major office development, deregulation of planning control in the 1980s led to significant locational shifts which worked to the detriment of access by public transport in general and local rail networks in particular. It can be concluded from this research that the selection of locations where out-of-centre office development has taken place has been largely market driven and opportunistic, with little regard to the traffic generating implications or the potential for access by the public transport network. The footloose nature of B1 developments has worked very much to strengthen the hand of developers and weaken the ability of planning authorities to defend their policies:

> Faced with an option of accepting 'an IBM' in an out-of-town business park or 'no IBM at all' leaves the local planning authority with no realistic choice but to bow to the desire of the developers (Foster and Eastman, 1993, 489).

The failure to integrate the development of such major growth poles as the Docks or the developments around Manchester Airport with the local rail network is symptomatic of the 1980s approach. In the case of the Airport, this is despite the introduction of new rolling stock and the opening of a new rail link and station at the Airport. In addition, notwithstanding the major shifts in national planning policy in the early 1990s, it is notable that the Use Classes Order remains unchanged despite an investigation of its impacts concluding that:

> Sufficient has been learnt during the period in which the new Order has been in practice for worthwhile change in their application and practical effect now to be envisaged. This could seek to enhance the achievement of the original and desirable objectives, whilst perhaps reining back some of the significant land use changes that have been set in progress. (DOE, 1991, 56).

Whereas the research in Manchester was concerned with relationships between land use planning and a mature local rail network, which included the conversion of part of it to the Metrolink light rail system, the research in Sheffield was specifically focused around the issue of the regenerative impacts of the introduction of a new light rail system. Light rail is relatively new to Britain but has been around much longer in other countries. In the USA in particular there is a well established research thread on the impact of such schemes on urban development (Cervero, 1984). Cervero and Landis reflect on this experience and emphasise that:

> urban transport and land development have had a rather tenuous relationship in the US over the postwar period. In isolated settings, rail transit seems to have produced value gains and induced growth. However the circumstances that brought about these changes are not easily generalisable (1995).

A crucial factor seems to be the role of planning authorities:

> rail transit can have a strong influence on the location, intensity and timing of new development, especially when it is supported by positive development incentives and co-ordinated land use/transit planning (Orski, 1980).

This view is supported in the UK by Hurdle: 'clearly transport can stimulate activity, if planned to do so' (1992, 9). This seems to be the point. Although there is a great deal of caution in the literature as to the potential regenerative impacts of transport development (Grieco,1994), the situation in Sheffield was one whereby neither of the two land use planning authorities was single mindedly pursuing the initiation of development around the route. Sheffield City Council was operating under the same deregulated planning regime as Manchester City Council. The SDC, which was given a much more interventionist role and resources to back it up, was market driven and focused its development initiatives on the road network and securing the provision of low density development with generous provision of car parking. Also, as in all cases of rail network development in Britain, the SYPTE which was introducing Supertram, was only vested with powers to construct the system and could not acquire land around the route for associated development.

Data from the 1991 Census shows that the modal split for use of rail for the journey to work in Greater Manchester was 2.5 per cent and 0.8 per cent in South Yorkshire, the situation having changed little since 1981, although use of the car increased substantially in each area with a commensurate reduction in bus patronage (Beatty and Haywood,1997). However experience since then in Manchester shows that light rail, in particular, can be attractive to passengers, including car users, and ridership has surpassed expectations (GMPTE, 1996). But, the overall conclusion of this paper must be that from the land use planning point of view, notwithstanding the major shifts in planning policy which have taken place in Britain in recent years, a much more interventionist and regulatory approach is needed if significant shifts in the patterns of land use development are to take place in ways which favour the rail mode. This could include the identification in development plans of locations having high rail accessibility, measures of which can be quantified, to which major employee and visitor activity generators could be steered. Some British planning authorities are already experimenting with such measures (Kerrigan and Bull, 1992) and they are the basis of Dutch 'A-B-C' locational policy, which involves clarifying all potential development/redevelopment sites according to their accessibilty characteristics and all potential developments by their mobility (needs) profiles, and then steering new developments into appropriate sites (Sturt, 1992, 50).

Such locational control also implies the refusal of planning applications for high employee and visitor activity generators in locations which do not have the prescribed accessibilty characteristics. In developing such locational policies there is a need to review some of the sacred cows of British planning, especially green belt policies, the transport implications of which have been overlooked until recently. This might result in the granting of permission for development on green field sites in rail corridors in areas which under previous policy regimes could not have been developed. Such matters are very much on the planning agenda in Britain at present as both Central Government and local planning authorities wrestle with the projected increases in housing demand in the context of the new agenda of sustainable development (Hall and Breheny, 1996) (Secretary of State for the Environment, 1996).

Despite the fact that the ownership and operation of railways in Britain is now largely in the hands of the private sector, the operation and development of the network are still heavily dependent on public subsidy. In order to secure more sustainable patterns of urban form and the associated improvement of rail services, there is a need to build closer links between the public bodies responsible for planning policy and the private rail companies. This could involve empowering railway companies to get involved in riparian land development, or the creation of new bodies, perhaps joint public/private ones, which can intervene in the land development process in ways which produce more sustainable patterns of urban form and provide funds for rail investment. An alternative way of linking land development with the funding of rail infrastructure might be to reintroduce direct taxation of betterment or to introduce American style 'impact fees' (Goodchild, et al, 1996). The latter involves assessing the infrastructural impacts of new development and securing funds from developers to pay for them. Any of these innovations, or a combination of them, could secure development of land accessible by rail and divert some of the possible improvement to rail infrastructure development to ensure high quality services. Identification of the most effective way of proceeding will necessitate a thorough review of the interface between local town planning and transport planning authorities, railway companies, and the private interests involved in the land development process.

Finally, many commentators on the transport crisis in Britain and elsewhere express reservations as to how effective 'carrot' measures can be with regard to persuading drivers to use public transport, and see the need for the regulatory 'stick', with measures ranging from more stringent parking restrictions to full-blown road pricing. Notwithstanding these arguments, with a new Labour Government committed to addressing environmental problems and a transport White Paper scheduled for early 1998, there is a major opportunity to use the tools of the statutory planning system, which are still there despite their lack of use for many years, as part of an integrated transport strategy, particularly if they are associated with the use of betterment for securing the necessary investment in public transport infrastructure.

References

P. Abercrombie (1944) *Greater London Plan*, HMSO, London.

P. Ambrose (1986) *Whatever Happened to Planning?*, Methuen Andover.

W. Ashworth (1954) *The Genesis of Modern British Town Planning*, London, Routledge and Kegan Paul Ltd.

C. Beatty and R. Haywood (1997) 'Changes in travel behaviour in the English Passenger Transport Executives' areas: 1981-91' *Transport Geography* vol. 5 no. 1 pp 61-72.

British Railways Board (1963) *The Reshaping of British Railways*, HMSO, London.

S. Brownill (1990) Developing London's Docklands: Another Great Planning Disaster, Paul Chapman Publishing, London.

C. Buchanan (1963) Traffic in Towns: A study of the long term problems of traffic in urban areas, HMSO, London.

M. Castells (1989) *The Informational City*, Basil Blackwell, Oxford.

R. Cervero (1984) 'Light Rail Transit and Urban Development', *Journal of the American Planning Association*, vol. 50 no 2 pp. 133-147.

R. Cervero (1986) *Suburban Gridlock*, Centre for Urban Policy Research, Rutgers University, New Brunswick NJ.

B. Cervero and J. Landis (1995) 'Development Impacts of Urban Transport: A US Perspective' in D. Banister (ed), *Transport and Urban Development*, , E & FN Spon, London.

G.E. Cherry (1974) *The Evolution of British Town Planning*, Leonard Hill, London.

City of Manchester (1961) Manchester Development Plan, Written Statement.

C. Clark (1958) 'Transport - Maker and Breaker of Cities', *Town Planning Review*, vol. 28 no. 4 (January) pp. 237 - 250.

B. Cullingworth (1994) 'Fifty years of Post-War Planning', *Town Planning Review*, vol. 65 no. 3 pp. 277-303.

P.W. Daniels (1977) 'Office Locations in the British Conurbations: Trends and Strategies' *Urban Studies*, vol. 14 pp. 261 - 274.

P. Damesick (1979) 'Office Location and Planning in the Manchester Conurbation' *Town Planning Review*, vol.50 no.3 pp. 346-366.

Department of the Environment (1972) *Town and Country Planning (Use Classes) Order 1972*, HMSO, London.

Department of the Environment (1980) Circular 22/80, Development Control - Policy and Practice, HMSO, London.

Department of the Environment (1985) *Cmnd 9571, Lifting the Burden*, HMSO, London.

Department of the Environment (1986) *Building Businesses Not Barriers*, HMSO, London.

Department of the Environment (1987) *Town and Country Planning (Use Classes) Order 1987,* HMSO, London.

Department of the Environment (1989a) Regional Policy Guidance Note No.4, Strategic Guidance for Greater Manchester, HMSO, London.

Department of the Environment (1989b) Regional Policy Guidance Note No.5, Strategic Guidance for South Yorkshire, HMSO, London.

Department of the Environment (1990) *Cm1200 This Common Inheritance*, HMSO, London.

Department of the Environment (1991) *An Examination of the Effects of the Use Classes Order1987 and the General Development Order 1988*, HMSO, London.

Department of the Environment (1993a) *Reducing Transport Emissions Through Planning*, HMSO, London.

Department of the Environment (1993b) *The Effectiveness of Green Belts*, HMSO London.

Department of the Environment (1994a) *Cm2426 Sustainable Development:The UK Strategy,* HMSO, London.

Department of the Environment (1994b) *Planning Policy Guidance Note 13:Transport*, HMSO. London.

Department of Transport (1989a) National Road Traffic Forecasts (Great Britain) 1989, HMSO, London.

Department of Transport (1989b) *Cm 693,Roads For Prosperity*, HMSO, London.

Department of Transport (1989c) Circular 3/89: Section 56 Grant for Public Transport, HMSO, London.

H.D. Dickinson (1964) 'Is Public Transport a Social Service?', *New Society, 12 March*, pp. 6-7.

H.J. Forshaw and P. Abercrombie (1943) *County of London Plan*, Macmillan and Co., London.

M.S. Foster and C.R. Eastman (1993) 'Parking and Public Transport - the Effect on Mode Choice: A Study of B1 Developments', *Traffic Engineering and Control* 34 pp. 480-83.

P.E. Garbutt (1985) *London Transport and the Politicians*, Ian Allan, London.

B. Goodchild, C. Booth and J. Henneberry (1996) 'Impact Fees: A Review of Alternatives and their Implications for Planning Practice in Britain', *Town Planning Review*, vol. 67 (2) pp. 161-181.

P. Goodwin et al (1991) *Transport:The New Realism*, Transport Studies Unit, University of Oxford.

Greater Manchester Council (1986) *Greater Manchester Structure Plan*, GMC, Manchester.

Greater Manchester Passenger Transport Executive (1996) *Trends and Statistics 1986-96*, GMPTE, Manchester.

M. Grieco (1994) The Impact of Transport Investment Projects upon the Inner City, Avebury, Aldershot.

P. Hall, et al (1973) *The Containment of Urban England, Vol. 1,* George Allen and Unwin, London.

P. Hall (1992) Urban and Regional Planning, 3rd edition, Routledge, London.

P. Hall and M. Breheny (1996) *The People: Where Will They Go?,* Town and Country Planning Association, London.

R. Haywood (1996) 'More Flexible Office Location Controls and Public Transport Considerations: A Case Study of the City of Manchester', *Town Planning Review* vol. 67 no.1 pp. 65-86.

R. Haywood (1997) Railways, Urban Form and Town Planning in London: 1900-1947, *Planning Perspectives* vol. 12 no. 1 pp, 37- 69.

J. Henneberry (1988) Conflict in the Industrial Property Market, *Town Planning Review* vol. 59 no. 3 pp. 241 - 262.

J. Hillman (ed), (1971) *Planning for London*, Penguin, Harmondsworth.

R. Home (1992) 'The Evolution of the Use Classes Order', *Town Planning Review* 63, no. 2, pp. 187-201.

D. Hurdle (1992) 'Does Transport Investment Stimulate Economic Activity?', *The Planner* vol. 78 no. 9 pp. 7-9.

R. Imrie and H. Thomas (1993) *British Urban Policy and Urban Development Corporations*, Paul Chapman Publishing, London.

M. Kerrigan and D. Bull (1992) *Measuring Accessibility - A Public Transport Accessibility Index,* PTRC Summer Conference Paper, PTRC, London, pp. 245 -56.

T. Kitchen (1996) 'The Future of Development Plans: Reflections on Manchester's Experiences 1945-1995', *Town Planning Review* vol. 67 no. 3, pp. 331-353.

C.M. Law (1991) *Office Development in Greater Manchester*, Metrolink Impact Study, Working Paper 3, Department of Geography, University of Salford.

C. M. Law and D. Dundon-Smith (1994) *Metrolink and the Greater Manchester Office Market:An Appraisal,* Metrolink Impact Study, Working Paper 13, Department of Geography, University of Salford.

P. Lawless and G. Dabinett (1995) 'Urban Regeneration and Transport Investment: A Research Agenda', *Environment and Planning A,* vol. 27, pp. 1029-1048.

U. A. Lehrer (1994) 'Images of the Periphery: the Architecture of Flexspace in Switzerland', *Environment and Planning D: Society and Space,* vol. 12, pp. 187 - 205.

Manchester City Council (1984) *City Centre Local Plan* , Manchester City Council, Manchester.

Manchester City Council (1993) *Transport Policies and Programme, Submission for 1994/95,*, Manchester City Council, Manchester.

O. Marriott (1967) *The Property Boom*, Pan Books Ltd, London.

Martin and Voorhees Associates; Colin Buchanan and Partners, local authority staff (1976) *Sheffield and Rotherham: A Transportation Plan for the 1980s*, Martin and Voorhees Associates, London.

Ministry of Housing and Local Government (1955) *Circular 42: Green Belts,* HMSO, London.

Ministry of Transport (1966) *Transport Policy:Cmnd 3057*, HMSO, London.

Modern Railways (1997) 'Altram Signs for Salford and Eccles', *Modern Railways*, vol. 54 no. 585, p. 326.

C. Orski (1980) 'The Federal Rail Transit Policy: Rhetoric or Reality, *Transportation*, vol. 9, pp. 57-65, quoted in I. Dickens (1992) 'Transport Investment, Economic Development and Strategic Planning : the Example of Light Rail Transit', *Planning Practice and Research*, vol. 7 (2), pp. 9-12.

R. Player, R. Haywood (1996) Land Use Planning Along the Strategic Road Network and Supertram Lines, SIS Paper no.29, CRESR, Sheffield,.

J. Raper, D. Rhind, and J. Shepard (1992) *Postcodes: the New Geography*, Longman, Harlow.

Royal Commission (1940) Cmd.6153: Royal Commission on the Geographical Distribution of the Industrial Population, HMSO, London.

Salford City Council and Trafford Metropolitan Borough Council (1981)*Salford Docks/Trafford Park Enterprise Zones.*

Salford City Council (1985) *Salford Quays:The Development Plan for Salford Docks*, Salford City Council, Salford.

Secretary of State for the Environment (1996) *Household Growth: where shall we live?: Cm 3471*, HMSO, London. Sheffield City Council (1991) *Draft Unitary Development Plan,*, Sheffield City Council, Sheffield.

Sheffield Corporation (1969) *Mosborough Master Plan*, Sheffield Corporation, Sheffield,.

Sheffield Development Corporation (1990) A Vision of the Lower Don Valley: Design Principles for Development, SDC, Sheffield.

Sheffield Development Corporation (1992) *A New Valley: Made in Sheffield*, SDC, Sheffield.

M. Smith (1990) British Politics, Society and the State Since the Late Nineteenth Century, Macmillan, London.

A. Sturt (1992) 'Going Dutch', *Town and Country Planning*, 61, February, pp. 48-51.

SYPTE (1978)*Transport Development Plan*, SYPTE, Exchange Street, Sheffield.

A. Thornley (1992) *Urban Planning Under Thatcherism*, Routledge, London.

A. F. Williams (1985) *Rapid Transit Systems in the U.K.: Problems and Prospects*, Institute of British Geographers, London.

Notes

1. The research in Manchester has been supported by the Centre for Regional Economic and Social Research (CRESR), a research institute of Sheffield Hallam University. The research in Sheffield has been carried out through CRESR and supported by the South Yorkshire Passenger Transport Executive; the Department of Transport; the Economic and Social Research Council Transport; and the Enviroment Research Programme Contract No L119251020. Karl Dalgleish (now of Ecotec Ltd) and Tony Gore were members of the CRESR research team involved in the research into planning applications. Also the advice and support is acknowledged which was received form F. Robinson (University of Durham) and C. Law (University of Salford), also from T. Worsley (Department of Transport) and P. Haywood (SYPTE). Thanks are due to Sue Hewinson of Sheffield Hallam University for preparation of the maps.
2. The PTAs are elected member level organisations, the PTEs are their officer level counterparts.
3. For dwellings a major development constitutes one where the number of proposed dwellings is 10 or more or a site is over 0.5 hectares where the number of dwellings is not stated (DOE, 1994, 1). For all other uses a major development is a scheme where the proposed floorspace is 1000 square metres or more, or where the site is one hectare or more.

Part 4

Freight Industry

CHAPTER 14

A comparison of deregulation in the road haulage markets of the UK, EU and USA

Michael Browne and Julian Allen

Introduction

The first section of this paper outlines the key trends occurring in freight transport (such as total quantity of goods moved, modal split) together with other general statistics for the United Kingdom (UK), European Union (EU) and the USA. This data helps to construct a picture of freight transport and establish the main similarities and differences between the three regions being studied. The second section addresses the deregulation of freight transport in each market, documenting both the legislation and rationale for deregulation in each case and concluding with a summary of the impacts of deregulation in each region. The third section examines the freight industry and major developments in freight transport operations and services in each of the three regions/countries. This consists of a review of each marketplace together with a discussion of the main sectors in the freight transport market. The paper ends by considering a comparison between financial performance and profitability in the USA and the UK. This is followed by a short discussion of some of the important external influences that will be felt by all concerned with freight transport services.

Freight trends and comparisons

Before going on to discuss in detail the nature of trends in the different markets it is interesting to compare some of the more general statistics on freight transport and market size for the UK, the European Union (EU) and the USA. Tables 14.1 to 14.4 contain some information on area and population together with the key features of freight transport and modal split for each market.

Table 14.1 Area and population in 1993

	Population (millions)	Area (000 sq km)	Population density (per sq km)
UK	58	244	239
EU (15)	369	3243	114
USA	258	9373	27

Source: Department of Transport, 1996

Table 14.2 Road and rail infrastructure in 1993

	All roads (000 km)	All roads per 1000 sq km (km)	Rail network in operation (000 km)	Rail network per 1000 sq km (km)
UK	389	1592	16.8	69
EU (15)	3501	1470	160.7	57
USA	6278	670	177.7	19

Source: Department of Transport, 1996

It may be worth noting that although the total length of all roads in the USA is almost twice that of the EU (15), when compared per unit of area, the concentration of roads is greater in the EU (15) than in the USA. The total rail network in the USA is only slightly greater than that in the EU but again the density of the EU rail network is of course much higher.

Table 14.3 Road freight transport comparison

	Road freight moved (billion tonne km)		Freight moved per person (000 tonne km)	Goods vehicle km (billion km)		Goods vehicle km per person (000 km)
	1983	**1993**	**1993**	**1983**	**1993**	**1993**
UK	94	132	2275	45	65	1116
EU (15)	665	920	2493	179	260	705
USA	802	1285	4980	657	872	3378

Source: Department of Transport, 1996

Some striking differences can be seen from Table 14.3:

- The far larger geographical area of the USA in comparison with the UK means that the average haul distance in the USA is far greater than in the UK.
- Road freight moved in the USA increased by 60 per cent between 1983 and 1993. In the EU (15) it increased by 38 per cent, and in the UK it increased by 40 per cent.
- Goods vehicles kilometres performed increased by 33 per cent in the USA between 1983 and 1993, while in the EU (15) there was an increase of 45 per cent over the same period.
- Freight activity for road movements in the USA is much greater than in the EU (15) although the population is only approximately two-thirds.
- Tonne kilometres of freight moved per person are far greater in the USA than in the EU (15).

Table 14.4 highlights the dominant position of road in the UK and the EU in terms of tonne kilometres performed. In the USA, rail performs more tonne kilometres than road. Whilst rail freight has declined significantly in the UK and EU over the last 30 years, it has not diminished to anything like the same extent in the USA and is widely used in long distance movements, especially of bulk goods and in the movement of containers.

Table 14.4 Modal split for national goods transported by national vehicles in 1993

	UK			EU (15)			USA		
	Road	Water	Rail	Road	Water	Rail	Road	Water	Rail
Bn Tkm	131	0.2	14	920	106	222	1285	416	1620
%	90%	0%	10%	74%	8%	18%	39%	13%	48%

N.B. 'Bn Tkm' is billion tonne kilometres
'Water' refers to inland waterway i.e. it excludes coastal shipping

Source: Department of Transport, 1996

Deregulation of freight markets

Deregulation of industry has been an important theme for national governments for the past 20 years. The freight markets in each of the countries and regions studied in this paper (the UK, the European Union and the USA) have undergone some degree of deregulation. The impact of deregulation in each case is discussed below. However, it is important to note that the term `deregulation' does not have the same meaning in each of these freight markets. In the UK, for instance, deregulation of national road haulage took place following the 1968 Transport Act. In the EU as a whole, by contrast, the earliest steps towards deregulation concerned international transport. Only much more recently have some of the mainland European countries sought to deregulate their national markets. In the USA it refers to inter-state freight and intra-state transport; however, until very recently intra-state transport remained regulated - with each state supervising rates and capacities for movements wholly within the state.

It is also important to note that the nature of the regulations that existed prior to deregulation were very different in each region and, therefore, deregulation has brought about different changes in each case. For instance, in the USA, deregulation has brought about relaxation of freight rate controls. In the UK, freight rates were not controlled under a regulated market and the impact of deregulation was therefore rather different since the changes related mainly to issues of haulage capacity and market entry. It follows that the meaning of deregulation in any one country or region is highly dependent upon two factors:

- the characteristics of past regulations
- the aims of new policy measures.

Cooper (1991) argued that a number of consequences can occur as a result of deregulation, depending upon the circumstances. Some of these consequences, listed below, are put forward as arguments in favour of deregulation, whilst others are held up by believers in regulation as potential negative impacts of deregulation:

- reduction in freight rates
- greater efficiency of freight transport operation
- loss of rail traffic
- increasing concentration of ownership in the haulage sector
- worse services for rural areas
- increases in haulage bankruptcy
- lower wages for haulage employees
- poorer safety record in road freight operations.

It is often difficult to prove that economic deregulation is directly and solely responsible for subsequent events in the haulage sector. This is because separate but related action (such as enforcement of safety regulations) can be used to prevent unwanted events arising from deregulation [for a full discussion of these impacts see Cooper, 1991 and Cooper, Browne and Peters, 1994].

Deregulation in the USA [1]

Deregulation in the USA was brought into effect in 1980, under the Motor Carrier Act. It was designed to remove a variety of economic controls on inter-state trucking, applied through the Interstate Commerce Commission (ICC). The powers of the ICC were considerable prior to 1980. Applicants wishing to enter the inter-state trucking sector had to justify their services on the grounds of `public convenience and necessity'. In many cases the ICC would then restrict carriers according to the geographical area of their operation and types of commodity that they could carry. But, once established, carriers could organise themselves into rate bureaux which would formulate freight rates, then publish and file them with the ICC. The additional role of the ICC, once it had accepted the rates, would be to ensure strict adherence to them.

Deregulation through the Motor Carrier Act removed any of the powers of the rate bureaux and the ICC, particularly with respect to market entry and rates control. However, the Act did not apply to intra-state trucking which was not deregulated until 1995. Prior to deregulation, there was considerable variation between states in their freight regulation. Two of the larger states, California and Texas, had stricter regulation than elsewhere. Texas limits both haulage capacity and rates, while California controlled only rates and is currently leaning towards a degree of deregulation. One study estimated rates in Texas to be 30 per cent higher than inter-state rates for comparable distances (Canny and Rastatter, 1989).

Critics of intra-state regulation argued that it resulted in inefficient and expensive trucking services and that deregulation would be beneficial to the economy and the coherence of US-wide trucking policy. The Supreme Court decision to free Federal Express from all regulations in California encouraged intra-state deregulation, as other trucking firms lobbied Congress to give them similar rights to Federal Express (Corsi, 1994). Intra-state trucking deregulation finally occurred at the beginning of 1995. It is too early to assess the effects of intra-state deregulation, but there is already some evidence of falling freight rates and carriers becoming casualties of the increased level of competition.

The rationale behind deregulation in the USA was based on recognition in the late 1970s that inter-state regulation, coupled with strong trucking unions, was impairing the efficiency of the US economy. Deregulation was used to remove unnecessary economic privileges that had reduced competitiveness in trucking. However, despite the 1980 Act and the reduction in economic regulation in the haulage industry, there has been an increased concern about regulations to do with safety and the environment. Congestion levels in urban areas have led to increasing research into policies and approaches to controlling traffic levels and recycling and waste management are also receiving greater attention.

Deregulation in the UK

UK freight deregulation was brought about through the 1968 Transport Act. The provisions of the Act applied to all freight in the UK, which does not have a federal system of government as exists in the USA. Deregulation was prompted by the Geddes Committee in 1965, which concluded that the systems of road freight licensing then in operation did not serve the public interest, particularly in the area of safety. Under the 1968 Transport Act, action was therefore taken to remove capacity restrictions applied to the haulage sector and to introduce new safety controls on road freight operation.

In the UK, by the 1960s, the machinery of regulation set up in the 1930s had become almost irrelevant, because periodic reforms of the regulations had amounted to a process of liberalisation. What the 1968 Transport Act effectively did was to dismantle a redundant administration for regulation and refocus new regulations towards achieving safety objectives. As a result, deregulation in the UK apparently produced no major achievements but brought no major upsets either. One impact that was not necessarily anticipated was that it stimulated the larger road transport companies to develop services that took advantage of their scale of operation in a market where there was no longer any quantitative control over entry. It could be argued that deregulation encouraged the development of more sophisticated distribution services combining transport, storage and order processing operations.

Deregulation in the EU

For many decades, markets for road freight services in the EU were characterised by strong economic regulation, affecting haulage capacity and/or the prices charged for services (tariffs) both within and between

Member States. However, important changes have now taken place that are having a significant impact on road freight markets and the supply of services.

To a great extent these changes were initiated by the European Commission as part of its plan to complete the Single European Market by the end of 1992. Before 1985 the position on economic regulation in EU freight markets can be summarised as shown in Table 14.5.

An important point illustrated by Table 14.5 is the extent to which haulage markets were controlled by government prior to 1985, both with respect to haulage capacity and prices charged. Only in the UK and Luxembourg could free-market conditions be said to apply, with hauliers deciding how many trucks to operate (subject to safety standards) and hauliers and their customers together deciding prices for services.

During the 1980s the position began to change, with consideration of deregulation of international haulage and cabotage at an EU level and domestic deregulation by several countries at a national level. The driving force behind deregulation in the EU relates to the Treaty of Rome, and the right of an individual from any Member State to provide goods and services elsewhere within the Union. The European Commission argued that the system of bilateral permits which controlled international haulage in the mid 1980s amounted to cargo reservation.

There are now no quantitative entry restrictions to the international road haulage industry and any operator registered in a Member State can now apply for an international road haulage licence. Cabotage operations have also been subject to significant changes. In the road freight industry the term 'cabotage' refers to domestic work carried out wholly within one country by a foreign haulier. Prior to 1989, cabotage operations were not allowed within the EU. However, since this date, cabotage operations have become legal for hauliers from Member States. At present over 30,000 quotas are available for EU hauliers wishing to undertake cabotage and this quota will be increased by 30 per cent each year until 1 July 1998 when the system will be abolished and cabotage operations will no longer require permits. However, it will still be necessary for foreign hauliers to meet any specific qualitative operating requirements within the domestic market in which they are performing cabotage.

Weights and dimensions and certain other technical characteristics of vehicles participating in international traffic are regulated, and more recently these technical aspects have been extended to include safety aspects such as speed limiters. Although a maximum gross vehicle weight of 40 tonnes has been agreed for international intra-Union movements, EU domestic markets have maximum limits ranging from 38 tonnes (in the

UK) to 60 tonnes (in Sweden). In 1992 and 1993 agreements were reached on fuel and minimum vehicle taxes respectively, which are intended to lead to a reduction in the disparities in these taxes between Member States.

Table 14.5 Economic regulation of road haulage in the EU before 1985

Road freight market	Economic regulatory provision	Comments
1) International haulage (that is, the transport of goods by a third-party haulier between Member States of the Union)	Bilateral and multilateral permits, reference tariffs	Permits frequently in short supply; prices mostly market-driven
2) Cabotage (that is, the transport of goods within a country by a non-resident international haulier)	Forbidden in all Member States	Some illegal cabotage undertaken
3)Domestic haulage	Permit and/or tariff control in Germany, France, Italy, Ireland, Spain, Portugal, Denmark, Netherlands, Belgium, Greece; liberal markets in Luxembourg and the UK	Severity of control over haulage markets, particularly enforcement, very variable

Source: Cooper, Browne and Peters, 1994

However, despite the trend towards harmonisation there is a strong degree of concern at the extent of differences between markets and at the way that national governments treat the road freight industry. There is widespread feeling among UK hauliers and the trade associations that represent them, that recent changes in freight vehicle charging systems in a number of EU countries have unfairly discriminated against UK hauliers with international operations. The basis for this feeling is that in several instances in which road user charging schemes have been introduced, national governments have reduced vehicle taxation accordingly for national hauliers. However, UK-based international hauliers are having to pay these road user charges in addition to unchanged levels of vehicle excise duty.

The tax scheme introduced in Germany, Belgium, Denmark, the Netherlands and Luxembourg in 1995 (known as the Euro-vignette system) has resulted in charges on either a daily, weekly, monthly or annual basis for operating vehicles of 12 tonnes or more on motorways in these countries. Annual charges are 1500 DM for vehicles with up to three axles and 2500 DM for vehicles with four or more axles. This is payable by hauliers from any country including domestic hauliers. However upon the introduction of this tax, vehicle excise duty was reduced by an amount similar to the cost of the vignette in order to offset these costs for domestic hauliers. For UK-based international hauliers however, vehicle excise duties have remained unchanged and must be paid in addition to these new road user charges; some UK-based international operators have argued that this amounts to unfair competition. This point of view was given some support by the EU Transport Commissioner who suggested that the UK should join the Euro-vignette system and that UK hauliers could be given a corresponding reduction in vehicle excise duty (Brignall, 1996).

The European Commission has no power to change the rules of domestic freight operation in any Member State, except if they are in breach of the Treaty of Rome. However, it is clear that the Commission has had an indirect impact on domestic regulation as a result of successfully promoting change in regulations governing international haulage and cabotage. National governments have had to consider the road haulage sectors in their own countries, particularly with respect to future competition.

Comparison of USA, UK and EU freight markets

USA freight industry

The revenue earned by the USA trucking industry was valued at $345 billion in 1994; representing five per cent of GDP (Standard and Poor, 1995). It is a very large market with many sectors. Prior to deregulation, US trucking companies were classified as either common, contract, private or special carriers. Common carriers served the general public, contract carriers operated for hire services under specific contracts, private carriers transported their own goods and special carriers had exemption from federal economic regulations, for example, fish and livestock haulage (Button and Chow, 1983). However, classification in this manner became unnecessary after the 1980 Act and the distinctions between common, contract and private carriers has become blurred in recent years. The key sectors in the American industry are:

- parcels and packages,
- regional and national less than truckload (LTL),
- local trucking
- truckload services (TL).

LTL services involve the carriage of a number of customers goods on a hub and spoke network, whilst TL is more concerned with point to point of vehicles carrying one customer's products. The relative importance of each of these sectors is shown in Table 14.6. To a great extent these sectors are also becoming less clearly defined over time as services offered by companies from very different backgrounds begin to overlap.

Table 14.6 Breakdown of trucking revenues by sector in the USA in 1994

Sector	**Per cent of total US trucking industry revenue**
Truckload (ICC)	18.8
Non-ICC trucking	33.0
Long haul LTL	2.6
Major regional LTL	2.2
Local trucking	37.7
UPS	5.7

Source: in Standard and Poor, 1995 (compiled by Merrill Lynch & Co. Inc.- Interstate Commerce Commission)

Private fleets (i.e. own account hauliers) are still of importance in the US freight market. According to Brown and Greenlee (1995) they haul approximately 60 per cent of intercity traffic and operate about 6 million vehicles (although many of these are likely to be small vehicles for local collection and delivery work).

The degree of competition introduced by deregulation has been significant with many new companies entering the industry and therefore expanding capacity, leading to lower freight prices. This has meant that operating and financial results are under constant pressure. At the same time the degree of concentration in the industry has increased since deregulation (i.e. an increasing proportion of the total market is accounted for by a small number of large firms). This began to occur in the LTL sector in the 1980s and in the TL sector more recently, and is expected to continue throughout the current decade. However, because of the different nature of the two types of service, LTL is more suited to consolidation, given the need for terminal networks in this sector. The extent of the increase in concentration in the LTL sector is shown in Table 14.7.

Since deregulation, bankruptcies and financial losses in the LTL sector have been widespread, and the sector still does not appear to have reached equilibrium. Three of the fifty largest LTL carriers of 1992 declared bankruptcy and left the industry in 1993. Major mergers also occurred in 1993. However at the same time some smaller carriers have been successful. Rakowski (1994) noted that of the top 50 LTL carriers in 1979:

- 24 had ceased operations by 1994,
- Another eight were involved in mergers with partners that eventually closed,
- Four more firms left the LTL industry.

Therefore, in total, 36 of the top 50 LTL firms in 1979 were no longer in the sector by 1994.

Transport costs in the LTL sector have declined in real terms since deregulation as a result of both efficiency improvements and reductions in labour rates. However, at the same time, prices have been falling at a faster rate than cost reductions and this has affected profitability (Corsi, 1994). Since 1989 revenue per ton[2] for long-haul LTL carriers has increased by less than two per cent annually. Costs have increased by at least this amount, thereby reducing carriers' margins (Standard and Poor, 1995). This is related to the recurring problem of overcapacity and the commodity nature of freight services.

Table 14.7 The 50 largest LTL companies in the USA; 1992 versus 1979; percentage shares of operating revenue; group share of top 50 revenue

Group	Share of revenue (%)		Relative change
	1979	1992	
Top 3 firms	25.7	41.0	+59%
Top 5 firms	36.0	51.9	+44%
Top 10 firms	49.6	66.4	+34%
2nd 10 firms	18.9	15.5	-18%
3rd 10 firms	13.8	8.8	-36%
4th 10 firms	10.1	5.6	-45%
5th 10 firms	7.7	3.7	-51%

Source: Rakowski, 1994.

The cost structures of the LTL and TL carriers differ significantly. Total labour costs, including wages and fringe benefits, account for approximately 65 per cent of total revenue for major LTL carriers, with fuel costs representing four per cent of total revenue. Meanwhile, labour costs for TL carriers only account for approximately 40 per cent of total revenue and fuel costs for 15 per cent (Standard and Poor, 1995). This difference is explained by the unionised nature of long haul LTL services. In addition, LTL carriers tend to also have higher fixed costs than TL carriers because of their need for extensive terminal facilities.

Unionisation among those working in the LTL sector is a significant factor in both the cost structure and reliability of the sector. The strike in April 1994, involving 75,000 members of the Teamster's Union affected approximately 60 to 70 per cent of LTL business. The strike continued for 24 days and was estimated to have cost the firms involved approximately $1.0 billion (Standard and Poor, 1995).

Increased levels of competition in the TL sector has forced firms to seek innovation in order to ensure survival. Corsi (1994) notes that: ‘an entire class of carriers commonly referred to as *advanced truckload* firms, emerged with a cost structure that clearly outpaced industry competitors. The advanced TL firms focused on high-density, long-haul corridor traffic for their primary business and they were well rewarded for their efforts'.

Competition between TL and railroad services has been intense in the USA. Rail freight began to offer double-stack container train services in the 1980s at very competitive rates. This attracted some TL customers to use these services. These advanced truckload firms have responded by introducing a number of strategic initiatives: forging intermodal partnership agreements with railroads, committing to containerisation, and shifting to shorter-haul markets with shipper-carrier partnerships to minimise shippers' total logistics costs (Corsi, 1994).

Concentration has also occurred in the railroad sector. By the early 1990s fourteen independent railroads together accounted for more than 90 per cent of total rail freight revenues. Railroad services have focused on bulk and commodity haulage and have achieved increased market share in coal and chemical movements (Bowersox and Cooper, 1992).

Intermodal transport companies have also emerged since deregulation which have joint ownership of railroad, trucking and water-borne operations. Some have come about through acquisition, some by railroads being allowed to form trucking subsidiaries and others by strategic alliances between railroads and trucking companies to jointly improve their competitive position and service portfolio. This sector of the market has been growing at a rate of more than three per cent per year (Bowersox and Cooper, 1992).

UK freight industry

Over the past decade the UK road haulage industry has, along with other UK industries, faced varying economic conditions. Levels of business activity have fluctuated from the highs of the economic boom of the late 1980s to the lows of the recession experienced during the early 1990s. This is illustrated in Figure 14.1 which shows the annual change in goods lifted and goods moved, the two key measures of output in the road haulage industry. Both of these measures can be seen to have risen extremely sharply between 1986 and 1989, increasing annually at a faster rate than gross domestic product over this period. However, goods lifted and goods moved began to fall after 1989 and continued to do so up until 1992. It is important to note that there was an increase in both goods lifted and moved in 1993, and this suggests that, in general, economic conditions were beginning to improve slowly.

Figure 14.1 Road freight transport in the UK 1980 - 1986

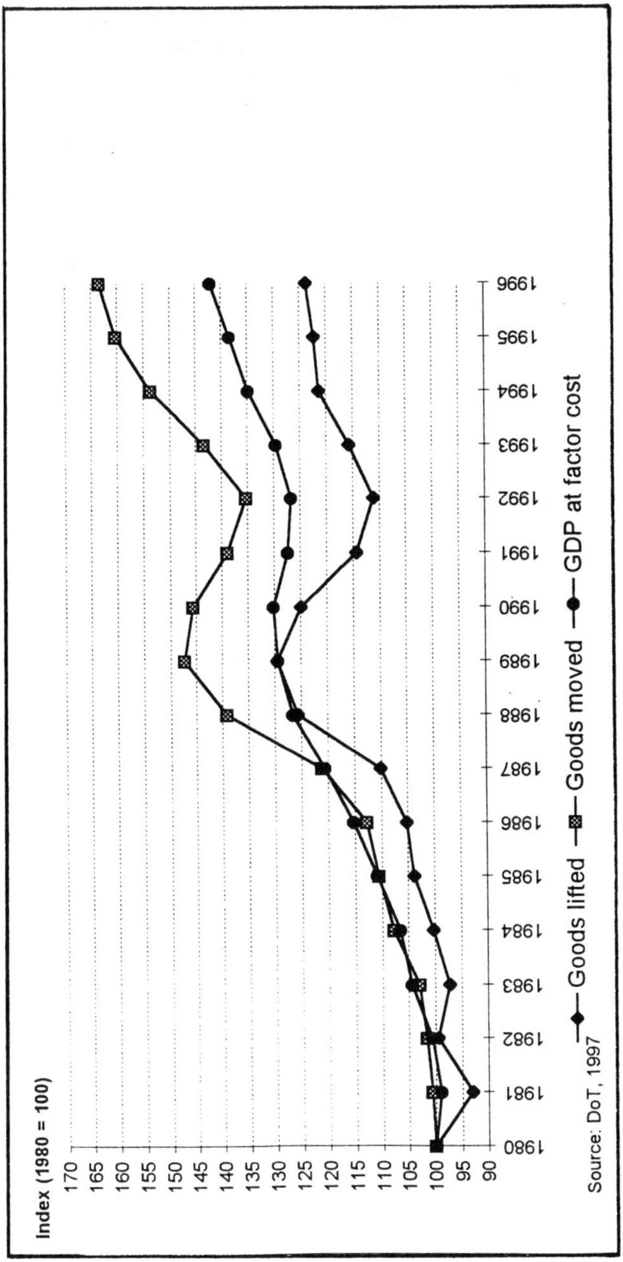

It is evident from the figure that goods lifted declined more rapidly than goods moved during the recession and this is explained by the fact that goods are in general being moved over ever increasing distances (this being the logical outcome of rationalisation of distribution networks). Both of these measures of freight activity began to decline after 1989, whereas gross domestic product (GDP) did not begin to fall until after 1990. Both declined at a faster rate than UK GDP and also rose at a faster rate than GDP in 1993. These two points indicate the sensitivity of the road haulage industry to changes in economic and business conditions.

Users' total expenditure on road freight rose sharply between 1987 and 1989, but remained relatively unchanged until 1992, after which it began to increase again (Department of Transport, 1993). When this total spend is contrasted with expenditure per tonne kilometre performed, an important pattern can be identified. During the earlier period of rising freight activity, and hence increasing total expenditure, the expenditure per tonne kilometre fell by approximately 10 per cent in real terms. This reduction can be linked to the use of larger vehicles and improved utilisation levels of these vehicles.

Two of the most important costs associated with road haulage are fuel and drivers' wages. These two components combined typically represent approximately 50 per cent of total vehicle operating costs. Fuel bills have fallen in real terms over recent years as a result of pump prices and improvements in vehicle engineering and technology. However, the UK road haulage industry has experienced difficulties in controlling labour costs, and they have often risen faster than the retail price index (RPI) over the past decade. Even in the recession of 1991/92 labour costs in the industry rose by 7.5 per cent compared with a rise in RPI of 4 per cent (Institute of Logistics, 1993).

A clear trend in the UK road haulage industry in the last twenty years or more has been the move away from own-account operations towards the use of third party operators. This trend has continued during the 1980s and 1990s with third party haulage responsible for over 74 per cent of all freight moved in the UK in 1996 compared with 61 per cent in 1980 (Department of Transport, 1997). A variety of reasons can be advanced for this shift. It is often asserted that using third-party services enables a company (manufacturer or retailer) to cope more effectively with problems associated with peaks in demand. It is also true that third-party providers have specific distribution and management skills that may enable their customers to benefit from higher levels of service or lower costs (or a combination of the two). A survey of food manufacturers found that the four most important ways that third-party distribution companies could add value were (i) helping to improve customer service, (ii) increasing efficiency (iii) reducing costs and (iv) providing expertise and specialised skills (Browne and Allen, 1997).

A further reason for contracting-out the transport function is that many large companies wish to focus on what they define as their core business. This appears to have been one of the motivating factors in the grocery retail sector in the UK in the early 1980s. This core business philosophy was combined with concerns about the industrial relations problems associated with a strongly unionised environment in own-account transport operations. By contracting out, the retailers were able to release financial assets for store expansion, gain the benefits of innovation from the service providers and avoid some of the management complexities of dealing with difficult industrial relations problems. Clearly, the relative importance of these various reasons will change over time and will, in any event, be rather different form one industry sector to another.

The UK hire or reward (third-party) haulage sector is predominantly made up of relatively small operators, typically with a fleet of fewer than five vehicles. But although these small companies dominate in terms of numbers within the hire or reward sector it is the larger companies which dominate the market in terms of the total vehicle fleet and, therefore, capacity. In the UK seven per cent of hire or reward companies operating in domestic markets own 54 per cent of the vehicles (see Table 14.8).

Table 14.8 United Kingdom domestic road hauliers, 1995

Fleet size	Percentage of firms	Percentage of total vehicle fleet
1-5 vehicles	85	33
6-10 vehicles	8	13
11 vehicles and more	7	54

Source: Adapted from data provided by Directorate of Statistics, Department of Transport, 1995

The UK hire or reward haulage sector shows a similar degree of concentration to that which exists in the LTL sector in the USA, with the five largest firms by turnover accounting for approximately half of the turnover generated by the top fifty companies (see Table 14.9).

Table 14.9 The 50 largest freight transport companies in the UK; 1995 percentage share of top 50 turnover

Group	Share of turnover (%) in 1995
Top 3 firms	36.6
Top 5 firms	50.4
Top 10 firms	67.9
2nd 10 firms	15.4
3rd 10 firms	8.0
4th 10 firms	4.8
5th 10 firms	3.7

Source: based on data in Motor Transport Top 100, 1996.

Work carried out by the authors in 1994 examined the size and performance of eight key sectors in the UK road haulage market (Browne et al, 1994). Table 14.10 shows the average turnover of companies in 1992 in the samples for each of the sectors.

During the past fifteen years the types of services offered by hauliers have tended to become increasingly specialised with some of the larger firms moving away from general haulage and instead establishing contract distribution and in some cases dedicated services. There are a variety of reasons for this trend but among the most important is the desire among larger companies to develop services that have less pressure on profit margins and where scale is important. It is difficult to gauge accurately the profit margins within the road haulage industry since there are so many companies providing many different services. However, it is generally considered to be a very competitive industry - an analysis of company accounts for 374 companies from 1982-1992 showed that average profit margins varied from three per cent to six per cent during the ten year period. Profitability was generally better for the larger companies (Browne et al, 1993).

Table 14.10 Average turnover of UK road freight companies in eight key sectors

Sector	Average Turnover in 1992 (£ million)
Agriculture and livestock	2
Express and Parcels	40
General Haulage	5
International	11
Tankers	13
Temperature Controlled	23
Tipping	3
Warehousing and Distribution	76

Source: Browne et al, 1994.

EU freight market for international transport

On 1 January 1995, Austria, Finland and Sweden joined the EU, increasing the number of Member States to 15. This development was one of a series of political initiatives that have major implications for transport services throughout Europe. Two years earlier, on the 1 January 1993, border controls and customs arrangements within the EU were lifted following the creation of the Single European Market (SEM) under the Treaty of Maastricht. Central to the logic of creating the European Union free from unnecessary trading restrictions has been the desire to encourage the development of European companies able to compete on a global basis. It is often claimed that fragmented national economies within Europe have resulted in too many small companies in certain important industrial sectors. Dismantling barriers to trade and opening up new market opportunities allows companies to grow and become more competitive. Inevitably this is also likely to result in the relocation of certain economic activities as some companies become larger and others fail.

An enlarged and more integrated Europe will certainly influence demand for transport. Indeed, markets for goods and services in Europe have become much less fragmented over the past ten years. Through the latter half of the 1980s and the early 1990s there has been a discernible shift away from a mainly national approach to a more unified European strategy. In general, companies will increasingly regard the EU as their home market rather than having their trading horizons restricted to a single country. This in turn has important implications for logistics services. For example, increased trade between Member States will create new demands for logistics services such as transport and warehousing. New and improved transport infrastructure is helping to promote international freight transport - in some cases this may in turn lead to concerns about the increased length of haul of goods movement.

Despite the steady shift towards a more integrated Europe absolute volumes of freight moved are still greater for national flows than they are for cross-border flows. The growth in cross-border flows is encouraging for international hauliers and the market potential for this traffic looks promising. But hauliers from some countries can be considered much more international than others. For example, German operators are only responsible for about one-third of the international intra-Union traffic loaded and unloaded in Germany, compared with the Dutch who have a two-thirds share of traffic to and from the Netherlands. Indeed, transport operators from the Netherlands have traditionally provided a significant proportion of all European cross-border capacity. To some extent this is a reflection of the role of Rotterdam as a gateway port for Europe. The result of this gateway role is that Dutch transport operators have very strong outward traffic flows on which to base their business strategies. This has been backed up by the development of sales networks in other European countries. However, this is not confined to Dutch operators and many transport companies in the UK have also established extensive European sales networks.

One way in which transport companies can enter into foreign markets is through the establishment of operating centres in other countries and gradually increasing their networks. However, rather than follow this evolutionary and somewhat slow route to growth in foreign markets, some firms prefer the prospect of mergers, takeovers or strategic trading alliances with operators based in other European countries as a means to gaining a foothold abroad. For instance, a number of large, well known transport firms registered and operating in the UK are owned by overseas companies. To give some idea of the scale and importance of this foreign ownership the following firms are all among the top 100 UK transport companies (based on annual turnover): Kuehne and Nagel, Danzas, Gefco, Schenkers, Inter Forward, Hoyers and Norbert Dentressangle. A number of UK transport firms also own foreign haulage companies or have European subsidiaries and operations; some of the largest and best known include Exel Logistics, Christian Salvesen, Transport Development Group, United Transport and P&O.

It is also important to note that a number of firms use subcontracting as a means to entering the European road haulage market. This tends to involve large transport companies taking on work in other countries and then subcontracting this to smaller firms domiciled in these countries. Subcontracting also obviously occurs regularly within domestic markets, either on a one-off basis or sometimes as part of a long term contract between transport companies. In many cases it appears that smaller companies are increasingly dependent upon such subcontracts for their work.

Cabotage

Each EU member state is issued with a proportion of the total EU quota for cabotage permits. In the UK the demand for cabotage permits has generally been lower than the supply. Reasons for the low uptake of permits within the UK include:

- In most years the UK is a net importer and therefore operators tend to be able to obtain return loads and do not need to seek domestic work abroad.
- If the haulage company does not have a subsidiary in the country in which it is performing cabotage it will have to purchase the services of a financial representative to charge and account for VAT on their behalf.
- The cultural barriers that exist such as different languages and different business approaches.

A useful distinction can be made between what has been referred to as 'casual' and 'network' cabotage. Casual cabotage is cabotage undertaken in an opportunistic fashion by an international haulier, and not part of a planned, coordinated operating system. In contrast network cabotage is undertaken on a regular basis as part of a planned system by an operator who performs both international and national services. In the longer term it is the latter that could represent a threat to domestic hauliers if offered by large European transport companies with extensive transport networks. Yet in some ways the more any given operation performs cabotage the more likely it becomes that its operating costs will tend towards the same level as the national haulier and therefore any cost advantage will be eroded.

The right to perform cabotage operations has received considerable attention. Yet the data on the levels of cabotage carried out suggest that its importance has been somewhat exaggerated - certainly for most Member states of the EU. During 1995 the UK haulage industry had a small positive balance in cabotage (Croner, 1997). The British (i.e. vehicles registered in Britain) performed 46 million tonne kilometres in other countries, compared with 41 million tonne-kilometres carried by foreign hauliers between points in Britain. This amounts in both cases to an increase of 12 per cent on the previous year. But despite a 12 per cent increase, the total represents only 0.04 per cent of total hire and reward traffic in Britain (one of the lowest proportions within the EU). By contrast the German haulage market was more attractive for hauliers from neighbouring countries. Almost 1.5 per cent of all tonne kilometres performed in Germany was carried out by foreign hauliers, mainly from the Netherlands, Belgium and Luxembourg. This does however represent a small increase from the figure of just under one per cent reported by Crawford (1994).

Conclusions

Comparing performance and financial margins

USA freight market

As previously noted, deregulation has had the effect of reducing transport costs as a result of competition leading to lower prices, and the increase in ability for carriers to negotiate rates with shippers. Transport costs for both TL and LTL have decreased significantly after deregulation. In the LTL sector research has indicated that operating costs fell by 27 per cent in real terms between 1977 and 1987. The greatest reduction among key cost components during this period was in labour costs, which fell by 38 per cent (Corsi, 1994).

Improved utilisation of vehicles also helped to reduce unit operating costs. Average length of haul increased by 48 per cent between 1977 and 1987 in the LTL market; this was due to LTL carriers being able to expand their geographic coverage post-deregulation. This helped the average vehicle miles per vehicle to increase by 32 per cent between 1977 and 1987 (Corsi, 1994).

Profit margins in the US LTL sector tend to be relatively low. Of the top 50 LTL firms (by revenue) in 1992, four ceased operations by 1994, eight showed net losses in 1992 and six had profit margins of less than one percent. This reflects the competitive nature of the sector.

UK freight market

A study based on an analysis of published company accounts for a sample of 374 UK road haulage firms in 1993 provides a useful comparison with the US performance data above (Browne et al, 1993). Turnover increased by approximately 60 per cent over the period from 1983 to 1991 among the sample of companies studied. Return on total assets (ROTA) rose from approximately five per cent to eight per cent during the boom in the economy, peaking in 1988. ROTA declined from this eight per cent level to approximately two and a half per cent in 1991, during which time economic conditions were worsening. While it is relatively easy for companies to achieve reasonably high levels of ROTA when the economy is performing well and growth is occurring, this becomes far harder to achieve during a recession.

The pre-tax profit margin followed a similar cyclical pattern, moving between three per cent and six per cent over the period. A fall in the pre-tax profit margin between 1989 and 1991 is partly responsible for the reduction in ROTA at this time (the other factor is poor asset utilisation). Such levels of profit margin are typical of a highly competitive industry. Road haulage companies may not be in a position to influence prices in the industry on their own but need to reduce asset levels or costs in order to improve ROTA when economic conditions are difficult. Some UK hauliers have undoubtedly already taken this action.

Working capital as a percentage of turnover exhibited a typical level of plus or minus two per cent over the period. This shows that the industry can operate on extremely low levels of working capital in comparison with most other industries. This is due to the nature of road haulage; companies in this sector tend not to have many creditors (the main creditors being leases, fixed finance, diesel and wages).

Financial comparison between top USA and UK carriers

Data available on the financial performance of the top LTL carriers in USA in 1992 (Rakowski, 1994) and the performance of the top UK haulage companies for this same period (Motor Transport, 1994) makes a direct comparison of the largest operators in two markets possible. This is shown in Table 14.11.

Table 14.11 Turnover and profitability of top 5 UK and US LTL freight companies in 1992

	Turnover in 1992	Pre-tax profit as a % of turnover
Top 5 UK freight companies		
NFC plc	£1901.6 M	5.5
P & O Industrial Services Division	£1560.7 M	5.5
Lep Group plc	£1346.6 M	-0.3
Ocean Group plc	£1202.5 M	0.8
Transport Development Group	£599.5 M	6
Top 5 US freight companies		
Yellow Freight System	$2239.4 M	0.7
Roadway Express	$2155.8 M	2.7
CF Motor Freight	$2079.6 M	-1.7
Overnite Trans.	$872.7 M	6.9
ABF Freight System	$843.1 M	1.2

N.B. All the above financial data refers to year end December 1992 with the exception of NFC plc for whom the data refers to year end October 1993.

Source: Rakowski, 1994 and Motor Transport, 1994.

In the case of all the markets reviewed it is apparent that the haulier is often in a weak bargaining position when it comes to negotiating price increases especially for the more basic transport services. Not surprisingly pressure on profit margins leads to a focus on cost control or reduction. In section 3.2 it was noted that fuel costs and drivers' pay together could amount to 50 per cent of total operating costs. Being able to reduce these costs could therefore yield an important competitive advantage to a transport operator. In the past few years there has been growing concern about the practice of using drivers from low wage economies to drive commercial vehicles registered in high cost countries. As yet there has been limited analysis of the problem and therefore hard evidence is scarce - although it has been recorded that a number of Netherlands based companies employ drivers from Hungary and Russia (Sheppard, 1994). In north America the same phenomenon is said to occur with drivers from Mexico being employed by firms operating in the USA. It is important to stress that there is not necessarily anything illegal about this practice and the main safety issues relate to observance of drivers hours regulations, health and safety issues and observance of road traffic regulations and vehicle condition. It is clear that in any regional market there may often be considerable differences between wage rates prevailing in one county and those of another. Over time these differences may become less. In the short term this trend may cause concerns for employees of transport companies and there are, of course, issues of labour exploitation and equity that need to be appreciated and considered.

External factors

Concern about the use of energy, in transport as well as other areas could lead to increased taxes on fuel. Road haulage prices would rise if fuel taxes were sharply increased, as fuel makes up a significant component of operating costs. This in turn could have an impact on logistics strategy. The commercial vehicle of the future will have to be more fuel efficient and cleaner in terms of reduced emissions of carbon monoxide, nitrogen oxides and particulates. Legislation setting standards for the "clean lorry" will be important in ensuring that the next generation of commercial vehicles is able to meet tougher environmental standards.

Traffic congestion will also be more important in future. Forecasts in most developed countries suggest that demand for transport will rise and could continue to outstrip the ability of governments to supply sufficient infrastructure capacity. The effects of this mismatch between demand and supply could influence logistics strategies. If congestion increases then road transport will become less reliable, less productive and more expensive. The use of rail and intermodal freight services will become increasingly attractive in these circumstances.

There is little doubt that there is increasing pressure on road transport users in many countries (both car users and truck operators) to cover the external costs they generate. This pressure comes from a variety of sources and for a variety of reasons. Among the most important has been a general growth in concern about the environment, especially the increase in pollutants emitted by road transport as traffic levels have continued to rise. In the case of road haulage there have also been concerns about safety. Within the UK the Select Committee's Report on the 'Adequacy and enforcement of regulations governing heavy goods vehicles, buses and coaches' made numerous recommendations about how enforcement could be improved (Transport Committee, 1996). The regulations examined concerned mainly those governing the design, construction, use, maintenance and operation of vehicles and their conduct on the road. A key conclusion reached by the Committee was that:

> While the rules governing HGVs, buses and coaches are in the main adequate, the chance of even the most serious violators being caught is too low and the punishments for those who are detected are very often lenient and an ineffective deterrent; for enforcement to be effective either offenders should have a high chance of being caught or punishment should be exemplary. [para 96 page xxxii].

In the absence of economic regulation over capacity, market access and tariffs it is clear that regulations and enforcement relating to safe operations will be viewed as having a special importance. Despite the removal of economic regulations in freight markets in both the USA and Europe there has been an increase in freight transport regulation concerning safety and protection of the environment and this trend is likely to continue.

In the case of the liberalisation of cabotage, the concern was initially related to the possibility for low cost operators to gain market share at the expense of higher cost operators (who would probably claim that they operated to higher safety standards). So far there is no conclusive evidence that cabotage freedom has led to significant market distortion. In some ways it seems useful that the issue of cabotage has if anything turned attention towards the need for regulations governing safe operation of goods vehicles and the enforcement of the regulations. At the European level there would appear to be scope to make enforcement strategies rather more similar across the Member States of the EU. By doing this there will be a steady reduction in the problems caused when standards appear the same but are enforced in different ways.

Regulation has implications for modal share within freight transport. For example, legislation about vehicle sizes and weights can be argued to have some influence on the likely share of traffic that can be carried by rail. Enforcement of regulations on drivers' hours and overloading could also have an influence on the price per tonne-kilometre prevailing in some sectors of the road haulage industry. Although with the greater sophistication in the way transport is viewed as part of the supply chain and with the development by more firms of what may be called a logistics approach, the impact of small changes in price are likely to be diminished. If road haulage costs were to rise substantially as a result of, say, an increase in taxation or the introduction of road user charges then it is possible that modal share would change. The importance of this is difficult to gauge - what seems clear is that there is growing pressure from local, national and European level to ensure that all modes meet their external costs and this will lead to an increase in real terms of the cost of road haulage. Many of the larger road transport organisations are aware of the need to continue to improve their efficiency and develop innovative approaches to meeting the requirements of their customers within ever tighter operating standards and restrictions set by government.

Despite growing concerns about environmental issues and the rise in awareness of sustainability it seems unlikely that there will be a return to regulation of the haulage industry through government control of capacity and price - this applies to all the markets reviewed in the paper. However, in the case of all three markets discussed it seems inevitable that there will be:

- an increased emphasis on regulating goods movements in cities and urban areas - by means of bans and lorry controls;
- an increased emphasis on internalising the external costs of transport;
- some increases in enforcement regimes in order to overcome what are perceived to be very different but nonetheless important abuses of regulations such as illegal operation and drivers' hours limits.

These changes will be regarded as additional costs and constraints by some operators. However, the more responsible companies will benefit from appropriate regulations introduced and then rigorously enforced. In this way competition can be determined through normal commercial practices and there are no advantages to those organisations and individuals prepared to ignore the regulations concerned with safety and the environment.

References

D. Bowersox and M. Cooper (1992) *Strategic Marketing Channel Management*, McGraw Hill, New York.

M. Brignall (1996) 'Kinnock wants UK in vignette scheme', *Commercial Motor*, 4-10 July, p.5.

T. Brown and J. Greenlee (1995) 'Private trucking after deregulation: managers' perceptions', *Transportation Journal*, 35 (1), pp. 5-14.

M. Browne and J. Allen (1997) *Logistics and distribution trends in food manufacturing: A report on the second University of Westminster survey*. University of Westminster, London.

M. Browne, J. Allen, S. Cohen and B. John (1993) *Performance Trends in Road Transport*, Lloyds Bowmaker, Bournemouth.

M. Browne, L. Albanese, J. Allen and B. John (1994) *Road Transport Industry - Financial Trends*, Lloyds Bowmaker, Bournemouth.

K. Button and G. Chow (1983) 'Road haulage regulation: a comparison of the Canadian, British and American approaches', *Transport Reviews*, 3 (3), pp.237-264.

J. Cooper (1991) 'Lessons for Europe from freight deregulation in Australia, the United Kingdom and the United States of America', *Transport Reviews* 11(1) 85-104

J. Cooper, M. Browne and M. Peters (1994) *European Logistics: Markets, Management and Strategy*, Blackwell, London. Second edition.

T. Corsi (1994) *Motor Carrier Industry Structure and Operations*, proceedings of International Symposium on Motor Carrier Transportation, Williamsburg, Virginia, 31 May - 4 June 1993, National Academy Press, Washington.

K. Crawford (1994) *Introduction of road cabotage in the European Community*, Proceedings of Seminar A PTRC Summer Annual Meeting, pp.153-162.

Croner (1997) *Road Transport Operation*, Eurobriefing note, Issue no. 127 (page 2). Croner Publications Ltd, Surrey.

Department of Transport (1994) *Transport Statistics Great Britain 1993*, HMSO, London

Department of Transport (1996) *Transport Statistics Great Britain 1995*, HMSO, London.

Department of Transport (1997) *Transport of Goods by Road in Great*

Britain 1995, HMSO, London.

Institute of Logistics (1993) *Distribution Cost Survey*, Institute of Logistics, Corby.

Motor Transport (1994) *Motor Transport Top 50*, Motor Transport and Lloyds Bowmaker, Surrey, March.

Motor Transport (1996) *Motor Transport Top 100*, Motor Transport and Lloyds Bowmaker, Surrey, June.

J. Rakowski (1994) 'The Continuing Structural Transformation of the US Less Than Truckload Motor Carrier Industry', *Transportation Journal*, 34 (1), pp.5-14.

G. Sheppard (1994) 'Flags of convenience', *Commercial Motor*, 14-20 July, pp. 42-43.

Standard and Poor (1995) *Hard Landing for Truckers?*, Credit Analysis Reference Service, London.

Transport Committee (1996) The Select Committee's Fifth Report *The adequacy and enforcement of regulations governing heavy goods vehicles, buses and coaches*. 356-1. London: HMSO

Notes

1. The summary from 'Deregulation in the USA' to 'Comparison of USA, UK and EU freight markets' is based on Chapter 8 in Cooper, Browne and Peters, 1994 .
2. One US (short) ton = 907 kg.

Part 5

Airlines

CHAPTER 15

The ownership and regulation of international airlines

Kenneth Button

Abstract

Trends towards liberalising domestic transport markets have extended pressures to deregulate international transport. Developments within the European Union are one manifestation, but the movement is much wider. This paper explores specific issues surrounding changes in the regulatory regimes confronting international air transport services suppliers. It looks at the nature of recent trends in policies related to ownership and access and pays particular attention to the transatlantic market. The main focus is on airlines' reaction to the changing regulatory environment in which they operate and the longer term implications on new market entry and airline users. Particular attention is paid to airline alliances' roles in these gradually liberalising markets.

Introduction

International air transport has undergone considerable change over the past decade. The European market has become a focal point for the liberialisation of international services, and the US's 'Open Skies' policy has contributed to longer distance routes as well. Demand has expanded as both business and leisure users seek to travel more. Technology has brought down the price of operations, along with the commercial pressures that accompany more liberal markets. Airlines have also reacted to the new circumstances, both in terms of the way activities are managed and the growing number of alliances fostered between carriers.

This new situation raises a number of important legal, economic and management issues. The focus of this paper will be limited to initially offering a short assessment of the key trends affecting the international air transport model. It will outline significant institutional reforms affecting the way the market is developing before looking at initiatives between domestic and international markets. The core will concern itself with carriers' response to these developments, particularly the structure of airline alliances. Finally, this discussion will link with the development of transport policy in the EU.

Aviation trends

Globalization and internationalization are two major industrial trends of the late twentieth century. This is reflected in part in the significant growth of trade taking place in the 1990s, with real export growth in industrialized countries making up the Organisation for Economic Cooperation and Development at over 7 per cent per annum. Equally, a significant rise in foreign ownership of assets is now estimated to total about $1.7 trillion. It is too early to judge whether these trends represent passing fads or genuine long term adjustments in the way production and trade is conducted. Preliminary indications are, however, that they are more than just transient trends.

All this has been taking place at a time when the institutional structure of provided air transport services has seen significant developments. The 1978 US deregulation of its domestic passenger market, combined with its commitment to an 'Open Skies' approach to international aviation in 1979, have been instrumental in changing not only the way US policy is conducted but also, through both demonstration effects and direct knock-on effects, the ways in which many other air transport markets are now regulated.

In particular, the intra-European market is moving rapidly towards a situation akin to that in the United States. While many European countries have unilaterally liberalised their domestic markets, the EU, since 1988 through a succession of 'packages', has moved to a position that has left air transport within the Union largely free from economic regulation.[1]

Outside of Europe and North America, the majority of national markets in South America have been liberalised with different types of extensive privatisation programs. Australia and New Zealand have been deregulated as well. Additionally, establishment of the World Trade Organisation has also brought into play (albeit in an extremely small way) a new and geographically wider policy-making institution to supplement roles already played by those such as the International Civil Aviation Organisation and the International Air Transport Association. And aviation issues are on the agenda of new regional groupings such as the Asian-Pacific Economic Council.

This combination of market trends and institutional reforms, combined with rising incomes and increased leisure time, have contributed to the steady growth in demand that has taken place. Additionally, technology advances in aircraft efficiency have risen, and air traffic control systems can handle greater volumes of traffic. All this has exerted positive effects on the cost side of the international air transport equation.

As a result, air passenger traffic since 1960 has grown worldwide at an annual average of 9 per cent, freight by some 11 per cent and mail traffic by 7 per cent. In 1995, some 1.3 billion passengers were carried by the world's airlines. Civil aviation has become a major service industry contributing to both the domestic and international transport systems. It facilitates wider business communications and has been a key component in the growth of tourism which is now one of the world's major employment sectors.[2]

All the indications are that this trend towards expansion will continue into the foreseeable future (albeit at differential rates) in various geographical sub-markets. While forecasting remains an art rather than a science, it seems likely that passenger traffic will grow at a rate of between 5 and 7 per cent in the foreseeable future, much of it in the Asian-Pacific region (up to 9 per cent a year). Forecasts also indicate a slower growth in the more mature US and European markets.

In line with other sectors, aviation has experienced significant moves towards internationalisation in terms of market structure. Indeed, it has been the stated objective of the UK carrier, British Airways, that it intends to become 'global'. In pursuit of wider market coverage and to enhance internal efficiency, airlines have followed a number of courses. The recent development of alliances is perhaps the most controversial of these.[3] Equally, cooperation between scheduled airlines to develop computer reservation systems represents another course of action, with the Amadeus system in Europe an example of this.[4]

Institutional developments

US domestic aviation was heavily regulated in the late 1920s. Responsibility for the economic regulation of interstate airline services was exercised by the Civil Aeronautics Board (CAB) which controlled market access and fares through the right of disapproval. A combination of forces brought regulatory change in the late 1970s.[5]

The subsequent 1978 Airline Deregulation Act constituted a Big Bang in the sense that a single act radically changed the way the domestic aviation market was regulated.[6] The move did not immediately establish a free market but rather, a time schedule was established for the relaxation of price and entry regulations. By January 1983, all fare and entry regulations were eliminated except the provision that carriers must be fit, willing and able. The CAB was abolished in 1985 since its residual functions over such things as international aviation and mergers were transferred to other agencies.[7]

Policy reform in the EU has been more complicated and protracted. While an EU Common Transport Policy has existed since the signing of the Rome Treaty in 1957, aviation was initially excluded.[8] Countries regulated their own domestic aviation, and a bilateral system governed international air transport within and outside the Union. Air service agreements were generally restrictive and often allowed only one airline from each country to operate a route. While this may be similar to CAB entry policy on US interstate routes, the situation was more stringent in that over 90 per cent of bilateral agreements involved a controlled capacity with an obligatory 50:50 revenue pooling. Further, some 900 agreements excluded fifth freedom rights.[9] The fares charged were agreed by both countries and excluded competitive pricing. Finally, many were 'flag' carriers largely state owned and often in receipt of substantial state subsidies.

Change within has come slowly and taken a number of forms.

Initially, there were domestic aviation policies within a number of EU members. Some, as in the UK, were *de facto* changes which did not entirely free the market but instead saw the national regulatory agency being more liberal in the allocation of licenses and fare flexibility. Other countries such as France, Spain, Italy and Germany, were less inclined towards liberalisation. Reforms have been accompanied by greater private sector involvement in the provision of services. In some instances (such as British Airways) complete privatisation of former state companies occurred at an early stage. More common instead (in Germany and the Netherlands) was a gradual selling of the airlines.

From the mid-1980s, moves were made to liberalise bilateral agreements between Members. This started between the UK and the Netherlands in 1984 with relaxed rules on market entry and later on tariffs. Subsequent agreements (particularly involving the UK, Germany, France and Spain) all embodied various degrees of liberalization. This was not EU-wide but confined to markets where Members were of a less regulatory orientation.[10]

Of particular importance external to the EU was the US 'Open Skies' policy. Since 1979, it has attempted to develop liberal agreements with individual EU states, although the only long standing liberal agreement has involved the Netherlands. Since 1994, the US has had agreements with a number of other European countries, although none with major international airlines. In 1996, an interim arrangement with Germany was reached that amounted to an Open Skies agreement.[11]

The EU's direct involvement in developing a common aviation policy emerged in the late 1980s as a result of several key legal cases.[12] These involved European Court of Justice rulings between 1974 and 1978 which determined that articles in the Treaty of Rome relating to state aid and competition policy did apply to aviation. The Commission itself was also not adverse to the idea of reform, as it had published a series of proposals in the late 1970s and 1980s on how liberalisation might be implemented.[13]

The *Nouvelles Frontières* case in 1985 indicated that the Commission's power to act on fares was stronger than had been supposed. To pre-empt the actions of the Commission, the Council introduced a more radical package of measures in 1988. The rules of competition apply permanently and directly to aviation, with block exemptions in place for joint planning of capacity, revenue sharing, consultation on tariffs and aircraft parking allocations, computer reservation systems (CRS), and ground handling services. A directive adopted allowed limited pricing freedom so that fare applications had to be accepted by a Member if it reasonably related to costs. A phased relaxation over the normal 50-50 division of traffic was implemented on international routes so that after three years, it could split by up to 60-40. Multiple designation for routes was gradually brought in as fifth freedom rights spread.[14]

This 'First Package' of reforms was modest yet insufficient to meet the commitments associated with moving to a Single Market.[15] A 'Second Package' was thus initiated in 1989 that removed government-to-government capacity sharing arrangements; introduced the notion of double disapproval of fares (whereby the governments of both countries on a route must object to fares set by the operators); prevented governments discriminating against airlines, and provided that technical and safety standards were met. At the same time, ownership regulations were reformed which made foreign participation easier and fifth freedom operating rights automatic.[16]

The final thrust came in 1993 which initiated a phased move that resulted, by 1997, in a regulatory framework for the EU similar to that in US domestic aviation, although leaving open questions regarding the EU's role with respect to international aviation outside the Union. From January 1993, any EU airline could fly between Member States without restriction, and within Member States other than their own, they would be subject to some controls on fares and capacity. Since April 1997, full cabotage has been permitted, and fares are generally unregulated.[17]

Foreign ownership among Union carriers is permitted, and for EU internal purposes, they have become European airlines. This change does not apply to extra-Union agreements where national bilateral arrangements still dominate the market. As a result of these changes, a considerable increase has occurred in cross-share holdings and the expanding number of code sharing and similar alliances among airlines within the Union.[18]

Domestic and international markets

The main focus is on internal markets for air services, but it would be inappropriate to ignore the wider international environment where air transport is supplied. For example, important features extend beyond the boundaries of the EU. The nature of modern networks means that the EU notion of interoperability and interconnectivity must embrace this wider spatial dimension.

This link between the internal EU market and external markets is a feature not unique to aviation but extends to all EU network industries. Indeed, it is one of many factors that make policy formulation regarding EU aviation and similar sectors so difficult. These are:

- Europe represents an important transit area for traffic originating outside of the region and destined for locations beyond it.
- European airports often cater for traffic with origins or destinations outside of the EU as well as for internal EU traffic.
- European air control and navigation infrastructure is jointly used by internal EU traffic, those with origins or destinations outside of the EU and traffic overflying EU countries.
- A large number European airlines have extensive operating networks extending well beyond EU boundaries with plans to manage their activities and tie them in with intra-EU services.
- Many EU airlines have equity involvements with those outside of EU.
- The rapid growth in alliances over recent years involving EU and non-EU carriers blurs operational boundaries.

- Many companies outsourcing EU air carriers also serve non-EU airlines and they plan their commercial activities in this wider framework.
- Many European airlines receive government subsidies, and no distinction exists as to whether they are spent on intra-EU services or on external routes.
- Air transport policy is a comparatively new EU concern, and most national policies have long been based on much wider international considerations that cannot easily be disregarded.

These links mean that the air transport policy of the EU has not grown in geographical isolation but has instead entwined with developments elsewhere. Future policies will inevitably need to take these linkages into account. Equally, the EU has been influenced by the effects of events elsewhere. United States deregulatory aviation policies have exercised significant influences on EU member countries who, as a result, adopt similar policies. These are tailored to meet local and national objectives, and enjoy the benefit of hindsight in circumventing problems associated with US reforms.

European developments have not attracted the same analytical attention as the US Airline Deregulation Act.[19] This relative dearth of work is due in part to a slower and phased pace in Europe which makes quantitative assessments of each policy reform difficult. Unlike the US, deploying econometric analysis is compounded because reform has been across a range of diverse regulatory systems. While lessons may be learned from developing aviation markets, there are also many features unique to Europe.

Airlines' reaction functions

The notion of alliances is one recently under public scrutiny in the wake of several publicised efforts by major international airlines to link their operations. The nature of ties differs between groupings, like the success of partners in gaining official ratification and the way they operate and manage their alliances.

The US domestic market first saw these emerge in 1967 as a way for jet and commuter operators to jointly develop markets in an era of tight economic regulation. The late 1980s and early 1990s saw the growth of new forms which embraced somewhat different characteristics and served different purposes. They were less institutionalised in that they were formed by privately owned commercial airlines outside of government or inter-governmental agency initiatives. This was also reflected in international alliances.

These alliances are in a continual state of flux as well. According to the *Airline Business* survey, for instance, the Spanish carrier Iberia reduced its alliances from 27 in 1995 to 13 by May 1996.[20] Over the same period, Austrian Airlines cancelled six agreements and added four new ones, and Swissair added six and dropped three. Similar patterns emerge for a number of others. These changes are part of a tidying-up process that carriers formulate towards more coherent network strategies.

The exact number of alliances is unclear, not only because of the dynamic nature of the arrangements involved, but also the term 'alliance' is generic. In a strict legal sense, it may mean some degree of equity ownership of one carrier by another. More often, it is interpreted to embrace such things as code-sharing agreements, interchangeable frequent flier programs and coordinated service scheduling. Equally, airlines are often involved in many different kinds of alliance, sometimes embracing a single partner but also with several others. A more recent feature is that several major carriers link their activities in so-called 'galaxies'.

An annual survey by *Airline Business* attempts to track alliances involving major carriers and reports changes in some of the main features (Table 15.1). Growth is immediately obvious as is the relatively small number of alliances which involve an equity stake and the slow growth in their numbers.[21]

The North Atlantic market embraces a number of strategic alliances where airlines are involved in code-sharing and other forms of cooperation across a large number of routes to strategically linked networks. This type dates back to the Global Excellence alliance formed by Swissair, Singapore International Airlines and Delta in 1989. Other alliances, such as Continental and Alitalia and United and British Midland, involve code sharing between specific regions. The vast majority of 'point-specific' alliances, are relatively minor targeted affairs that generate few controversies.

The type of features characterising an alliance can be divided into distinct categories:

- Full mergers of domestic airlines were a feature of the US following deregulation, as the initial period of instability moved into one of consolidation and rationalisation. There have also been significant merger activities within the Canadian market and an important number within Europe in recent years.[22] Mergers of this type are the most extreme form of an alliance and a traditional way carriers coordinate operations and other activities. They enjoy the advantage that complete control of a carrier is in the hands of a single board, and resources can be allocated more effectively.[23] In practice, however, mergers are not always successful. Generally, mergers of overlapping networks in any transport industry offers fewer economies than combining interfacing networks. In some instances, problems also arise because those involved have miscalculated the costs of transition.[24]

Table 15.1 Airline alliances 1994 - 1996

	1996	1995	1994	% change
Number of alliances	389	324	280	38.9
With equity stakes	62	58	58	6.9
Without equity stakes	327	266	222	47.3
New alliances	71	50	-	-
Number of alliances	171	153	136	25.7

Note: New alliances are those entered into since around May of the previous year and not then listed as planned. Alliances restricted to frequent flier cooperation were included in 1994, but excluded in 1995-6. The actual number of alliances in 1994, the first year Airline Business compiled information, was marginally higher than stated as some alliances went unreported. However, some domestic regional operators owned by majors were included in 1994, but excluded in 1995-6.

- There has been a recent tend for equity level holdings to fall short of a full merger, particularly when airlines from two separate countries are involved and national laws limit the extent of foreign ownership.[25] Such forms of investment have grown considerably in recent years. Table 15.2 provides details of some major examples. What the table does not show, however, is the degree of control equity holdings can afford an airline; particularly, voting rights are often less than the relative amount of capital involvement.

Table 15.2 Foreign ownership of major airlines

Country	Airline	%	Country	Airline	%
	EUROPE		NORTH AMERICA		
Austria	Austrian	20	USA	America West	20
	Lauda	40		Continental	20
Belgium	Sabena	49		Delta	10
France	TAT	49		Hawaiian	33
Germany	Deutsche BA	49		Northwest	24
Hungary	Malev	30		USAir	22
Luxembourg	Luxair	13	Canada	Canadian	33
Russia	Air Russia	31			
UK	Air UK	45	AUSTRALASIA		
	BMA	40	Australia	Quantas	29
			New Zealand	Ansett NZ	100

Source: Airline Business December 1995

Equity holdings have the advantage over looser alliance arrangements in showing stronger commitment on the part of carriers for longer term cooperation. They also act as a device to prevent other carriers from forming alliances with those involved. It also proves to be a prudent investment if the synergy effects anticipated meet or exceed expectations. This appears to be the case with the KLM equity stake in Northwest Airlines. A difficulty with such holdings is that they represent a commitment that can prove costly to reverse if the alliance fails to deliver the anticipated benefits. The second Continental airlines bankruptcy in 1991 was expensive for its equity holding partner, SAS. In many cases, regulations over airlines' control means that a foreign carrier cannot obtain the level of governance normally commensurate with its equity holding.

- Code-shares can be across a wide range of services (as with major strategic alliances) but more often, they involve a single or a small network of services.[26] Networks can be attractive to smaller carriers wishing to expand but whose own system does not ideally dovetail with other single airlines. The tendency, however, is that such arrangements are short lived as new coalitions emerge in a dynamic marketplace.[27]
- Hub-and-spoke operations, particularly the 'banking' of flights (a concomitant of effective operations), can be more efficient if carriers coordinate flight patterns. By allowing traffic to be consolidated and transshipped between flights, hub-and-spoke operations can enhance load factors and let airlines reap any benefits of economies of existing scope and scale. By agreeing to coordinate schedules, two airlines can increase the potential amount of traffic that on-lines across combined networks.
- Franchising is traditional in sectors such as fast food and clothing. Its appeal in aviation is allowing major carriers to spread their brand name and generate revenues on thin routes without the necessary major capital investments. It can also help funnel traffic to a major carrier at slot-constrained airports where its own frequencies cannot be increased. The franchisee gains from increased market recognition as well. Franchises are growing in popularity throughout international markets, especially in Europe. In 1996, British Airways had six arrangements that carried a combined estimated total of 3.4 million passengers and generated more than $77.7 million in revenue. British Airways' franchises now extend beyond Europe and include, since October 1996, the South African Airline Comair.[28]

The economics of alliances

Neither management science nor economic theory provides an uncontroversial single reason why airlines unite and form alliances. Undertakings may occur for a variety of reasons.

A traditional concern of anti-trust authorities is that alliances may be created to concentrate market power and exploit any monopoly advantages that exist. The result may be a reduction in supply and higher prices for users. The judgment here is difficult, because in situations where economies of scale, density or scope still exist, where market instability is emerging due to excess competition, or where there is a restrictive institutional regime, merger or cooperative behaviour can often increase efficiency.

In terms of basic economics, an alliance can have an impact on both the demand for a carrier's services and its cost structure. The ultimate effect on customers will also depend on the nature of the market in which the alliance operates. For example, if it is only a small part, the incentive for carriers to maximize efficiency and keep prices low is likely to produce considerably different results than where alliance carriers dominate a market and can erect barriers to new entry.

Because of the diverse forms airline alliances take and the ways in which potential elements can be combined, it is dangerous to generalise. Some broad points can be made in a simple diagram of a specific air transport market.

In Figure 15.1, D_1 and C_1 represent the demand and cost curves confronting a carrier on a particular route. These are linear, and the cost curve is horizontal simply for presentational clarity. If the airline joins with one or more of its competitors on this route in an alliance, then one would expect that.[29]

- by virtue of joint marketing, code-sharing, common frequent flyer programs, coordinated scheduling and so on, demand for the airline's services would move out (e.g. to D_2) and
- because of the potential for rationalizing services, scope for spreading ground costs and so on, costs would fall (e.g. to C_2).

The extent and nature of these shifts in demand and the cost schedules depend not only on the details of the alliance but also on carriers' response outside of it. If the alliance resulted in other, non-partner carriers becoming more cost conscious, cutting fares or improving services, then the demand curve could not shift or slightly move out. Equally, a pure code-sharing alliance may have minimal effects on costs. For purposes of exposition, simple cases are treated here.

Figure 15.1 Fares and passenger miles associated with an airline prior to making an alliance and after making an alliance

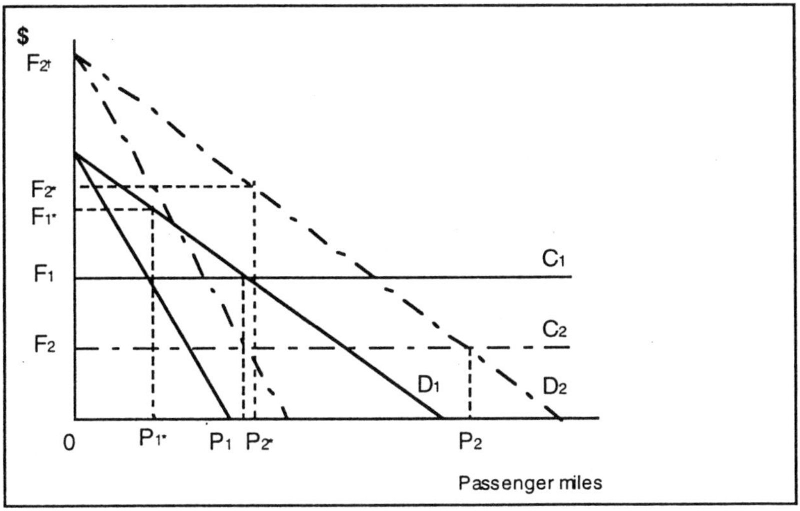

A number of possible outcomes emerge from this framework, each dependent on the nature of the market involved.

- If it is initially competitive and the alliance has no significant effect on conditions (e.g. involving two small players), then the outcome will be that full benefits of any cost reductions and service improvements will be passed to consumers (reflected diagrammatically in enhanced demands). Fares will fall from the old level of F_1 to the new level of F_2, and passenger seat miles will rise from P_1 to P_2.
- If the market is initially monopolised and the alliance brings new players into the game (such as through a blocked space arrangement whereby a carrier buys blocks of seats on another airline's planes to resell), then fares may fall and passenger miles increase. Initially, the fare level would be F_1*, if no price discrimination existed but would fall to F_2.
- If the alliance gives partners monopoly power on a route formerly competitive (this could be associated with control over capacity in an international bilateral agreement, landing slots, dominant frequent flyer program, etc.) and carriers seek to maximize profits, then the outcome will depend on the adopted strategy. If a single fare approach is used, then conventional theory indicates that fares charged by the airline

under review will change to F2* and passenger numbers to P2*. The ultimate fare level may be higher or lower than the original, depending on the extent to which the alliance results in cost savings. Equally, the number of passengers may be higher or lower than P_1 depending on the elasticities of demand that are involved.

- Widespread yield management practice, however, enables carriers 'to price down demand curves' rather than simply charging a single uniform fare. If there is perfect price discrimination, the alliance airline would price differentially down where demand meets cost, to a level where it carries P_2 passengers. (This gives the same number of passenger miles as the competitive final outcome, but the consumer surplus enjoyed by users with competition is translated into profits for the airline.) Fares in this case would vary between F2† and F2. The airline benefits by generating additional revenues equal to the area under the demand curve above C2 and to the left of P2.
- Alliance carriers may decide not to maximize monopoly profits from the arrangement but rather, managerial slack emerges (so-called X-inefficiency). In this case, the illustrated carrier's cost function may not fall entirely to C2 and the resultant pricing decision may not exploit the full price discrimination potential. The result would be less passenger miles than under the previous scenario and less revenues to the carrier.
- While enjoying short term monopoly power, the alliance may be concerned that earning excessive profits could make new entry difficult to combat. Alternatively, fear that high profits may bring a policy response from the government may influence their actions. In this case, fares may be set below those associated with either the single price monopoly or yield management. The extent to which this deviation occurs depends on the degree of risk aversion felt by alliance carriers.

The above is clearly a gross simplification but illustrates problems associated with assessing the economic implications of alliances, especially from an anti-trust perspective. In practice, the situation is more complex; aviation is a network industry, and any effect on one link may impact elsewhere on the network.[30]

Figure 15.2 offers a simple illustration of the issue types involved in assessing the economic effects of an alliance. If carriers operate between routes A and B and result in partners having a monopoly position, then the actual degree they can exercise power depends on a variety of factors. If there are indirect services between A and B via C and D, or even via E, for example, then the demand for services will be conditional, dependent on these alternatives.

If the link between A and B forms only part of many longer trips (say involving travel between A and F), then the number of options open to travellers is even larger. They may avoid the link by taking a direct flight (A to F) or other alternative indirect routes via C or D. If the services between A and B are frequently used in more complex travel patterns involving several interchanges (say as one link in a trip from E to F), then the number of possible competitive options is even larger.

Figure 15.2 Network implications of airline alliances

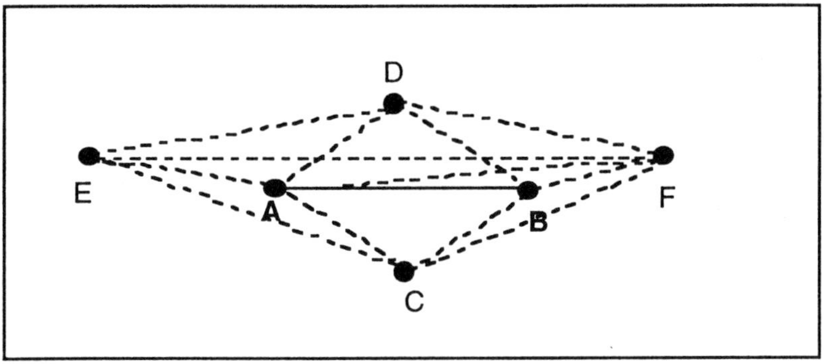

The situation depicted in Figure 15.2 essentially relates to a point specific route alliance (namely between A and B). A strategic alliance poses more complex problems in defining relevant markets. By coordinating activities across a network, the airlines involved may be in a position to influence fares and service quality on indirect routes that potentially compete for traffic with the A to B service. In the above illustration, this could be true if alliance partners were dominant hub carriers at A and B.

Whether this is against public interest depends on the extent to which efficiency gains outweigh efforts to extract profits beyond an investment's normal rate of return. If the specific details are of a blocked space variety with the prospect that each carrier would behave in a monopolistic way in its own markets, then expectations are that competition within the aircraft would keep fares down to at least a workable level.

Strategic alliance arrangements involving high market segmentation are less likely to provide consumer benefits where, despite prior hub dominance by each partner, a degree of competition existed before. This will occur unless a coordination of activities by the airlines reduces costs or the countervailing powers of such things as potential new market entry or the threat of government regulations temper fare setting strategies.[31].

Elements of an alliance involving coordination between carriers' activities can lead to cost savings. This was the driving force behind aircraft maintenance alliances in Europe in the late 1960s and subsequent formations of consortia to develop CRSs. Cost savings are also a major factor in securing ties with airlines from the west and those in other parts of the developing world (e.g. Lufthansa's links with Air China).

Alliances motivated by cost reduction considerations can be viewed as a way for carriers to reap economies of scale, density and scope without the rigidities and uncertainties accompanying a full merger. Savings may result from common ground facilities and improved scheduling.[32] Many smaller airlines have excess baggage handling and check-in facilities that can be utilised by a larger operation. Blocked space arrangements can lead to lower unit costs by making better use of an aircraft's seating capacity or by rationalising the number of flights required to carry a given number of passengers. And scheduling changes can reduce the number of flights required to provide a certain service level.[33]

More difficult trade-offs are made when considering instead mergers of looser forms of airline alliances. While mergers offer potential efficiency gains from reduced managerial overheads and tighter controls over duplication of costs, initial transaction costs of such things as redundancy pay to staff can be high. They can also take up liquidity in payments for equity holdings, weaken balance sheets and pose problems if the sector moves into recession.

On the demand side, alliances are viewed as extremely important. Economies of market presence are a strong contributor to the revenues most airlines enjoy. A traditional argument is that revenues tend to increase faster than the market share enjoyed by a carrier. By combining (and thus enjoying a larger market share), partners may anticipate a rise in revenue.

Additional traffic resulting from lower fares may accompany lower costs. Northwest and KLM have claimed this of their alliance. A wider range of options, both in terms of flights and final destinations offered by networks, provides potential consumers with a large number of products to choose from. Coordination of scheduling, *de facto* on-lining, and the seamless service image this creates enhances this effect as does the availability of common airport lounge facilities and joint frequent flyer programmes.[34]

Predictions of future demand patterns suggest that alliances may serve a second market penetration purpose by allowing airlines in geographical markets where likely growth is forecast to be slow (mainly North America and Western Europe) to expand into faster growing areas. The notion of 'global reach' becomes important in this context, and alliances permit this opportunity to be pursued in a flexible manner.[35]

A number of major alliances have come about when one or more partners experienced financial problems, thus providing a mechanism for a financial injection. This has recently occurred with the formation of the KLM/Northwest and the British Airways/USAir alliances, with the European carriers putting assets into their respective US partners.[36] Within Europe, Swissair has done the same for the Belgium carrier, Sabena, and within North America, American Airlines has put money into Canadian Airlines. In some cases, the injection came as part of the carrier's privatisation, such as the 25 per cent stake British Airways took in the Australian carrier Qantus and the 30 per cent Iberia took in Aerolineas Argentinas.

Airport congestion is a major problem that is set to become worse in the foreseeable future.[37] While technological developments and investments in new and expanded facilities will increase the physical capacity of the available infrastructure, environmental and safety concerns mean that overall capacity will likely grow slower than demand for transport services. It seems unlikely, for example, that the 50 per cent increase by the end of the century envisaged in the Outline Plan of the Trans European Airport Network will materialise.[38]

The provision of extensive scheduled air services is essentially a network industry. As such, it exhibits a variety of features suggesting it might be inherently prone to instability and, at the very worst, to under supply.[39] Once service has been scheduled, there is every incentive to fill those seats if the marginal cost of carrying extra passengers is exceeded by the revenue generated. This can mean that if insufficient funds contribute to the long term costs of the aircraft and other assets, services will eventually be withdrawn. Since carriers would have perfect foresight in an extreme case, they would anticipate this situation and not even bother entering the market. Insufficient capacity would then be offered. Economists refer to this as the empty core case.[40] To remain in network markets where potential instability exists or even enter, carriers may seek to insulate themselves from competition extremes. They may justify trying to retain customers through frequent flyer programs or corporate discounts. But by uniting with other carriers through alliances, problems of excessive competition may be reduced further. With extending networks and service attributes and the introduction of code sharing, capacity can be filled more completely and with greater certainty. Blocked space arrangements also offer greater flexibility on the supply side.

Isolating the types of environment where alliances form for reasons of combating market instabilities (rather than market exploitation) is difficult, and to date, analysis has relied on indicative and almost anecdotal evidence.

Empirical evidence

Considering the scale of controversy often surrounding the formation of major international strategic airline alliances, little rigorous empirical analysis has been completed in looking at real world implications.

What has been done focuses on the North Atlantic, short term implications of the carriers involved and their immediate competitors. An early assessment is in Oster and Pickrell,[41] but the main body of work includes two major US studies,[42] analysis by the UK authorities,[43] and several academic papers.[44]

Much of the analysis is US in origin and reflects both the country's tradition of undertaking quantitative analysis as part of its policy making processes as well as explicit concern about its traditionally dominant aviation sector position. Their carriers' performance in the main trans-Atlantic markets has been variable since the adoption of an Open Skies philosophy (see Table 15.3).

The vast bulk of empirical information relates to strategic alliances. The more common, point specific ones have often been ignored, although for individual markets they may be important. Equally, airlines with a number of alliances over a range of routes could potentially have more market control than a carrier with a single alliance. This type of arrangement is less transparent and, for this reason, has received even less attention than large, strategic airline alliances.

Table 15.3 US airlines' share of European-US traffic

	1980	1984	1988	1992
United Kingdom	55%	61%	56%	46%
Germany	46%	48%	53%	57%
France	48%	54%	62%	68%
Netherlands	7%	7%	11%	21%

One reason for an overall lack of empirical analysis is that assessing is seldom simple. Alliances have multi-dimensional, wide-ranging spatial and temporal effects that impact in different ways on many groups. While aggregate assessment of economic implications can be important, it is the incidence of costs and benefits on particular groups or geographical areas that impacts most on public policy. Policy concern may revolve around relative benefits and costs affecting consumers as opposed to airlines, but equally, it may also be influenced by the combined implications for a nation's own carriers and passengers.

The earliest US studies undertook by academics explored the implications of alliances on the structure of the industry and, to a more limited extent, sought out elements which led to successful alliances.[45] They tested a number of hypotheses, including the extent to which a partner's ability to adjust international route structures to emphasize hub-to-hub routes strengthened an alliance. They isolated a number of differing features that characterized each alliance (see Table 15.4) and found those associated with Northwest/KLM generated the greatest advantages in terms of increasing load factors and market share. This alliance saw the greatest realignment of strategies.

The Gellman Research Associates study looked in detail at USAir/British Airways and Northwest/KLM. Their approach involved an econometric model designed to reflect the way consumers select an airline. The parameters of the model were estimated for the first quarter of 1994. This snapshot was taken at a time when the Northwest/KLM alliance had matured but also when the implications of the USAir/British Airways alliance was still evolving.

The results estimated the impact on revenue, costs and profits of the carriers involved and others serving common routes (see Table 15.5). It concluded that the two alliances generated benefits for both the airlines and passengers involved. USAir/British Airways and Northwest/KLM increased their market shares on routes where code-sharing was practiced by 8 per cent and 10 per cent points respectively.[46] While benefits in terms of lower fares and improved services were the result of efficiency improvements, distributional implications occurred with other carriers losing traffic and revenue.

The US General Accounting Office study took a different approach by examining accountancy and other data over relevant sub-markets and combined this with interview information. The approach was broader but lacked the technical detail of the earlier analysis. It also showed that airlines participating in alliances benefit, albeit to varying degrees. Some

Table 15.4 The impacts of the early strategic international airline alliances

	Continental/SAS	Delta/Swissair	Northwest/KLM
Carriers realigned their route networks to take advantage of alliance partners' hubs	Yes	No	Yes
The traffic increase for the alliance was greater than the average North Atlantic traffic growth	No	Yes	Yes
The traffic increase for the alliance on routes between the US and the alliance partner's country was greater than the overall market growth between the US and the alliance partner's country	No	Yes	No
Load factors increased for both alliance carriers overall on trans-Atlantic routes	No	No	Yes

Source: Martin Dresner, Sue Flipcop and Robert Windle, 'Trans-Atlantic airline alliances: a preliminary evaluation', *Journal of the Transportation Research Forum*, vol. 35, 1995, pp. 13-25

Table 15.5 Estimated implications of the USAir/British Airways and Northwest/KLM code sharing arrangements in 1994 ($ million)

	Revenue	Cost	NetProfit	Consumer Benefit	Social Benefit
USAir/BA alliance					
UsAir	7.9	-2.3	5.6		
Other US carriers	-41.7	14.9	-26.7		
US total	-33.8	12.6	-21.1	4.9	-16.2
British Airways	45.8	-18.6	27.2		
Other non-US carriers	-1.3	0.5	-0.8		
Non-US total	44.5	-18.1	26.4	5.4	31.8
Grand total	10.7	-5.5	5.3	10.3	15.0
Northwest/KLM alliance					
Northwest	24.6	-8.5	16.1		
Other US carriers	-25.6	9.9	-15.7		
US total	-1.0	1.4	0.4	13.0	13.4
KLM	18.6	-8.0	10.6		
Other non/US carriers	-16.5	7.9	-8.6		
Non-US total	2.1	-0.1	2.0	14.1	16.1
Grand total	1.1	1.3	2.4	27.1	29.5

Source: Gellman Research Associates, A Study of International Airline Code Sharing, Office of Aviation and International Economics, Office of the Secretary of US Department of Transportation, Washington, 1994.

of the gains come from generated traffic but a significant amount come from transfers of non-alliance carriers.In the case of Northwest/KLM, Continental Airlines estimated that it lost about $1 million in revenue in 1994 as a result of competing with this alliance. The study also points to a number of new (or reintroduced) international services that resulted from alliances involving US carriers.[47]

Conclusions

The past twenty years have witnessed considerable changes in how the air transport market is perceived. The traditional notion that strong regulation, along with a high degree of state ownership was necessary if economic and social goals were to be attained, has faded. This reflects changes in the way society views air transport, as well as a clearer understanding of how the sector functions in a more market-orientated system.

Recent trends have seen a movement away from regulating prices, market entry and services. State ownership has declined as privitisation takes place and, in many cases, it has extended to infrastructure as well. The situation is not, however, an entirely free market; at the international level, bilateral arrangements still dominate most major markets. Airlines, however, have been inventive in this brave new world. While restraints still remain on behaviour, they have circumvented institutional impediments through the creation of various forms of alliances. Clearly, evidence exists that these alliances have not been perfected - they are still too volatile for that. Rather, they have provided a mechansim for carriers to evolve in changing market conditions.

Notes

1. K.J. Button (1996) 'Aviation deregulation in European Union: do actors learn in the regulation game?', *Contemporary Economic Policy*, 14: 70–80.
2. In addition to passenger transport, aviation is also an important form of freight transport. Some estimates suggest that it carries up to 60 per cent of world trade by value, and it is forecast to rise 80 per cent by 2014.

3. Most airline alliances are concerned with scheduled services, but some involve charter airlines, see *Avmark Aviation Economist* (1994) 'Is the Anglo-German alliance succeeding?', May: 8-9.
4. B. Humphreys (1994) *New Developments in CRSs*, ITA Documents and Reports 32, Paris.
5. A series of academic studies emerged and unfavourably contrasted the performance of regulated interstate aviation with more liberal intrastate services. Fresh intellectual ideas concerning regulatory capture and market structures brought a new questioning of the underlying rationale for controls. Demonstration effects from other network industries, such as UK trucking, indicated that liberal markets could be competitive and stable. At the macroeconomic level, Keynesian policies to reduce perceived cost push inflationary pressures sought ways of lowering prices.
6. Air freight transport had been the subject of liberalizing reforms in the previous year.
7. S. Borenstein (1992) 'The evolution of U.S. airline competition', *Journal of Economic Perspectives*, 6: 45-73.
8. EEC Commission (1961) *Memorandum on the General Lines of a Common Transport Policy*, EEC, Brussels.
9. Fifth freedom rights allow an airline of one country to carry traffic between two countries outside of its own registry as long as the flight originates or terminates in its own country of registry.
10. What they did was provide early evidence of the implications of change. Irish-UK reforms resulted in lower fare levels, with benefits estimated in 1989 at £24.9 million for the 994,000 passengers already using the route at the time of regulatory change and £16.2 million for 1.3 million additional passengers generated post liberalisation. See Barrett, S.D. (1990) 'Deregulating European aviation – A case study', *Transportation*, 16: 311-27.
11. The longer standing issue in the context of an EU air transport policy is whether it is still appropriate for individual Member States to engage in such bilateral arrangements or whether a common EU negotiating position should be taken. In 1996, EU states gave some soft-negotiating rights to the Commission, see Association of European Airlines (1995) *EU External Relations*, AEA, Brussels.
12. K.J. Button and D. Swann (1989) 'European Community airlines - deregulation and its problems', *Journal of Common Market Studies*, 27: 259-282.

13. Commission of the European Communities (1979) Contribution of the European Communities to the Development of Air Transport Services, *Bulletin of the European Communities*, Supplement 5, Brussels; Commission of the European Communities (1984) *Towards the Development of a Community Air Transport Policy, Civil Aviation Memorandum No. 2*, COM(84) 72 Final Brussels.
14. For more details see K.J. Button and D. Swann (1992) 'Transatlantic lessons in aviation deregulation: EEC and US experiences', *Antitrust Bulletin,* 37: 207-255; K.J. Button, (1992) 'The liberalisation of transport services', in D. Swann (ed.), *1992 and Beyond*, Routledge, London; D. Vincent, and D. Stasinopoulos (1990) 'The aviation policy of the European community', *Journal of Transport Economics and Policy*, 24: 95–100, 1990; D. Stasinopoulos (1992) 'The second aviation package of the European community', *Journal of Transport Economics and Policy*, 26: 83–87; D. Stasinopoulos (1993) 'The third phase of liberalisation in community aviation and the need for supplementary measures', *Journal of Transport Economics and Policy*, 27: 323–328; A. Midttun, (1992) 'The European market for aviation: a sociological inquiry into the political economy of a complexly organized market', *Journal of Economic Issues*, December: 1063-1094.
15. Commission of the European Communities (1985) *Completing the Common Market*, COM(85)310 Final, Brussels.
16. A further key event at this time was the *Ahmed Saeed* case which had implications for defining the relevant market area in aviation for the application of EU competition policy. It meant competition laws apply to domestic aviation within Member States and to services between a Member State and a non-Member of the EU.
17. In practice, however, there have been difficulties in implementation e.g. the European Court of Justice ruling on non-French access to Orly-Paris Airport.
18. The Commission has also found that routes flown within the EU rose from 490 to 520 between 1993 and 1995, 30 per cent of Union routes are now served by two operators and 6 per cent by three operators, 80 new airlines have been created while only 60 have disappeared, fares have fallen on routes where there are at least three operators and overall, 90-95 per cent of passengers are travelling at fares lower than prevailed in 1993. Commission of the Euoprean Communities (1996) *Impact of the Third Package of Air Transport Liberalization Measures*, COM(96) 415 Final.

19. S. Borenstein (1992) 'The evolution of US airline competition', *Journal of Economic Perspectives,* April: 45-73.
20. J. Gallacher (1996) 'A clearer direction', *Airline Business*, June: 22-52.
21. The data presented is not, however, definitive and one finds, for instance, *The Economist* produced slightly different figures claiming that there were 401 alliances in 1995, double the number it estimated four years earlier.
22. Pat Hanlon (1996) *Global Airline Competition in a Transnational Industry,* Butterworth, London.
23. D. Carlton, W. Landes, W. and R. Posner (1980) 'Benefits and costs of airline mergers: a case study', *Bell Journal of Economics,* 11: 65-83.
24. Mergers are often controlled by national governments, although the degree of control can vary. Individually, European countries have taken a variety of positions as has the EU in recent years. In many cases, mergers (such as those between British Airways and British Caledonia and Air France and UTA) have only gained approval by the airlines relinquishing routes or slots.
25. The relative importance of airline alliances involving equity stakes has been declining, with *Airline Business* recording less than 16 per cent of agreements in May 1996 compared with 18 per cent in 1995 and 21 per cent in 1994. This does not mean there has not been a large increase in absolute numbers, and other surveys indicate that from 1992, ownership stakes of above 20 per cent have predominated, see J. Lindquist (1996) 'Marriages made in heaven', *Avmark Aviation Economist*, January/ February: 12-13.
26. Technically, a code-share is a marketing arrangement between two carriers that allows them to sell seats on each other's flights under their own designator code. In the case of connecting flights of two or more code-sharing carriers, the whole flight is displayed as a single carrier service on a CRS.
27. A stronger form of code sharing involves blocked space arrangements. In this case, one carrier buys space on another airline's aircraft which it then sells in its own right using its own designator code.
28. L. Jones (1996) 'Keeping up appearances', *Airline Business*, 12: 38-42.
29. In this framework, it is assumed that the market is stable and viable prior to any alliance being formed. In practice an alliance may be

struck to keep loss making carriers in a market but this is not dealt with here.

30. Linked to this, there are problems in defining the relevant market. This issue was avoided in the diagrammatic presentation, but from an anti-trust policy perspective is a matter of crucial importance. In many cases an individual route may only have the alliance carriers providing services but may still not enjoy any significant monopoly advantage. There may be competition, for instance, from a variety of alternative, indirect services or the service itself may mainly act as only one element in a wider complex network of services where there are many different routings.
31. Even if a threat of entry remains, cooperative undertakings may contain any competitive threat when their actions reduce costs by adopting limited price strategies. They set prices below what small, independent suppliers can offer but still in excess of their own costs. In other cases, cooperative undertakings may set prices to attain a reasonable return for shareholders yet not maximize them; essentially, management will aim to create conditions under which it is subject to less competitive stress.
32. Savings through bulk purchasing by the partners in the Global Excellence Alliance (Singapore Airlines, Delta and Swissair) are estimated to generate savings up to $21 million a year for the airlines.
33. United Airlines and Lufthansa consolidated trans-Atlantic services into fewer flights after forming their 1996 alliance.
34. Strategic alliances benefit from network value simply because of the range of options they offer potential travellers. USAir/British Airways created a network that, in theory, could serve 17,000 city pairs and Northwest/KLM created 36,450 city pairs. Singapore/Delta/Swissair serves over 300 cities.
35. There is also the fear that not being a member of a global alliance will leave an individual carrier isolated and at a competitive disadvantage.
36. The introduction of foreign capital into US carriers had the advantage of circumventing problems that may have arisen if other financial plans had been pursued. If either KLM or British Airways had been US carriers, there would have been domestic US anti-trust laws to consider.
37. A study of European airports by the Stanford Research Institute forecasts that by 2010, there will be capacity problems at about half of Europe's airports.

38. Commission of the European Communities (1993) *Outline Plan of the Trans-European Airport Network*, Draft Communication VII-C4/Com 2/, Brussels.
39. K.J. Button (1996) 'Liberalising European aviation: is there an empty core problem', *Journal of Transport Economics and Policy*, 30: 275–291.
40. T.K. Smith (1995) 'Why air travel doesn't work', *Fortune*, April 3: 26–36.
41. C. Oster and D. Pickrell (1986) 'Marketing alliances and competitive strategy in the aviation industry', *Logistics and Transportation Review*, 22: 371-87.
42. Gellman Research Associates (1994) *A Study of International Airline Code Sharing, Office of Aviation and International Economics*, Office of the Secretary of US Department of Transportation, Washington; US General Accounting Office (1995) *International Aviation: Airline Alliances Produce Benefits but Effect on Competition is Uncertain*, GAO/RCED-95-99, Washington, DC.
43. UK Civil Aviation Authority (1994) *Airline Competition on European Long Haul Routes*, CAP 639, CAA, London.
44. M. Pustay (1992) 'Towards a global airline industry: prospects and impediments', *Logistics and Transportation Review*, 28: 103-28.; Dresner, M. and Windle, R. (1996) 'Alliances and code-sharing in the international airline industry', *Built Environment*, 22: 201-11; Dresner, M., Flipcop, S. and Windle, R. (1995) 'Trans-Atlantic airline alliances: a preliminary evaluation', *Journal of the Transportation Research Forum*, 35: 13-25; Youssef, W. (1992) *Causes and Effects of International Airline Equity Alliances*, Ph.D. Dissertation Series UCB-ITS-DS-92-1, Institute of Transportation Studies, University of California at Berkeley; Youssef, W. and Hansen, M. (1994) 'Consequences of strategic alliances between international airlines: the case of Swissair and SAS', *Transportation Research*, 28A: 415-31.
45. Desner *et al, op cit*. focused on the Continental/SAS, Delta/Swissair and Northwest/KLM equity alliances up to 1989.
46. For British Airways, this represented $27.2 million of additional net revenue and for USAir, $5.6 million. Their strategic alliance was estimated to benefit Northwest by $16.1 million annually and KLM by $10.6 million.
47. These include non-stop services between Zurich and Cincinnati (Delta/Swissair); European services to Memphis (Northwest/KLM); non-stop services between Houston and Rome (Continental/Alitalia);

and direct services between Vienna and Washington (Delta/Austrian Airways). Some alliances, by coordinating services of member airlines, also offered more choice of carriers and routes - three alternatives available between Indianapolis and Lyon being an illustration.

CHAPTER 16

Development of airport slot allocation regulation in the European Community

Chris Castles

Background

At present there is a small number of highly congested airports in the European Community (including some of the major European hubs) with a larger number of airports experiencing congestion problems during certain periods. Current projections of air transport demand growth indicate that the imbalance between capacity and demand at Community airports is likely to become more severe over time.

The allocation of scarce airport capacity is of central importance to attempts to create a competitive single market in air transport in the Community. In particular, if there is to be effective competition in air services it will be important to ensure that there is a mechanism by which carriers can gain access to congested airports. Without such a mechanism, insurmountable barriers to entry will continue to protect incumbent carriers and there is a risk that competitive conditions in the market will fail to develop. Council Regulation 95/93 (the 'Regulation') 'on common rules for the allocation of slots at Community airports' was passed on 18 January 1993. The purpose of the regulation is to create a common framework for allocating capacity at congested airports in the Community on the basis of 'neutral, transparent and non-discriminatory rules'.

Traditionally, capacity at congested airports has been allocated by airlines using voluntary guidelines developed over a number of years under the auspices of the International Air Transport Association (IATA). The guidelines set out administrative procedures, centred around the biannual scheduling conferences, and include a set of rules and criteria to be applied to the allocation of airport capacity. The IATA system is characterised by a high degree of self-regulation by airlines with reliance placed on consensual agreement to resolve problems.

The guidelines themselves have their origin in attempts by air carriers to facilitate the coordination of schedules during an era of restrictive bilateral air service agreements between states. They were not developed with the objective of facilitating air competition in mind. In response to pressure from policy-makers the guidelines were amended in 1991 to give some priority to new entrants in the allocation of available slots.

The 1993 Regulation was enacted in part to give legal force to existing custom and practice in slot allocation in the Community,[1] and it retains many key features of the IATA system: in particular the reliance both on self-regulation and on an administrative approach to the allocation of airport capacity. However, the Regulation goes beyond the IATA system in three important respects. First, it places a legal obligation on Member States to play a role in deciding on the need for slot allocation, and in the methodology of slot allocation more generally. Second, the Regulation establishes a set of institutional arrangements and a system of legally binding rules for the allocation of slots. Third, the Regulation makes further provision to facilitate new entrant competition than did the IATA guidelines.

The Regulation provides legal backing to the existing framework of administered slot allocation and does not seek any radical departure from this system. However, the limitations of the administrative system, which relies on procedures and rules for deciding the allocation of airport capacity between users, has been widely recognised. Specifically, incumbent carriers are favoured by being given protected 'grandfather' rights over previously allocated capacity and it fails to provide a market test that would ensure that capacity was allocated to the highest value in use. There has therefore been an active discussion of alternative methods, encompassing, in particular, market mechanisms as a means of achieving a more efficient allocation of capacity. Some such options have been tested in the USA including, for example, lotteries for the allocation of new capacity and provision for the secondary trading of slots.

Key provisions of the regulation

Under the provisions of the Regulation, Member States must ensure that opportunities for increasing capacity at congested airports are fully explored in consultation with airport users. Where a capacity analysis indicates that there is no short-term solution to a capacity shortage at an airport the responsible Member States must designate the airport 'fully coordinated'. This then triggers a number of other requirements. In particular, at fully coordinated airports, Member States must:

a) appoint a coordinator and ensure that he acts in an independent manner; and
b) establish a coordination committee (which may serve more than one airport).

The coordinator is responsible for allocating slots, monitoring the use of slots, and providing information on slot availability.

The coordination committee is an advisory body to the coordinator whose members are intended to be drawn from air carriers (and/or their representative organisations) using the airport concerned regularly, the airport authorities, and the air traffic control body.

At fully coordinated airports, the coordinator must allocate slots in accordance with Articles 8, 9, 10, and 11 of the Regulation. *Inter alia*, these require the co-ordinator to:

- recognise claims for historic slots, subject to a requirement that the slots were used the previous season;
- withdraw slots that have not been used;
- establish a pool of slots, comprising new capacity and withdrawn and surrendered slots. The coordinator is required to allocate 50 per cent of this pool to new entrants; and
- confirm the feasibility of slot exchanges between carriers or transfers between routes or types of service.

Weaknesses in the current system of slot allocation

The existing allocation system set out in the regulation has a number of inherent drawbacks including:

- the lack of clarity over what is being allocated (i.e. the rights and obligations that go with holding slots);
- the inefficiency of administered rules in allocating scarce capacity to highest value use;

- the potentially arbitary distinction being drawn between 'new entrants' and incumbents, if the balance of policy objectives remains undefined;
- the declining effectiveness of the 'new entrant' measures with shrinking slot pools at highly constrained airports;
- the unresolved conflict between the need for flexibility to reflect local conditions and the need coherently and consistently to pursue the objectives of the Single Market.

The regulation anticipates that the current system will be subject to review and revision, and recent debates in the industry have involved the discussion of a range of alternative slot allocation mechanisms to address the above weaknesses. In considering the merits of alternative allocation mechanisms it must be remembered that the process will continue to be subject to a number of important constraints, including the need for the neutral, non-discriminatory application of any mechanism, the need to comply with international law and to be consistent with the operational requirements of the international aviation industry, and the need to avoid destabilising existing commercial activities with unnecessarily precipitate action. These constraints effectively limit the practical options for fundamental change.

Indeed, the interdependencies inherent in the operation of the international aviation market mean that slot allocation initiatives will need to continue to be evaluated in conjunction with complementary policy developments (eg. on traffic rights, state aid and the role of national carriers). Hence, the policy intent should be that slot allocation should facilitate rather than drive or determine the development of the aviation markets served from EU airports.

Airline competition

A fundamental concern with the application of slot allocation at congested airports is its impact on airline competition. Capacity constraints at congested airports limit airlines' ability to expand services to meet demand. Hence they create the potential for incumbent airlines to exploit market power by pricing services above the level which would otherwise prevail if there were no restrictions on the supply of airport capacity.

A particular concern has been that shortages of slots prevent new entrants from competing on key routes. However, it should be noted that new entrant competition of itself does not alter the fundamental constraints on available capacity which creates the basic competition problem. Indeed it can be argued that new entrant competition may not always be the most effective means of increasing competition. For example, smaller new entrants may lack the scale and resources to mount effective competition against those network carriers able to buttress their route operating economics with transfer traffic. Also, where new entrants use smaller aircraft they may aggravate the congestion problem by using scarce airport capacity inefficiently. Confiscating slots from incumbents, and diverting them to new entrants, might enable incumbent carriers on the routes previously served by confiscated slots to raise their prices further as capacity on these routes is reduced.

In practice, it is not easy to predict the overall effect on competition of actions directed at one part of the air transport system. The structure, scale and geographic scope of the airline industry results in a diverse range of competitive behaviour which interacts dynamically at the route, hub and global network levels.

The basic aims of competition policy are to benefit the final users (the airlines' customers) by:

- providing incentives for airlines (and other air transport operators) to operate efficiently in the allocation and use of resources;
- constraining airlines' ability to exploit any market power through raising prices above the levels which would prevail under fully competitive conditions.

The efficacy of slot allocation mechanisms should be judged directly against these efficiency objectives rather than their ability to facilitate particular forms of competition. In this context, giving preferential access to slots for small new entrants may not be the most effective way of increasing competition at congested airports. Also the extent to which carriers will be influenced by the incentives to cut costs and prices efficiently will be driven to a large extent by their commercial freedom and by the extent to which they are insulated from competition. The future development of the structure of the airline industry in Europe and the policies being pursued in relation to deregulation, competition and the removal of state aid to airlines will therefore have a significant impact on the effectiveness of slot allocation policy.

In considering how slot allocation can be used to encourage competitive behaviour by airlines it is helpful to consider three interdependent dimensions of airline competition:

- at the route level;
- between hubs; and
- between global network carriers.

Airline users mainly experience the effects of competition at the route level through the range of routes served, the choices of carriers available and the prices and service levels offered. A key aim of policymakers is therefore to encourage the development of new services and ensure adequate competition between carriers on individual routes.

Experience suggests that carriers behave in a significantly less competitive manner on established duopoly and monopoly routes compared to those with three or more carriers. This has led some commentators (in particular the UK CAA) to suggest that a major objective of slot allocation policy should be to encourage the introduction of third carriers on community routes. The effect of this policy would be to concentrate the allocation of available new slots at congested airports on a limited number of 'thick' routes in Europe capable of sustaining three carrier competition.

However airlines can also compete through the operation of airport hubs and this dimension of competition should be taken into account in considering slot allocation alternatives. Airport hubs enable feeder services to support alternative routes through transfer traffic, which provides a number of benefits:

- carriers can exploit the efficiencies of economies of scale and scope by developing more extensive networks;
- competition on carriers' feeder routes can be stimulated while airlines build market share to sustain hubs;
- a wider range of route choice to passengers and increased service frequencies can arise from hub operations.

Nevertheless the development of hubs can also have a number of competitive drawbacks by:

- enabling some carriers to build up strong positions at hubs which are difficult to attack competitively, particularly if airport capacity is constrained;
- it can limit the development of direct, point-to-point, routes elsewhere and oblige passengers to make inconvenient transfers.

Closely related to hub competition is a similar form of competition between carriers seeking to develop global networks which feed traffic from one route and hub set to another. Global carrier operations are supported by alliances, customer loyalty programmes and similar operating and marketing devices aimed at achieving global coverage. These devices risk blunting competition in some segments of the market by introducing barriers to entry and potential use of cross subsidy for predatory pricing targeted at smaller competitors. They therefore pose dilemmas for EU policy makers. Global carriers need to exploit their networks to be able to compete effectively. Policy actions which weaken the ability of Community airlines to exploit the economies of global networks may inhibit their ability to compete with carriers based outside the Community.

It is difficult to judge the competition benefits of a market structure of hub-based and global airline network competition compared to a more diversified structure with many smaller carriers and niche operators in the congested European context. In practice, there are already a number of countervailing forces in the European air transport industry which are likely to result in a combination structure of point-to-point and hub or network based route choices, in contrast to the hub-oriented structure found in the USA. The role of airport slot allocation in these developments will remain crucial. Against the background of increasing airport congestion and changes in the competitive structure of the airline industry in Europe, it will become increasingly difficult for the current rules-driven slot allocation process to achieve policy objectives. The introduction of new mechanisms to enable market driven allocation of airport capacity and the opportunity for gaining access at congested airports would, if practical, relieve the strains on the current administrative mechanism for slot allocation.

Alternative allocation mechanisms

In this section alternative allocation mechanisms are reviewed, addressing:

- the initial allocation, i.e. different ways of allocating new (or released) capacity, considering in turn, auctions, 'first-come, first-served' approaches and lotteries;
- secondary allocation mechanisms, principally slot exchanges and transfers, distinguishing in particular between monetised trading and barter trade; and
- practical combinations of initial and secondary mechanisms.

Initial allocation

The requirements of non-discrimination and neutrality will generally mean that new or released scarce airport capacity must be made available to more than one potential user at the same time. There are three basic methods of allocating capacity on a non-exclusive basis. Each method achieves a rationing of scarce capacity, although the efficiency of the rationing method varies. The options are:

- Auctions. A perfect auction can, in theory, create an economically optimal allocation of capacity among users with differing valuations of that capacity. So long as all players bid their individual valuations, each unit of capacity will go to the player that places the highest value on that unit. Hence a perfect auction can, in principle, provide an optimal allocation of capacity, providing there are checks on uncompetitive behaviour;

- 'First-come first-served'. This approach is, for example, used to ration most road capacity. It can be supplemented by tariffs (eg. road tolls), and in principle, the tariffs could be set to ensure that capacity is used efficiently by those who place the highest value on it. The resulting pattern of demand and the prices paid would be the same as those set by a perfect auction;
- Lotteries. Lotteries are perhaps the least efficient approach to allocation. They can only achieve efficient outcomes accidentally, and are almost certain to generate windfall gains and opportunity losses. Although these effects can be reduced somewhat (by including entry fees, for example) lotteries are generally an unattractive way of allocating scarce capacity.

As noted above, slot auctions can, in theory, provide an optimal outcome to the slot allocation problem. In practice however, there are formidable difficulties in devising workable auction mechanisms for airport slots. In particular the value of a slot varies according to the user and the use to which other slots would be put, so that there are strong interdependencies between the values of different slots, which are only useful to an airline as a component of a viable schedule. The schedule establishes the relationships between connecting services and it is these relationships which determine the value of slots to individual airlines. These values cannot therefore be anticipated in advance of the schedule itself. The rules for an auction that took account of this interdependence would therefore be highly complex, in order to allow, for example, for multiple contingent bids, with different values bid depending on the complementary services provided under different outcomes of the overall auction. In practice, no such auctions have ever been implemented. Further complications arise if a schedule involves using the outcome of two independent allocation processes, for example, for a route between two highly congested airports.

This is not to say that all slot auctions would be impractical. For example, as the UK CAA has noted, long haul routes typically require lower frequencies, have more flexible timing requirements, and will often involve only one congested airport in the schedule. Under these circumstances slot auctions might be effective. Similarly, it might be possible to devise an auction to allocate a very limited pool of slots, within the framework of largely dedicated capacity. A large amount of capacity released at a core congested hub (such as that being sought by the EU from British Airways at Heathrow in the context of their proposed alliance with American Airlines) would however create more complex issues for an auction.

The mechanism of 'first-come, first-served' has traditionally predominated in airport slot allocation, being the basis on which grandfather rights have been acquired (typically before capacity becomes scarce), influenced to greater or lesser degree by the prevailing structure of posted airport tariffs. The primary benefit of first-come first-served is that it allows efficient use of capacity when there is no rationing requirement: the market can dictate the use of facilities without intervention of allocators (other than in setting posted prices). It remains the method by which capacity is allocated at non-co-ordinated airports in the EU, and at most US airports.

A first-come first-served approach to slot allocation means that additional aircraft movements at an airport are treated equally with 'historics'. They are accommodated within a given physical capacity by increasing the average delay experienced by all users. The approach is reasonably efficient if the marginal costs of additional users (in terms of the extra delay and other congestion costs they impose) are not great or sustained - i.e. at uncongested or easily expanded airports.

Where there is competition for scarce capacity some kind of pricing mechanism is required to reflect the marginal cost of an additional aircraft movement. An efficient outcome can only be secured if posted prices (eg. landing charges) can be set which perfectly anticipate the marginal value of capacity to users for the duration of their tenure.

In practice, there may be difficulties in estimating and setting accurately the structure and level of airport charges which would match supply and demand for airport capacity taking into account the peak profile of demand. Furthermore, the market clearing level of airport charges at many congested airports would be high, which has made this option unattractive to airlines. If all revenue were collected by airport operators it would generate large financial surpluses which may blunt the incentives for efficient behaviour, including carrying out the investment needed to relieve the constraint. The alternative of government congestion taxes would also be resisted by the industry. However there is scope for applying price-based demand management techniques at many airports through differential peak pricing. Airlines have been resistant to accepting the sharp differentials between peak and off-peak airport charges which would be needed to manage demand to relieve congestion. However this may be short sighted because the alternative is either increasing aircraft delays or premature airport investment, both of which are costly to airlines.

In the absence of universally acceptable alternatives, lotteries have been used to allocate new capacity at the US High Density airports. As explained above, lotteries on their own are most unlikely to achieve an efficient allocation of capacity, and in practice, it will be necessary to allow some kind of *ex post* transfer of slots, for example through bilateral or multilateral exchange. However, even with such a transfer system, lotteries will almost certainly generate windfall gains for some users at the expense of other users. Even if all users receive the same number of slots, the interrelationship between slots and their inherent heterogeneity are both likely to benefit some airlines and disadvantage others.

Initial allocation of slots in practice

The above observations are reflected by current practice: first-come first-served remains the basic method of non-exclusive allocation for new/released slots. With 'grandfather' rights, the slots that are allocated also confer usage rights of indefinite duration. Therefore if demand outstrips supply at the prevailing levels of airport charges, the opportunities available to promote competition at congested airports remain inherently limited.

In response, different forms of administered 'positive discrimination' in allocating available slots have been devised (of which the EC's Regulation '50 per cent to new entrants' rule is one). Administrative rules give priority to selected types of user in the queue for new slots, with the rules implicitly seeking to proxy the relative values of different types of use as perceived by the rule-maker. Inevitably, such rules involve judgement by the rule-makers, rather than the market, over these relativities, and must involve arbitrary criteria in their detailed application.

In practice, the complexity of the task of coordinating airline schedules at a congested airport makes it difficult for policy makers to develop rules which anticipate all circumstances and provide unambiguous guidance to coordinators. Airport coordinators must therefore use their discretion in applying rules set by policy makers. This requires the interpretation of policy makers' intentions and the need to achieve balanced judgements on outcomes which reflect any tensions between policy objectives. Therefore, at best, such rules introduce 'rough justice' and are imperfect modifications of first-come first-served outcomes. In addition, rules of this nature carry with them transactions costs for their implementation, and create 'winners and losers' in competitive markets. As a result, they have hitherto been treated with caution by the industry.

Secondary allocation

The arguments above suggest that there are practical difficulties in achieving an optimum allocation of slots through the initial allocation process by slot auctions or other mechanisms. Secondary allocation processes involving bilateral transfer of slots between airlines can however improve the outcome of the initial allocation. In addition, if carriers are granted grandfather rights to slots, then some mechanism for slot transfer is needed to allow flexibility in scheduling over time.

It is therefore appropriate to define usage rights for slots that permit transfer from one holder to another. Bilateral exchanges of these rights can either be monetised (sales) or non-monetised (barter). These mechanisms are available for reallocating existing capacity, where the usage rights concerned are already held exclusively by a given party (eg. an airline).

Conceptually, any form of initial allocation followed by subsequent monetised trading could lead to an identical final allocation to the efficient one expected from properly functioning auction mechanisms. This is because the slots would be traded to those users who valued them the most. Depending on the underlying structure of posted prices however, some or all of the scarcity rents of slots would remain with the traders (eg. airlines) and not the original holder of the rights (eg. the airport or state acting as 'auctioneer').

Secondary trading of airport slots is not explicitly permitted by the current EC Regulation covering EU airports, although it is known to take place on a covert basis. The EC Regulation permits barter exchange of slots. But barter trade (slot exchanges) is unlikely to result in efficient trading, given that slots have different values. Those users for whom the value of their slots was below the relevant efficient rationing price would have no mechanism for gaining the full value of their slots during exchanges for lower market value slots, unless the exchange was accompanied by monetary transfers. Such compensating monetary transfers for unequal trades currently take place covertly, but this is still an inefficient form of trading since it requires an airline to have an initial presence at the airport in order to have slots to barter. It does not enable new entrants to buy themselves access to the market.

In practice, the strong assumptions required to ensure efficient monetised trading are not fulfilled in the aviation market at present. Air traffic rights, the scarcity rents conferred by grandfather rights and network economies of scope (realised, for example by hubbing), all confer market power on particular airlines at congested airports.

There has therefore been a widespread fear that unrestricted monetised trading would allow such airlines to exercise their market power to buy up stocks of slots to enhance their market position at the expense of less powerful competitors. In principle, similar behaviour can occur with barter trading, by incentivising the use of 'paper airlines' and 'babysitting' of slots to build up portfolios beyond those required for efficient operations. Nevertheless, the heterogeneity of slots inhibits the effectiveness of such behaviour by barter, and monetised trading would conceptually offer a more direct means to exercise market power in this way.

Practical combinations

It will be apparent from the above arguments that some combination of initial and secondary allocation mechanisms will be required for slots at congested airports. In this sub-section some practical combinations are discussed.

The traditional approach to slot allocation has relied on:

- a first-come first-served approach to allocate initial (grandfather) rights; and
- non-monetised exchanges via airport scheduling committees run by the airlines for reallocation purposes.[2]

The current EC Regulation, by preserving grandfather rights and non-monetised trading, does not fundamentally change the traditional allocation mechanism, but rather puts administrative pressure on the continued efficient use of grandfather rights (through the slot withdrawal provisions), and stratifies the non-exclusive allocation of new/released capacity through the new entrants' quota.

Monetised trading has been permitted since 1985 at the four so-called 'high density airports' in the US. New/released capacity at these airports has been allocated by lottery and existing capacity has been grandfathered to incumbents. Grandfather rights are subject to a use-it-or-lose-it rule, but this is largely unused, since an airline will prefer to sell unwanted slots than to surrender them for lottery allocation. The level of trading has been significant, although influenced by the level of ongoing corporate restructuring in the domestic industry and more recently focused on short-term leasing which is subject to less intrusive anti-trust oversight. With few 'new' slots created, lotteries have all but ceased in recent years.

The development of a secondary trading mechanism for slots at the congested European airports would be a useful step to address the weaknesses in the current administered system of airport slot allocation. It would introduce a strong monetary incentive to encourage the release of increasingly scarce airport capacity at congested airports by those airlines less able than others to make efficient use of slots. This could result in airlines reviewing their schedules and route patterns to the benefit of the overall efficiency of the system. It would also provide a means for meeting the demands of those airlines currently prevented from gaining access or expanding their services at these airports. The European Commission is considering a revision of the current slot allocation regulation which will allow monetary trading in airport slots. It is therefore relevant to consider both the benefits and risks of this development as well as the supplementary measures needed to enable an efficient form of trading.

It is clearly difficult to evaluate conclusively the competitive implications of such an approach. However, it appears that the economic incentives faced by airlines would be sharpened, with potential short-term gains as carriers rationalised their slot holdings in a manner that would allocate capacity according to individual airlines' valuations of slots rather than historic rights.

There is a risk which has been debated both within the Commission and in the wider aviation community, that the market power of a few large airlines with strong positions at key hub airports could enable them to dominate slot trading. Over time, the incentive for smaller airlines to sell slots at congested airports to large carriers able to benefit from network economies could strengthen. There could be a tendency for competition to be concentrated at the hub or global carrier level, with major airlines dominating particular congested airports. This could result in the benefits of competition being unevenly spread. Without effective competition at the level of individual routes, the tendency for price discrimination and cross-subsidy would increase to the detriment of particular groups of users.

However such an outcome is not necessarily inevitable. Effective airline competition can be fostered at a number of levels, including niche operators focused on thick routes, competition between alternative hubs for transfer traffic, competition between alternative (less congested) non-hub airports which could build up services over time, and competition from alternative modes (such as high speed rail). The dynamics of airline competition are such that it is not easy to anticipate and evaluate the effectiveness of the emerging structure of competition in advance.

The benefit of a more transparent and accessible secondary trading mechanism for slots is that it would free up the market and allow the true valuations of slots to be revealed. Slot trading could promote the potential for new entry and hence competitive behaviour. Unilateral slot sales would allow new entrants with sufficient financial resources to enter the market in congested airports. Although there is a danger of anti-competitive behaviour, secondary trading would enable the policy makers to work with the tide of industry developments. Regulatory authorities will need to oversee the operation of the market and to develop tests of competitive behaviour to prevent abuses of dominant positions.

Historic rights

At present slot usage rights in Europe are typically of indefinite duration. Carriers can therefore protect their market position, and considerable market power can be conferred on incumbent carriers at highly congested airports, although it also provides security to carriers and incentives to invest in the development of their route structure. In the US, slot leases, with slots operated for limited durations (many valid for a season or less) are more common. Replacing indefinite rights with leases of limited duration in Europe would effectively imply their forced sacrifice by the incumbent airlines concerned. As these rights have considerable scarcity value at congested airports, such a proposal would meet with fierce opposition from most of the industry affected, and would probably, on balance, adversely affect the European flag carriers based at these airports relative to their overseas competitors. Yet some believe that without forced sacrifice, there will be little real opportunity for increased competition at the most congested airports.

There are a number of dangers in introducing mechanisms which combine forced sacrifice of slot rights, administered allocation favouring certain categories of users, and monetised secondary trading, particularly if this were to be introduced suddenly. The windfall gains and losses of forced sacrifice , combined with non-monetised initial allocation biased towards new entrants, would encourage gaming to obtain free slots in the initial allocation, which could then be resold. Such behaviour would require the development of further rules and monitoring procedures to prevent it.

Furthermore, the effect of the financial drain on incumbent carriers seeking to replace surrendered slots by re-purchasing them on the secondary market would need to be considered carefully. These costs would be borne disproportionately by European carriers and would disadvantage them in comparison with non-European carriers. There is a risk that in the long run this would lead to a fragmented and weakened European airline industry unable to compete with global network carriers.

The introduction of elements of forced sacrifice through time limitation of slot rights should therefore only be introduced with caution. In particular, it should only be considered in the context of the longer term development of a complete package of measures to introduce market mechanisms into the slot allocation process. Any measures designed to force carriers to relinquish slots will need to have regard to the financial commitments made by carriers in obtaining and using those slots.

If slots are tradable, the terms under which they may be withdrawn will need to be carefully defined. Indeed, these terms will influence the value of slots. It should be noted that under the present regulation, slot withdrawal is the primary sanction available to coordinators. If slots were to acquire tradable monetary values, and/or their use was to be more tightly specified, this 'all or nothing' sanction might become increasingly severe in its impact on users and incentivise inefficient behaviour by threatened carriers (increased disputes, transfers to 'baby-sitters', etc.). Hence monetary trading may increase the need for more graduated enforcement mechanisms including fines for abuse of slot use obligations, operating off slot times or failure to use slots.

Conclusions

The purpose of this paper is not to prescribe any particular solution to the problems of slot allocation at Europe's congested airports. However, the above discussion allows a number of conclusions to be drawn, which are likely to be of relevance in developing these solutions. Some of the key points of this paper are summarised below.

First, fundamental change to the allocation mechanism would probably require consistent changes to the definition of slots as usage rights being allocated, to the monitoring and enforcement of the results of the allocation process, and to the institutional framework supporting it. There are unlikely to be 'simple fixes'.

Next, there are almost certain to be periods at some airports during which demand exceeds the supply of capacity by a significant margin in future. This implies that in these airports the voluntary surrender of grandfather rights by incumbents will be limited. Such surrender would be necessary to allow the slot pool mechanism of the regulation to provide significant opportunities to certain types of new entrant and competition.

Faced with this problem, there are two basic ways of releasing more capacity: forcing incumbents to give it up, or allowing monetised trading of the capacity concerned. With perfect aviation markets, the latter would be preferable, as surrender choices would be market driven and hence efficient, and the confiscation of incumbents' scarcity rents could be avoided. Other distortions in the aviation market have, however, created doubts over whether the potential for carriers to exert market power via monetised trading could be adequately policed. (Informal unequal trades can and do occur if formal monetised trading is prohibited, and these are also susceptible to the abuse of a dominant position, and are arguably more difficult to police).

Potentially, the disadvantages of the two basic methods of releasing capacity might be minimised by combining features of both. For example, by tightening the times at which grandfather rights are valid for subsequent seasons, incumbents would have less incentive to hold on to their rights, while not being compelled to give them up. Alternatively, rolling sacrifice (for example, requiring surrender of a proportion of rights each year,) could be timed to be introduced at a future time, and with increasing percentages over time, so that the discounted value of the implied sacrifice to incumbents is minimised, but the size of the slot pool grows over time in line with a widening demand/supply gap.

Against such a background, pool slots could be made of limited duration, and subject to monetised trading. A secondary market could thus be allowed to grow gradually, and if incumbents were still thought to derive market power from their grandfather rights, (temporary) limits could be placed on their holdings of traded slots.

The method of allocating new/released pool slots is likely to continue to require administration, if the pool is still largely a residual product of grandfather rights, as it will be difficult to institute effective auctions for a heterogeneous mixture of 'products' (slots) to be made available to heterogeneous users (eg. 'new entrants'). Structured (and potentially discriminatory) forced sacrifice would be needed to create homogenous 'pools' for auction.

The choice for allocating new/released capacity is therefore likely to continue to be between first-come-first-served, within administered quotas or priorities, or lottery, or both. In general, the larger the pool as a proportion of the congested capacity, the less need to segment the users for rationing purposes; while the greater the excess demand, the more problematic the policing of first-come first-served is likely to become (places in the queue will be bought for short-term advantage) and the more attractive lotteries will be (particularly if secondary trading can then allow efficient reallocation, as in the US).

However, the details of pool slot allocation mechanisms are potentially less fundamental than the basic decisions over sacrifice and monetised trading[3]. These could have significant impacts on the development of aviation markets in Europe, and the role of individual carriers within this development. For example, sacrifice could, as a generality, favour non-EU carriers at the expense of European flag carriers. It follows that the political will and broad consensus of Member States will be important preconditions for the effective introduction of fundamental changes of the type discussed above. In addition, the institution of greater competition for scarce capacity would involve a steep increase in the transactions costs of its introduction and effective administration.

Taken together, these considerations suggest that fundamental changes to the existing system of slot allocation in Europe should probably be focused, if they are introduced at all, on those airports and periods where the problems of scarce capacity are most acute.

Notes

1. The regulation has legal force in each Member State without the need for additional enabling legislation.
2. In fact, anti-trust legislation prohibits exchanges in the US and the reallocation function is performed by the FAA, with only informal 'grandfather rights' associated with incumbents during their six-monthly schedule filings to the FAA.
3. Although, it should be noted again, there is some danger in favouring new entrants too strongly at the (financial) expense of incumbents, who would be obliged to pay substantial sums to remain in the market.

CHAPTER 17

Allocating airport slots: A role for the market?

David Starkie

Introduction

It is a growing feature of modern civil aviation that many of the world's major airports suffer from inadequate runway capacity. Particular airports have become popular places to fly to and from or change planes at, but as the demand for flights has grown the expansion of runway capacity has failed to keep pace. There are a number of reasons for this. The overwhelming majority of airports are still publicly controlled utilities subject to political whims and often having tight budgets, but even where such constraints apply less, (as in the UK) expansion has been hampered by environmental limitations and other planning controls. Building new runways or lengthening existing ones is not impossible as London City Airport and Manchester both illustrate, but it is not an easy task.

In consequence, delays at busy airports have increased and expansion of air services has had to rely upon improvements in air traffic control technology and the adoption of new procedures to squeeze more aircraft movements through existing runways. The ability to do so has been at times remarkable. Little more than a quarter of a century ago, the (Roskill) Commission investigating locations suitable for a new international airport for London adopted the working assumption that the estimated capacity of Heathrow and Gatwick combined was 440,000 annual movements: in 1997 Heathrow alone handled 429,000. Nor has the process of improving throughput slackened in recent times. Nevertheless, severe shortage of capacity remains a salient feature. At Heathrow the runways are operating at, or close to, declared capacity during much of the day.

The economics of runway congestion

The economic consequence of this can be illustrated by reference to demand and supply diagrams. The first of these - Figure 17.1a - illustrates that when the demand for runway use is within declared runway capacity, the price to charge will be one that covers the cost of supply including a satisfactory return on the capital invested. In Figure 17.1a, for ease of exposition, average costs are assumed to be constant across all levels of output. However, if demand continues to grow, and grows more rapidly than capacity can be expanded, maintaining the price shown in Figure 17.1a will eventually result in excess demand (Figure 17.1b); airlines will want to use runways more than there is capacity is available and, unless corrected, queues on the ground or in the air will develop. In these circumstances, the economically correct (or efficient) price is one that balances demand with the supply available (Figure 17.1c). This is referred to as a rationing or market clearing price which, it should be noted, produces what is referred to as an economic rent, the potential for earning a return in excess of the average cost of supplying runway services. In reality, at a number of airports, not only is the price charged to the airlines less than the market clearing price but it is also, on average, less than the cost of supply. The Monopolies and Mergers Commission (MMC, 1996) in a recent report on BAA's London airports, for example, suggested that if air-side assets (runways, aprons for parking aircraft, piers and aerobridges) were to earn a seven per cent real rate of return, revenue from charges would need to increase by more than a third.[1]

However, and importantly, the fact that landing charges at congested airports are less than the market clearing price (and sometimes the cost of supply) does not mean that air travellers necessarily enjoy cheaper air fares. If it did, airlines too, would have a problem of allocating an excess of demand (in this case, of passengers to seats). Instead, airlines will be inclined to charge what the market will bear and make excess profits on the use of the scarce slots. In effect, the economic rent shown in Figure 17.1c is transferred to the airlines.

For Heathrow we have one indication of the size of the transfer from analysis undertaken for the 1993 RUCATSE Report. The Report suggested that the fares premium charged by airlines might be equivalent to £20 per passenger. At Gatwick, where, in contrast to Heathrow, charter traffic is important, Bishop and Thompson (1992) found that the prices charged for inclusive tours using the airport were significantly higher than prices for the same 'package' at other UK airports, by a magnitude of broadly five per cent.[2]

Figure 17.1 Demand and supply for runway capacity

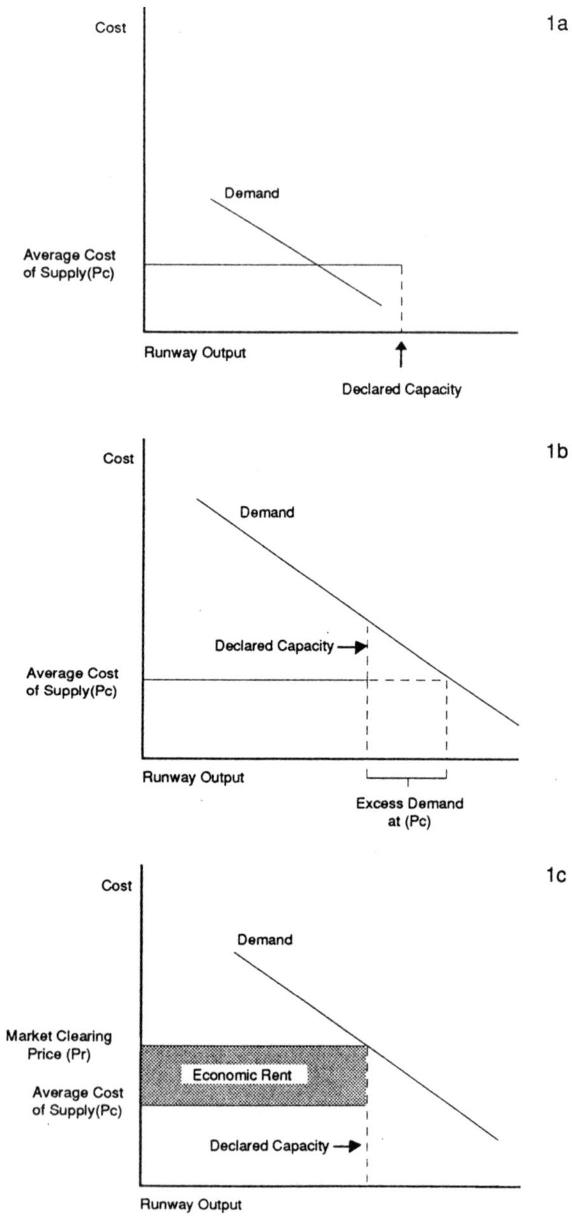

Administrative rationing

With the potential to charge such fare premia, naturally airlines are very keen to have slots at congested airports and, because price is not used to ration demand to the limited runway capacity available, some other rationing mechanism is needed. This mechanism used is an administrative process based on guidelines laid down by the International Air Transport Association (IATA), the airline trade association. These guidelines, first and foremost, recognise the historical use of 'slots' (the entitlement to use a runway at a particular time). An airline has a right to a slot if it has already made use of the runway at the same time during the preceding equivalent season.

These entitlements, known as 'grandfather rights', form the point of reference at bi-annual international conferences which take place to coordinate schedules at capacity restricted airports. At these conferences, airlines seek to modify their schedules by trading (or exchanging) their existing rights, or perhaps by trying to obtain additional slots that occasionally become available (although since 1990 the IATA rules have required a proportion of unclaimed slots to be set aside for use by new entrant carriers, defined as those with negligible or non-existent presence at the airport concerned). These basic guidelines were adopted into EU law, albeit it with some minor changes, by Regulation 95/93 early in 1993.

Because the basic rules encourage airlines to exchange slots and because airlines do not even have to use their existing slots for a specific route, established carriers dominate slot-constrained airports. For example, 95 per cent of the slots available at Heathrow for use in the Summer of 1994 went to those airlines that had made use of them in the corresponding season of the previous year. The few entrants that did commence services, did so on 'thin' routes to new destinations; there were no directly competing services introduced and those incumbent airlines that did obtain some additional slots used them to increase frequencies on their existing routes. Not surprisingly, this rule-based approach, although widely supported within the aviation industry, has been criticised by the anti-trust authorities.

The US approach

In the United States, for anti-trust reasons, the IATA based system does not apply; instead as a general rule airlines simply schedule their flights taking into account expected delays at the busier airports. At four major airports, however, (JFK and La Guardia in New York City, O'Hare in Chicago and Washington National), where demand for runway use is particularly high, the authorities stepped in many years ago to prescribe a limit on the number of flights during restricted hours and they have since permitted airlines to buy and sell their slot holdings at these four airports, albeit with conditions attached. The market is restricted to slots used for domestic services which in turn are divided into two groups: air carrier slots, and commuter slots (originally operated by aircraft with 56 or fewer seats). The latter category cannot be bought by the former and, in addition, slots used for subsidised 'essential air services' are excluded from the market. The regulations stipulate that any person is entitled to purchase, sell and mortgage a slot or to lease on a temporary basis.[3] However, slots not used for a stipulated minimum of time in a two-month period have to be returned to the FAA; that is to say, carriers must 'use or lose' their slot. Surrendered slots, or others becoming available, are assigned to a pool and reallocated using a lottery but with 25 per cent initially offered to new entrants.

These basic terms were introduced in April 1986 when airlines started to buy and sell those slots which they were holding as of 16 December 1995. More recently, small amendments have been made to the regulations. From January 1993 slots have had to be used for 80 per cent of the time in a two-month period (it was previously 65 per cent) and the definition of those entitled to slots from the reserved pool was broadened to include incumbent carriers with few slots. In addition, restrictions were introduced to prevent slots intended for new entrants being acquired by incumbents. Other amendments have adjusted the distinction between air carrier and commuter slots, a distinction which was originally introduced in order to strike a balance between maximising the economic use of runway resources and preserving services to smaller communities. The distinction is still maintained, but the aircraft size threshold for the use of commuter slots has been increased, particularly at O'Hare.

Data on how the market has worked in distinguishing between air carrier and commuter slots is available only for the first three years after its introduction. During the first six to nine months, there was an initial surge of activity as airlines acquired the slots they believed they could use best and disposed of those that could be sold profitably. Following this initial sorting out the number of outright sales declined but the number of leases grew, particularly short term leases reflecting the fact that some carriers require the use of a slot at limited times of the year only (Table 17.1). The distinction in the regulations between air carriers and commuters produced two separate markets and differences between them emerged. Commuter carriers have been more inclined to buy and sell, rather than lease, and some new commuter airlines took the opportunity to enter the market; as a general rule air carrier slots were traded between incumbents. An interesting feature that has emerged is that a significant number of slots is held by non-carriers. Nothing in the regulations prohibits communities from acquiring slots to enhance services to their region's airports and a number have done so. Some airlines have also mortgaged their slots to financial institutions.

Table 17.1 Summary of leases and sales at US high density airports 1986-88

	Air Carriers / *Commuters*							
	Leases < 6 months %		Leases > 6 months %		Sales %		Total Transactions	
1986	26.0	*34.0*	13.0	*1.0*	61.0	*65.0*	617	*159*
1987	79.0	*13.0*	1.0	-	20.0	*87.0*	774	*23*
1988	83.0	*17.0*	8.0	-	9.0	*83.0*	734	*99*

Predation and anti-competitive behaviour

With such an interesting example of a secondary market in airport slots now more than a decade old, it is to be expected that suggestions have been made for the introduction of a similar system in Europe. At the moment, Regulation 95/93 does not allow slots at European airports to be bought and sold, although it is less clear that this is strictly against the rules if money changes hands only when slots are exchanged (Bass, 1994, p.147). According to press reports, however, the Transport Directorate of the European Commission favours legitimising the monetarised trading of slots in a revised Regulation. Again according to press reports, this has been opposed by the Competition Directorate. The concerns of the latter appear to reflect those expressed by the US General Accounting Office (USGAO, 1996) and, more recently, the US Department of Transportation, in regard to the US secondary market. These concerns centre on the issue of dominance and predatory behaviour. Specifically, the argument is that established airlines with grandfathered rights will buy slots to keep entrants out of the market and, as a consequence, further increase their dominance at congested airports. In support of this view, both the GAO and the US DoT have pointed out that there had been few entrants into the four US airports which have a secondary market and that established airlines have increased their share of slots.[4]

This argument, that airlines with market power will engage in predatory bidding for slots, has been subject to close examination by McGowan and Seabright (1989). They accept that this is a serious objection but are of the view that, for established airlines, it is an expensive way to deter or drive out competitors. This is because at any one airport there are many slots each one of which has a large number of close or reasonably close substitutes (bearing in mind that slots are transferable between services). To keep a newcomer out of a particular market, an incumbent airline might, therefore, have to 'overbid' on a large number of slots. In these circumstances, McGowan and Seabright argue that it is more likely that an established airline will direct any predatory behaviour to the route (service) itself. This seems a reasonable argument, although the larger the share of slots held by the dominant airline(s) the less convincing it becomes. At an airport such as O'Hare Chicago, where over 80 per cent of all slots are in the hands of either American or United, the number of slots that these two incumbents would need to overbid on would be relatively small.[5] On the other hand, McGowan and Seabright's argument is reinforced even at airports such as O'Hare, once an additional factor is taken into account. Once a secondary market is introduced, slots, in effect, become tradable assets and, as a result, their value should be written into the balance sheet of the airlines. This value will reflect the capitalisation of the economic rent that they command, which suggests that slots will be bought when an airline is able to earn a satisfactory return on its investment in them and that they will be sold when it is unable to do so. The introduction of a secondary market, therefore, is likely to create strong pressures for under-utilised slots to be sold (even by the dominant airlines) or to be used more effectively.

Support for this view comes from an analysis of slot ownership and usage data at Chicago O'Hare (Kleit and Kobayashi, 1996). O'Hare has the most concentrated holding of slots of all the four US airports where slots are bought and sold, and it is also the one airport where airlines established before deregulation of the US domestic market have increased their slot holdings. The analysis, based on 1990 data, focused on the utilisation of slots at O'Hare, and specifically examined whether the two large dominant carriers (United and American) were using their slots more, or less, intensively than the smaller carriers: the usual market power argument would be that the dominant firm(s) have an incentive to reduce output. Slot utilisation was measured in three ways: the average rate at which slots were used; whether leased slots were used more or less intensively than owned and operated slots; and the average daily seat capacity per slot. The analysis indicated that the dominant carriers had a higher usage rate for their slots and that slots that they leased out were used at an equal or higher rate than the owned and operated slots. In other words, there was no indication that dominant carriers were hoarding poorly utilised slots, or were leasing slots to other airlines which would make little use of them. On the other hand, there was evidence that the dominant carriers were using on average smaller aircraft but it was suggested that this reflected the use of Chicago O'Hare as a regional hub by both United and American. The overall conclusion was that concentration in the slot market at O'Hare was not leading to anti-competitive behaviour. Instead, the evidence was more consistent with the hypothesis that efficiency considerations are generating concentration at O'Hare: large airlines with large networks are more likely to obtain additional value from use of the marginal slot.

Efficiency and equity

These processes would of course be more transparent and therefore more evident, if the price mechanism was used to balance demand with supply at popular airports. However, in the absence of the price mechanism being used in this way, the conclusion from the O'Hare data supports the case for having a secondary market in slots. Such a market is likely to increase the efficient use of slots at congested airports. The point can be shown analytically if we modify the argument illustrated in Figure 17.1b. This Figure showed that when prices are sub-optimal there is an excess of demand. In Figure 17.2 an extreme case is assumed where the whole of the tail of excess, but low value, demand displaces an equivalent quantity of high value demand (a feasible situation when the criteria for administrative rationing emphasises historic usage). In this situation it can be seen that there is a net loss of user surplus.[6] The introduction of a secondary market in these circumstances, enables those potential users of runway capacity with high added value, to compensate users with low added value through the purchase price; there is a net gain in welfare by, in effect, allowing the market to redistribute slots to those services from which airline passengers derive most benefit. There is, however, an interesting consequence of which Figure 17.2 is suggestive. At airports which introduce secondary markets, it is quite possible that the average fare yield will increase. This is because capacity is now utilised by flights from which passengers derive most benefit and airlines will be able to extract this additional economic rent in higher fare yields. This is not an adverse effect; it reflects the fact that at slot constrained airports, introducing a secondary market has added value to the network of services.

Figure 17.2 Potential loss of surplus from adminstrative rationing

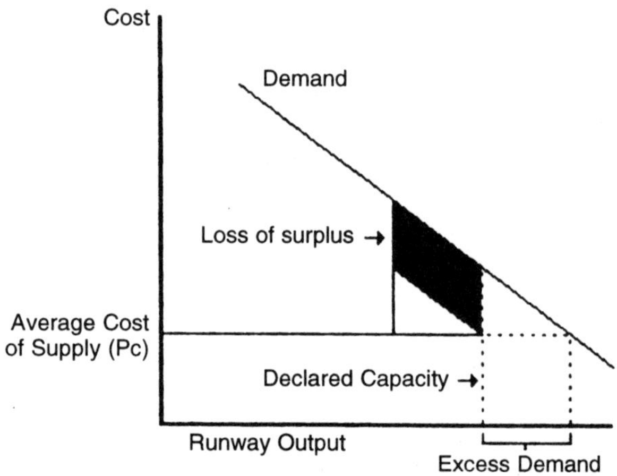

The taxing issue which follows from all this is: who should receive the economic rent? At the present time, with runway capacity priced sub-optimally, the economic rent goes to the incumbent airline which possibly obtained its slot(s) many years ago, for nothing.[7] Consequently, a secondary market in airline slots will mean that the incumbent receives a lump sum financial benefit when the slot is purchased, but this lump sum represents the present value of the scarcity rent associated with the slot.[8] The incumbent is, of course, currently receiving this scarcity value through the yield premium on fares charged to passengers. In this sense those with grandfathered rights already enjoy the 'windfall' and the introduction of a formal slot market will neither add to nor subtract from this.[9]

If this situation is considered inequitable then the way forward is not to confiscate slots from incumbents and redistribute them between other airlines on the basis of criteria such as the *assumed* competitive consequences. To do so merely emphasises the arbitrary process by which the rent is 'captured', and it will do little for effective competition in circumstances where there is a fundamental shortage of capacity: if one market sector, say the North Atlantic, receives a tranche of reallocated slots, this will generally be at the expense of competition in those markets from which the slots are withdrawn. A better way of approaching the problem is, clearly, to build more runways. Alternatively, if that is not possible, it is to increase the price charged for landing aircraft so that there is less of a divergence between the market clearing price and the price actually charged.[10] Increasing the price would have the effect of reducing the scarcity rent enjoyed by the incumbent airline, thus placing incumbent and entrant on a more equal footing. It does, of course, open up a range of additional issues that bear upon the regulation of airport charges such as the role of the 'single till'; the contribution made to airport finances by duty free tax concessions; and, indeed, whether there is scope for using the revenue from charges to compensate for environmental externalities. If increased charges were channelled into such payments there might be less resistance to the expansion of airports and building more runways might be more acceptable. But, these broader issues lie outside the compass of this paper.

References

T. Bass (1994) 'Infrastructure Constraints and the EC,' *Journal of Air Transport Management*, September, pp. 145-150.

M. Bishop and D. Thompson (1992) 'Peak-Load Pricing in Aviation: The Case of Charter Air Fares', *Journal of Transport Economics & Policy*, January, pp. 71-82.

A. Kleit and B. Kobayashi (1996) 'Market Failure or Market Efficiency? Evidence on Airport Slot Usage', in B. McMullen (ed.) *Research in Transportation Economics,* JAI Press, Connecticut.

F. McGowan and P. Seabright (1989), 'Deregulating European Airlines,' *Economic Policy*, October, pp. 283-344.

MMC (1996), *BAA plc, A Report on the Economic Regulation of the London Airport Companies*, MMC4, CAA, London.

D. Starkie (1994) 'Developments in Transport Policy: the US Market in Airport Slots,' *Journal of Transport Economics and Policy*, September, pp. 325-329.

D. Starkie (1992) *Slot Trading at United States Airports. A Report for DG VII*, City Publications, London.

D. Starkie and D. Thompson (1985) 'The Airports' Policy White Paper: Privatisation and Regulation', *Fiscal Studies*, pp. 30-42.

United States General Accounting Office (1996) *Airline Deregulation: Barriers to Entry Continue to Limit Competition in Several Key Domestic Markets, Report to US Senate.*

Notes

1. In spite of this conclusion, the industry Regulator (the CAA) decided that charges at Heathrow and Gatwick should fall still further in real terms over the five year period starting in April 1997. The process by which such a decision is arrived at is complex and hinges on what is called the 'single-till' approach. It is useful to note, however, that within the broader framework, BAA Plc does have a structure of landing fees which leads to significantly higher charges during defined peak periods. The company pioneered such a structure - still rare in world aviation -starting in the 1970s.
2. Part of the premium might however reflect the increased costs of operating at a capacity constrained airport.
3. For further details see Starkie (1992 and 1994).
4. Account should also be taken of the consolidation of the US airline industry as a whole during the 1980s and of the fact that unlike in Europe, gates at US airports are often on long term exclusive leases and are difficult to obtain (see Starkie 1992, 15-16).
5. At Heathrow, the largest airline (BA) holds less than 40 per cent of slots and therefore McGowan and Seabright's argument is more likely to apply.
6. A legitimate question is: why do not those airlines with historic rights serving low value markets switch services to more highly valued markets? In other words, is not the division of the market portrayed here, rather artificial? The division reflects both barriers to entry associated with international air service agreements and the absence of a market in slots; airlines do not have an incentive to earn a return on assets which are freely acquired and are not fully tradable.
7. A related issue is whether airlines enjoy any property rights to slots.

Opinion is varied but the legal view is inclined to argue that airlines do not have such rights. The rule governing the US market in slots states specifically that it does not give airlines any proprietary rights.

8. Estimates of the market price for a pair of slots at the four US airports vary from $1.0m off-peak to in excess of $4.0m at peak times. At Heathrow estimates vary enormously but generally start at $1.0m for a pair of peak period slots. These values might be contrasted with the book value for aircraft. The entrant would, of course, have to finance his purchase of a slot but raising finance against the security of a slot should not be at all difficult.
9. See Starkie & Thompson (1985) for elaboration.
10. If charges were raised to market clearing levels, a secondary market in slots would, of course, be unnecessary. However, given the practical difficulties of determining the market clearing price (which will fluctuate between seasons) it could be argued that it is more efficient to fix landing charges conservatively and allow a secondary market to fine-tune the situation.

CHAPTER 18

Whither airport regulation?

Stuart Condie

Introduction

This article presents a critique of today's economic regulation of UK airports. This is distinct from the many other forms of regulation faced by airports such as those pertaining to safety, noise levels or security standards. Before describing economic regulation an understanding of the industry structure and markets is required as well as of the nature of airport charges and the legal framework these sit within. After an overview of the present regulatory systems and associated topics such as slots and taxes, the present day difficulties will be outlined and some suggestions made on reform.

Structure of air transport

Before examining the nature and purpose of airport economic regulation, it is first necessary to understand the industry which airports are but a part of. Airports themselves are merely transport interchanges to enable passengers to switch from surface to air modes of transport. The product is air transport - or in some cases where air and rail compete this can be more widely defined. Figure 18.1 shows in diagrammatic form the multiplicity of parties involved in the supply of this product. Thus a package holiday to the Canary Islands for instance might involve the travelling public buying the holiday from a travel agent which arranges the contract between the buyer and the tour operator. The latter has contracted the travel component to an airline (which may or may not be vertically integrated with the tour operator and travel agent) which in turn contracts many of the

Figure 18.1 Multiplicity of parties involved in air transport

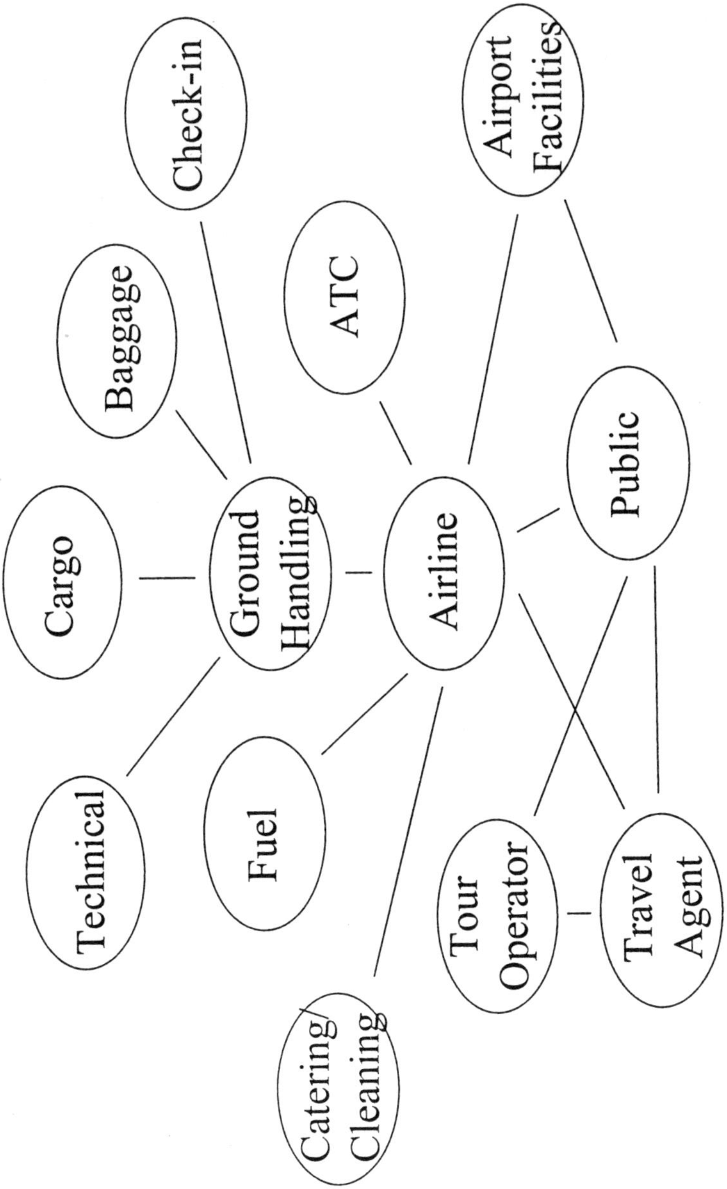

support services necessary to carry out the journey. The aircraft will need to be cleaned, fuelled and supplied with catering.

Once the passenger has reached the airport from his place of work or home, he will check in at the airline desk. This check in of passengers and their baggage is a complex process involving airline reservation systems (typically for scheduled carriers), computer systems for seat allocation and baggage tracking and reconciliation with passenger numbers. In many cases these check in processes and the manual handling and sorting of baggage together with delivery to aircraft are carried out by independent handling agents (e.g. Servisair or Gatwick Handling in the UK). These agents also perform other functions such as the handling of the aircraft on the apron (e.g. push back, first line engineering checks) provision of aircraft steps, control of the boarding process, loading and unloading of mail or cargo.

The airline also makes use of airport and air traffic control facilities. The former involves use of the terminal buildings by the passengers and airline staff plus the use of the runway and taxiway systems for aircraft take off or landing, and the apron, associated piers, satellites and gaterooms with linked airbridge walkways for loading and unloading of passengers and parking of the aircraft. Air traffic control is usually divided into local approach control (including aircraft ground movements) and en route ATC. Thus at any one airport there will be many agents involved in any one flight and this has to be multiplied several times around the globe as the aircraft flies several sectors and uses different states' airport and ATC services.

Particular features of the air transport industry

From the above brief description of the structure of the industry and the parties involved several features become apparent particularly when compared with other regulated companies.

1. It is a truly international industry. In the UK only some 20 per cent of passengers fly on domestic routes. It therefore follows that the supply of overseas airport and ATC facilities, ground handling services, fuel etc. is just as important as the situation in the UK.
2. The product is heterogeneous. This is primarily a result of the different purposes for travel - not just between business and leisure but also reflecting different niche markets such as inbound inclusive tours to the UK, visiting friends and relatives, those working for small businesses (or themselves) or attending conferences. The variety is

compounded by differences between nationalities. It is also further enlarged by the possibility of multiple ways of making certain journeys by changing planes at different points. Thus Frankfurt and Los Angeles can be served by direct flights or by transferring at European hubs (e.g. London, Paris, Amsterdam) or US hubs (New York, Atlanta, Dallas, Chicago, etc.). An illustration of this is shown in Figure 18.2 which shows the origin of all the passengers on a Gatwick-Dallas flight in 1996. Only 23 per cent started their journey in London. Scheduled carriers such as British Airways aim to attract high yield passengers by concentrating on high standards of service, sophisticated marketing, the Air Miles loyalty programme and strong brand image. Other new entrant, so-called low cost carriers often have direct telephone booking rather than complex computer reservation systems used by travel agents, no seat reservation, minimal catering or cargo carried and an emphasis on short haul, single aircraft type operation.

3. The market place tends to have oligopolistic characteristics - i.e. a relatively few number of players with much differentiation based both on price as well as quality and other branding/marketing characteristics. Thus the national base carrier tends to provide at least half the aircraft services from many European hubs (e.g. Paris, Amsterdam, Frankfurt). ATC tends to be a state owned monopoly supplier and consolidation in the travel agent/tour operator market has been investigated by anti-trust bodies in both Germany and the UK. Aircraft manufacture is largely in the hand of two companies, and engine manufacture is dominated by three firms worldwide.
4. The produce is discretionary; thus air travel is not a prerequisite for human life in the same way as water, heating, energy and communication.
5. The greatest power tends to rest with airlines. They tend to employ more staff, have tremendous trade union power bases, greater market capitalisation (Table 18.1 shows seven of Heathrow's customers have six times the market capitalisation of the whole of BAA) and very strong systems of political patronage. Airlines are also extremely litigious with legal action in the UK, European and US courts on wide ranging topics such as predatory pricing, unfair competition, noise quota rules, travel agent commissions, etc.

Figure 18.2 The importance of connecting passengers: BA 193 Gatwick-Dallas 12 July 1996

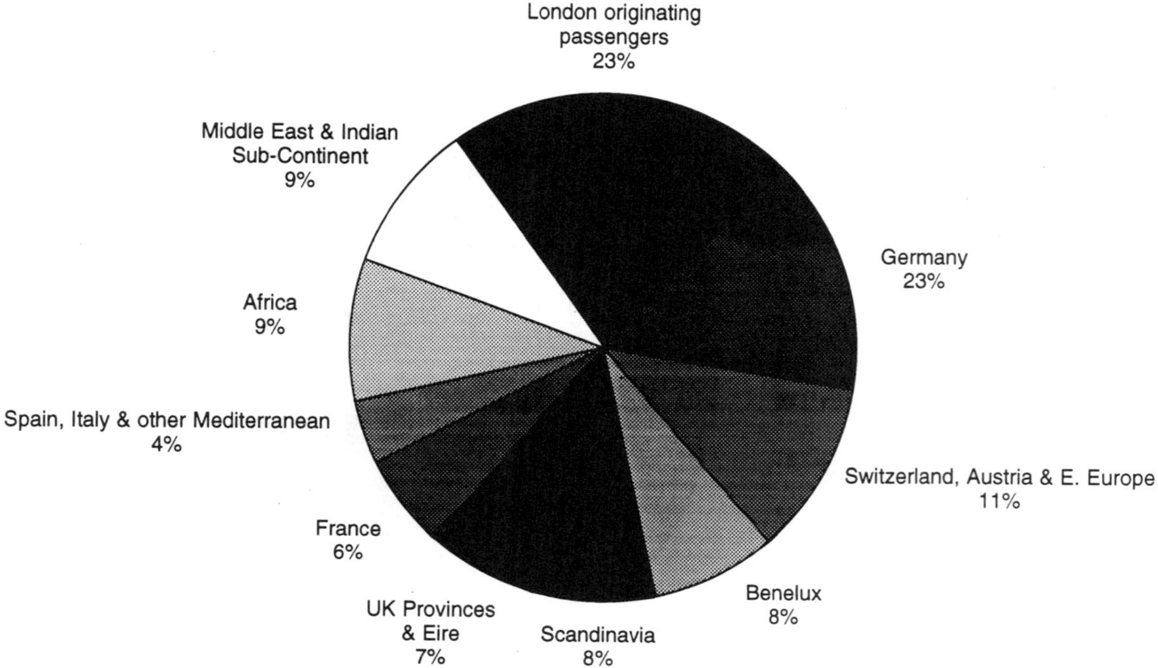

Table 18.1 BAA and its airline customers' market capitalisation

	£Bn
AMR	5.4
UAL	6.0
Delta	4.1
BA	6.9
Lufthansa	3.7
JAL	5.8
Cathay Pacific	3.1
BAA	5.8

Note: Market capitalisation as at March 1997

The airport market place

Many airports in the UK have been sold to the private sector starting with BAA plc in 1987. BAA accounts for some 75 per cent of air traffic in the UK and owns three London airports, three in Scotland and Southampton. The company also now has three management contracts in the USA (Pittsburgh, Indianapolis and Harrisburg), owns the contract to run Naples Airport and is a minority shareholder in Melbourne International Airport. Other UK airports sold off include Belfast Aldergrove and Cardiff (to the TBI Group), Bournemouth and East Midlands (to National Express) and Birmingham (to Aer Rianta/NatWest Ventures). Three other European airports (Rome, Copenhagen and Vienna) have local stock market quotations but are still majority owned by state or local government.

Despite this activity, most large scale airports are still run by organs of the local or central state (all the major hubs and international gateways in USA, Canada and Far East plus airports such as Paris, Frankfurt, Amsterdam, Madrid, Milan, Athens, Berlin, Munich, etc). In addition central governments retain influence over airports in several key areas irrespective of their ownership structure. Figure 18.3 lists some of these for airports in the UK.

Figure 18.3 Continuing government influence over UK airports

- Airport licensing
- Provision of Air Traffic Control
- Traffic distribution rules
- Limits on ATMs
- Security
- Noise
- Planning Permission
- Control Authorities

However, although many airports are still state owned or are heavily influenced by the state, competition between them has been increasing steadily over the last few years. This has been due to three factors:

1. Opening of high speed rail links; this has severely dented air traffic on routes such as those between London, Brussels and Paris, and from Paris to centres such as Lyon, Nantes, Bordeaux and Lille, and from Seville to Madrid.
2. Increasing use of airline schedules coordination and yield management systems to generate transfer traffic around hub airports. The proportion of transfer can be as high as 50 per cent of all passengers using the airport.
3. Growth in regional airports; the latter can generate a wide range of services by linking to contiguous hubs. Hence Air UK services to Amsterdam Schiphol from Norwich, Newcastle, Teeside, East Midlands, etc., provides a competitive product to traditional links to London.

Table 18.2 shows a breakdown of Heathrow's traffic by market segment and the competitors within each of these segments.

The business activities which airports are involved in also vary considerably (see Figure 18.1). Many provide approach ATC services for their airport (e.g. Bristol, East Midlands, Paris) or ground handling services (e.g. Paris, Frankfurt, Spanish airports) or aircraft catering (Rome) or airport retailing (BAA, Aer Rianta). BAA is comparatively rare in providing security services directly. On a micro scale there are many more products or services where the supplier may be the airport company, an airline, or a third party (e.g. cargo handling, fire and rescue services, stand allocation).

Table 18.2 Competition for Heathrow's traffic

Routes	Competitors	Passengers (mppa)
Paris, Brussels	Channel Tunnel	3.5
International/International Transfers	Continental Hubs	12.2
Domestic/International Transfers	Continental Hubs Regional UK Airports	6.7
Domestic (O/D)	Rail	4.0
International (O/D in North)	Regional UK Airports	3.4
International (O/D in South)	Continental Hubs Regional UK Airports	24.9
	Total	**54.8**

Notes:
a) O/D = Origin/Destination in North or South of UK
b) Passengers = millions per annum in 1996/7

Airport charges

Airports in Europe are remunerated by airlines for the use of their facilities by means of airport charges. Whilst these are not homogenous in level or structure they typically comprise four elements:

1. A landing charge to reflect use of the runway and taxiway systems.
2. A passenger charge to pay for use of the terminal buildings.
3. An aircraft parking charge to pay for the apron areas.
4. Miscellaneous charges such as for ground power, lighting, supply of air, use of air jetties, etc.

Recent research of European Airports (see Figure 18.4) shows that when airlines select which airports to use there is a wide range of factors which are taken into account. The primary factor is revenue generating potential (a combination of number of passengers, the yield or average ticket price paid and the likelihood of feed from other airlines). Ease of market access - in other words slot availability - is also important along with quality of baggage handling and check in.

Figure 18.4 International airlines selection criteria for European airports

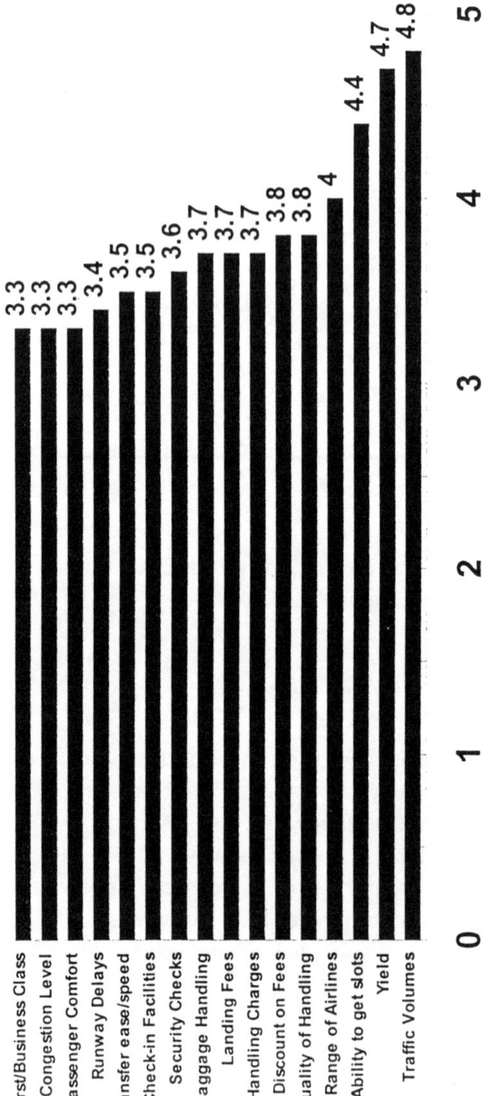

Source BAA
Sample of 30 International Airlines (BMSL) - October 1993
Scale of importance, 5 = extremely, 1 = not at all

Perhaps not surprisingly given the heterogeneity of the product and non price differentiation the research shows that the importance of fees (whether levied by the airport or by handling companies) is not paramount but features alongside other quality issues. This is repeated in the 1996 MMC report on BAA's London Airports (MMC4) which stated that 'airlines are concerned as much about the quality of airport facilities .... as about the airport charges they pay' (para 2.13).

Table 18.3 IATA Airlines 1994 - operating costs per ATK

Cost function/item	**Share of Total in 1994**
Flight deck crew	7.4
Fuel and oil	11.3
Flight equipment insurance	0.4
Maintenance and overhaul	9.1
Flight equipment depreciation	8.5
Rentals	3.6
Landing Charges	5.1
En-route charges	3.9
Direct operating costs	**49.5**
Station and ground	12.8
Cabin attendants	7.2
Passenger Service	6.5
Ticketing, sales and promotion	17.9
General and administrative	6.2
Indirect operating costs	**50.5**
TOTAL OPERATING COSTS	**100.0**

Note: ATK = Available Tonne Kilometre

Another reason for this state of affairs is that airport charges account for only a small proportion of airline costs. IATA 1994 cost data (see Table 18.3) suggests this is around 5 per cent, which is in line with a 4-6 per cent estimate supplied by the Comité des Sages to the European Commission in the same year. Higher figures have been suggested but these usually include ATC and handling costs which are normally charged for separately. An ITA study (Wrobel 1997) showed European airport charges as a proportion of ticket prices as follows:

	%
Domestic	5-8
Short Intra-EU	5-7
Long Intra-EU	2-3

However, on longer inter-continental routes the proportion is likely to be 1-2 per cent so that an overall average of 5 per cent seems reasonable. The IATA data in Table 18.3 also shows that other fees paid whilst at the airport - but for other services, such as fuel (11 per cent), ATC (4 per cent) and handling (possibly 7-10 per cent), are actually far more important than airport charges. The variation between airports in fuel prices alone (caused by government taxes and the degree of choice for fuel supply) shown in Table 18.4 can easily exceed the total cost of airport charges. Similar comment can be made of handling charges with significant variations being due to some markets being open to competition (e.g. London, Amsterdam) and others remaining in the hands of state owned airport monopolies (e.g. Frankfurt, Milan). This is described more fully in a 1996 Cranfield Institute of Technology report.

Table 18.4 Sample fuel prices

	Cents/USG		Cents/USG
London	64-66	Abu Dhabi	98-99
Rome	73-75	Jo'burg	101-104
Moscow	97-105	Lagos	125-126
New York	75-58	Singapore	79-81
Miami	59-61	Melbourne	90-91
Aruba	99-100	Beijing	119-130
Mexico	71-72	Hanoi	106-107
Quito	83-84	Hong Kong	82-83
Guatemala	113-114	Tokyo	96-103
Havana	115-116	Phnom Penh	118-121

Source: Airbus Industrie (March 1997)
USG = US Gallon

Airport economics

The most striking feature of the airport business is the success which these enterprises have had in generating other sources of revenue outside of airport charges. A profit centre report for Gatwick Airport (see Table 18.5) illustrates the extent of this. In this year airport charges accounted for barely 30 per cent of total revenue and made a £20M loss. Whilst these losses were counterbalanced by revenue from other services provided to airlines (mostly office and cargo rents) all the airports profits were accounted for by retail sales to passengers of which duty and tax free sales took over half. Without such retail sales and profits Gatwick would have lost £11M at a trading profit level.

Table 18.5 Gatwick Airport Limited: Profit centre reports 1994/95

	Income (£M)	**Operating Profit (£M)**
Airport Charges	73.6	(20.1)
Airline Services	45.0	9.1
Rents	*24.7*	*6.3*
Passenger Services	125.9	72.0
Duty/Tax Free	*78.6*	*39.6*
Total	**244.6**	**61.0**

The importance of non airport charges revenue has been increasing over the last five years (see Figure 18.5) as airports have been successful in better managing their retail product by more space, better merchandising, improved targeting of passenger groups and innovative products and services. It should also be mentioned that the market for airport retailing is a higher competitive one as even with duty and tax free products passengers have multiple opportunities to purchase - both outbound and at the destination airport on the return leg. Furthermore the current strong pound allied to low levels of excise duty in other states means that many overseas high street products are competitive.

Figure 18.5 Proportion of BAA London Airports revenue derived from airport charges (given as %)

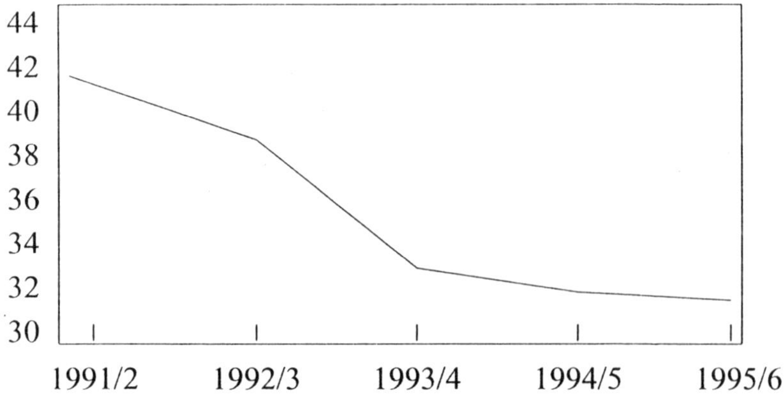

By contrast airport charges per passenger have been in decline for some years now. Even taking the period financial year 1980/81 to 1987/88 up to BAA's privatisation in 1987, Heathrow Airport charges per passenger had fallen by 60 per cent in real terms (equivalent to about RPI-7). The decline has continued since 1987 but the impact of regulation (see below) has been partially offset by increased costs due to new government security regulations following the downing of the Pan-Am jet over Lockerbie in 1988. As a result of this continual long term decline Heathrow and Gatwicks' airport charges are now on average amongst the lowest in Europe and about half their equivalent at airports in Germany, Austria and the Netherlands (see Figure 18.6). This point was recognised in MMC4 which stated that BAA's London 'airport charges are also relatively low compared with those at other international airports' (para 2.33).

Regulation of airport charges

The level and structure of airport charges are covered by a wide range of legislation (see Figure 18.7). The first of these is the Chicago Convention of 1944 which was an attempt to regulate the conditions under which commercial carriers could make international flights. The most important concept for airports contained in the convention is that of non discrimination and parity of treatment - that, for instance, a UK carrier should expect to pay the same charges for equivalent services in the USA

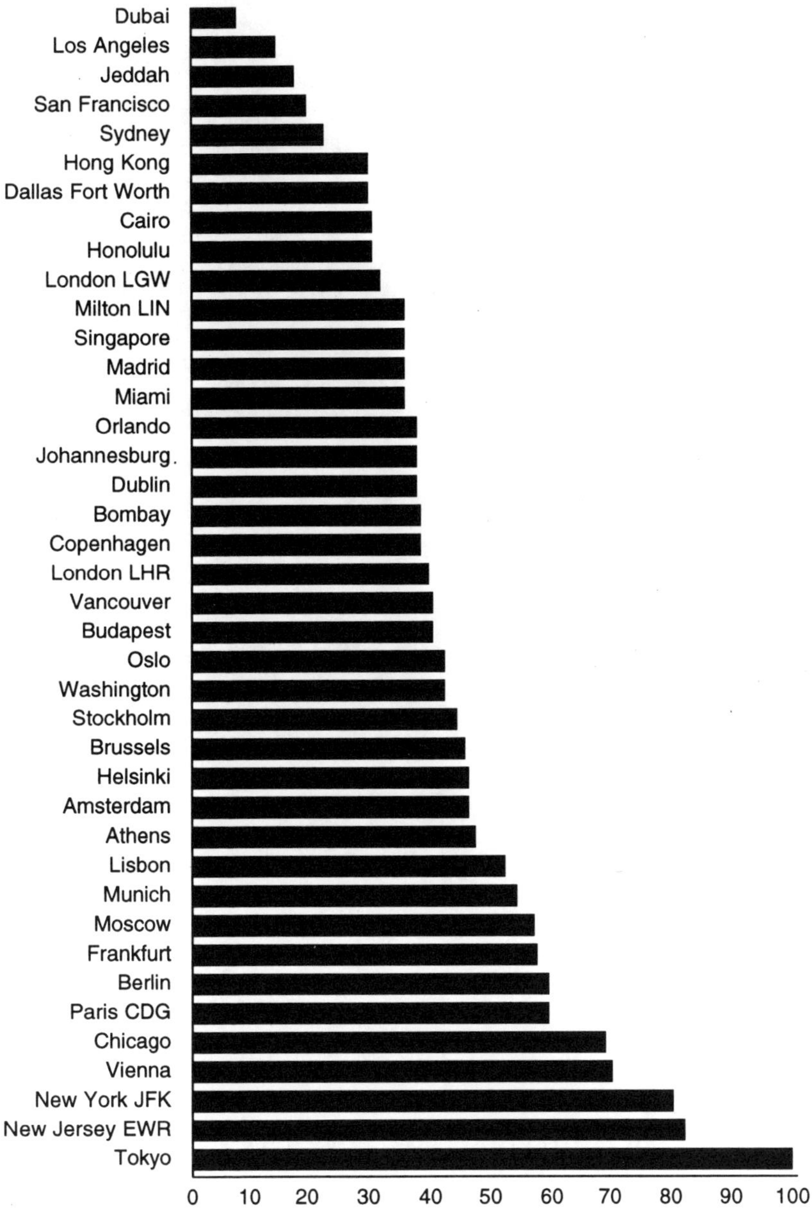

Figure 18.6 Symonds Travers Morgan - Airport charge index 1998

Rank	*Airport*	*SDR*	*Index*	*Rank*	*Airport*	*SDR*	*Index*
1	Tokyo	37862	100	21	London LHR	14857	39
2	New Jersey – EWR	31176	82	22	Copenhagen	14468	38
3	New York JFK	30440	80	23	Bombay	14215	38
4	Vienna	26325	70	24	Dublin	13900	37
5	Chicago	25980	69	25	Johannesburg	13881	37
6	Paris CDG	22484	59	26	Orlando	13823	37
7	Berlin	22393	59	27	Miami	13273	35
8	Frankfurt	21595	57	28	Madrid	13260	35
9	Moscow	21510	57	29	Singapore	13194	35
10	Munich	20356	54	30	Milan LIN	13168	35
11	Lisbon	19784	52	31	London LGW	11579	31
12	Athens	17770	47	32	Honolulu	11443	30
13	Amsterdam	17534	46	33	Cairo	11379	30
14	Helsinki	17283	46	34	Dallas Fort Worth	11150	29
15	Brussels	16951	45	35	Hong Kong	10932	29
16	Stockholm	16618	45	36	Sydney	8513	22
17	Washington	16091	42	37	San Francisco	7096	19
18	Oslo	16034	42	38	Jeddah	6297	17
19	Budapest	15199	40	39	Los Angeles	5436	14
20	Vancouver	14997	40	40	Dubai	2779	7

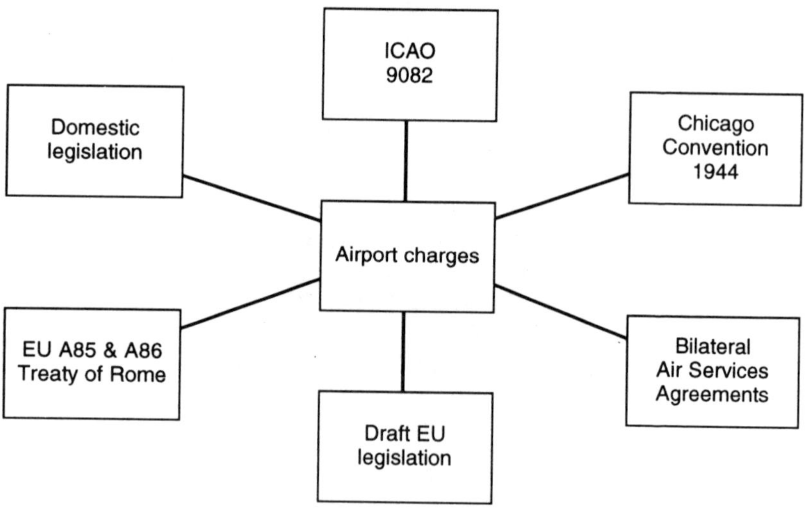

Figure 18.7 Airport charges: The legal framework

as a US domestic carrier and that the same would apply to US and UK carriers in the UK. Staying at the international level, there is also the International Civil Aviation Organisation (ICAO) 'Statement by the Council to Contracting States on Charges for Airports and Air Navigation Services' or Doc 9082 fifth edition, 1997. ICAO is a branch of the United Nations Organisation and Doc 9082/5 carries no legal weight but is more a statement of recommended best practice on the definition of charges, the activities they should cover, as well as some general principles such as the notion that airport charges should cover permissible costs plus an element for a reasonable rate of return.

It should also be mentioned that bilateral air services agreements between member states (these no longer apply to intra European flights) not only cover areas such as capacities, prices, number of carriers and gateway cities but, on occasion, also make statements about matters relating to airports such as the level and structure of airport charges and ground handling rights. There are particular provisions in the UK/US Bermuda 2 agreement which have been the source of long running disputes between the respective governments relating to airport charges for US carriers at Heathrow. This is described in more detail in MMC4.

At a European level there is another panoply of legislation. There is the general provision of the competition articles of the Treaty of Rome, in particular Article 86 relating to abuse of dominant position. Airlines have already been successful in using these general competition provisions (e.g. British Midland in a complaint about discounts on airport charges at Brussels Zaventem). In addition, there is specific legislation on ground handling (Directive 96/97/EC) which aims to open up the market, as well as draft legislation on airport charges (COM(97)154). The latter legislation introduces the three principles of cost relatedness of charges, non discrimination between users and transparency of charges.

Finally there is UK legislation on the economic regulation of airport charges. The basis for this is the Airports Act 1986 which was a prelude to the privatisation of the former British Airports Authority. This legislation is described in detail elsewhere (eg MMC4) but contains the following elements:

1. A two tier system of regulation. All licensed airports in the UK have transparency and accounts conditions as well as general prohibitions on discrimination, predatory or excessive pricing and abuse of dominant position. There is a procedure for complaints relating to these general trading conditions.

2. The second tier of airports have in addition a formal system of price control. This second tier currently includes Heathrow, Gatwick and Stansted (BAA plc London airports) and Manchester (owned by a consortium of Local Authorities).
3. Price control is of the classic RPI-X type with a review every five years by the MMC setting the formula for the next five years and establishing if any public interest issues have arisen.
4. The Civil Aviation Authority (CAA) acts as primary regulator (akin to OFWAT, OFGAS, OFTEL, etc) but does not carry out the price review itself and is obliged to implement any MMC findings on public interest issues. MMC proposals on price control are usually submitted for consultation to the air transport community before a final decision is taken by CAA.

Figure 18.8 CAA's responsibilities as per Airports' Act 1986

- To further the reasonable interests of UK airport users
- To promote the efficient, economic and profitable operations of such airports
- To encourage investment in new facilities at airports
- To impose the minimum restrictions that are consistent with the performance by the CAA of its functions

5. The CAA's responsibilities are also set out in the Airports Act and are reproduced in Figure 18.8. Unlike other regulators they are obliged to promote the economic and profitable operation of airports and to encourage investments in new facilities. The obligation to act in airlines' interests but also to encourage profitable investment has been at the core of airport regulation and is probably its most problematic aspect. As CAA's 1996 Report on London Airport Charges (CAP 664) states 'airlines do not want to put [BAA's incentive to maximise the investment that airlines want at the quality they want] at risk to achieve further relatively small reductions in charges which are declining anyway in real terms' (para 15.2).

6. The RPI-X price control is based on an average revenue yield per passenger (i.e. all the revenues from the different parts of airport charges tariff divided by the number of passengers). The value of X is set based on a rate of return on assets calculation which takes into account all the airports' non aeronautical revenue such as property and retailing (the so called 'single till' approach). There is also a '+S' component to the formula which allows the pass through of a proportion of additional costs caused by changes to government policy or standards on air transport security. The formulae imposed on BAA London Airports since 1987 are shown in Figure 18.9.

Taxes and slots: the hidden charges

There is much confusion over the different taxes collected by Governments in addition to airport charges. Many of these taxes are hypothesized into particular funds, e.g. noise taxes at the six largest French airports and to fund noise insulation schemes at Amsterdam Schiphol. Others are collected by the state for particular services provided but the degree of hypothesis is uncertain (e.g. security taxes at Germany, Austria and France). However, a third form of charge exists which is more akin to a conventional tax as the revenue goes into general state budgets. The prime example of this is the UK's air passenger duty (APD) currently set at £10 per departing passenger for EU flights and £20 for non EU. The magnitude of this should be compared with the average airport charge being around £4.75 per passenger (i.e. £9.50 per departing passenger) at Heathrow and about £3.60 (£7.20 per departing passenger) at Stansted and Gatwick. Not surprisingly the revenues received by HM Treasury from the APD are likely to exceed total UK airport charges by around 30-50 per cent. To put this point another way what is the public policy raison d'être for airport charges regulation when taxes are subsequently introduced which dwarf any potential consumer gain? Even if one assumes that the real reduction in airport charges at London airports since privatisation of about 25 per cent can all be ascribed to regulation (this in itself is a very bold assumption) the reduction in ticket price of £2 (using a London Airport average) pales into insignificance compared to ticket taxes of £10-20.

It will also be apparent from the discussions above that airport charges are not set at a rate to clear the market, but are based on a cost plus calculation using rates of return and asset valuations. What therefore happens to the difference between the economic value of a right to use a busy airport such as Heathrow and the equivalent airport charge?

Figure 18.9 Maximum permitted increases in airport charges per passenger at BAA London airports

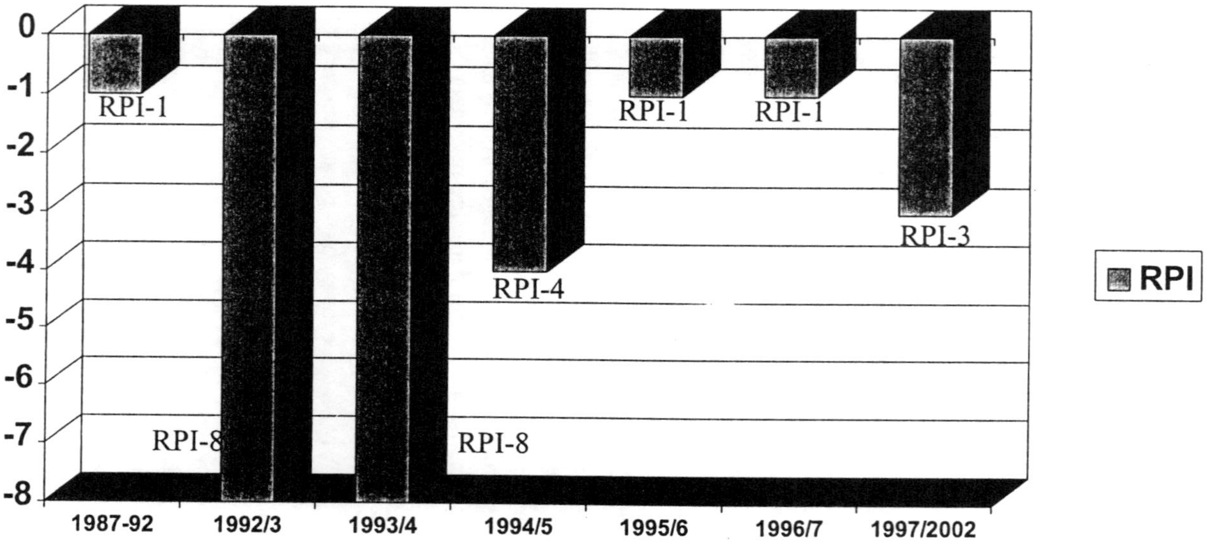

The answer to this question can be looked at in one of two ways. Firstly, what is the difference between Heathrow and Gatwick in air fares for equivalent points served? Taking Gatwick in the early 1990s as a relatively slot unconstrained airport the yield difference per passenger was reputed to be £50 on average for reasonably popular departure/arrival times.

Secondly, what are airlines prepared to pay for pairs of slots for convenient times at Heathrow? This is considered further in the chapter by David Starkie but the answer would appear to be anything up to £10-15M. If anything this could be an under valuation because of the grey nature of the market and its lack of a legal framework. When converted to a per passenger charge for a daily flight of a 747 size aircraft this equates to about £10 per passenger.

So, the conclusion can only be that the economic value of under pricing airport charges resides with airlines. Not only can they impose higher ticket prices but those airlines with grandfather rights for slots at Heathrow are also sitting on a substantial balance sheet value, namely the trading value of the rights to enter the market. Table 18.6 compares Heathrow average airport charges on a per passenger basis with the hidden charges of the APD and opportunity cost of the slots. The sum of APD and slot costs are between two and three times the average Heathrow airport charge.

Table 18.6 Heathrow airport charges vs other costs (1997/8)

	£ per departing passenger
Airport charges	9.50 (average)
Opportunity cost of slot	10
Airport tax	10-20
TOTAL	**30-40**

A critique of airport charges regulation

Why is regulation needed at all?

BAA's regulation is very unusual in that it concerns a product traded between companies and not between a company and the general public. In regulated industries RPI-X price caps almost always refer to the latter.

The main exception is Railtrack's track access fees to train operating companies but even in this case the two parties are free to negotiate different levels of fees for new investments such as Thameslink 2000.

Airlines at London airports are more akin to a bilateral oligopoly with BAA and BA having a degree of dominance. Given that these kind of market structures are normally subject only to orthodox domestic and EU competition law rather than price control, why should the same not be true of air transport? The imposition of formal price caps on this market structure merely redistributes wealth from the airport to the airline without any gain to consumers given the lack of market access to Heathrow. Furthermore, as described above, airlines have successfully used both the courts and political means to keep airport charges low. The other main beneficiary of regulation is HM Treasury through APD receipts. Consumer gains would appear to be insignificant.

The focus on price

Regulation was designed primarily for homogenous products delivered to the public to a common standard by utilities (e.g. water, telephone lines, gas, electricity). The air transport product is not homogenous and neither are airports - indeed what each airport chooses to provide varies dramatically with the common services between airports being very small compared to those which are supplied by third parties. From the above discussion it would appear that these third party services can cost far more than airport charges. Even when considering strictly airport activities airlines would also appear to regard quality issues and investment at least as highly as price. Is price regulation therefore missing the point?

Why the single till?

The single till feature is unique to airport regulation. Its rationale is difficult to understand when the markets concerned (retailing and property primarily) are both discretionary and competitive. There are some exceptions to this (e.g. terminal catering, short term car parks, airline offices within the terminals) but these account for only a minority of the revenue. The CAA is implicitly recognising this point by allowing BAA to reorganise some of its property activities into companies outside the single till. The only argument in favour of the single till would appear to be the custom and practice around the world of cross-subsidising operations from retail activities.

The 'cost plus' mentality

Airlines' continued insistence on cost plus charges (e.g. ICAO 9082, draft Directive 97/154/EC) and the rate of return on assets form of UK regulation have some bizarre effects on prices. Recently constructed terminal buildings with traffic building up have far higher charges than older airports with mature traffic levels as shown in Table 18.7.

Table 18.7 Cost comparison for hypothetical old and new airports

Airport	Capital Cost (£M)	Annual Deprec- iation (£M)	Other Charges (£M)	Pax (million per annum)	Charge per Pax (£)
A	200	10	5	5	3
B	100	5	5	10	1

The capital costs for airport A are higher for two reasons - firstly general construction costs tend to rise over time in real terms and secondly unit airport construction costs rise over time as easier sites are used first. Hence the lumpy nature of airport investment plus rising real unit costs leads to a saw tooth progression of charges if a strict cost plus regime is followed (see Figure 18.10). At BAA's London airports a combination of these factors plus regulation and the single till have led to falling charges at Heathrow (the most congested airport) and rising charges at Stansted (which is lightly utilised). This is shown in Figure 18.11 where the average charge at Heathrow has fallen by 13p whilst Stansted has risen by 73p and the price differential between them has almost been halved. Such a pricing regime shows scant regard for external costs such as noise, emissions, congestion, surface access difficulties, etc, which are obviously higher at busier airports. This feature is bitterly opposed by both environmental groups and those dedicated to improving regional airports (see MMC4 Chapter 13), and alienating local authorities and neighbours is probably against the long run interests of the air transport community.

Figure 18.10 The cost related charging saw tooth

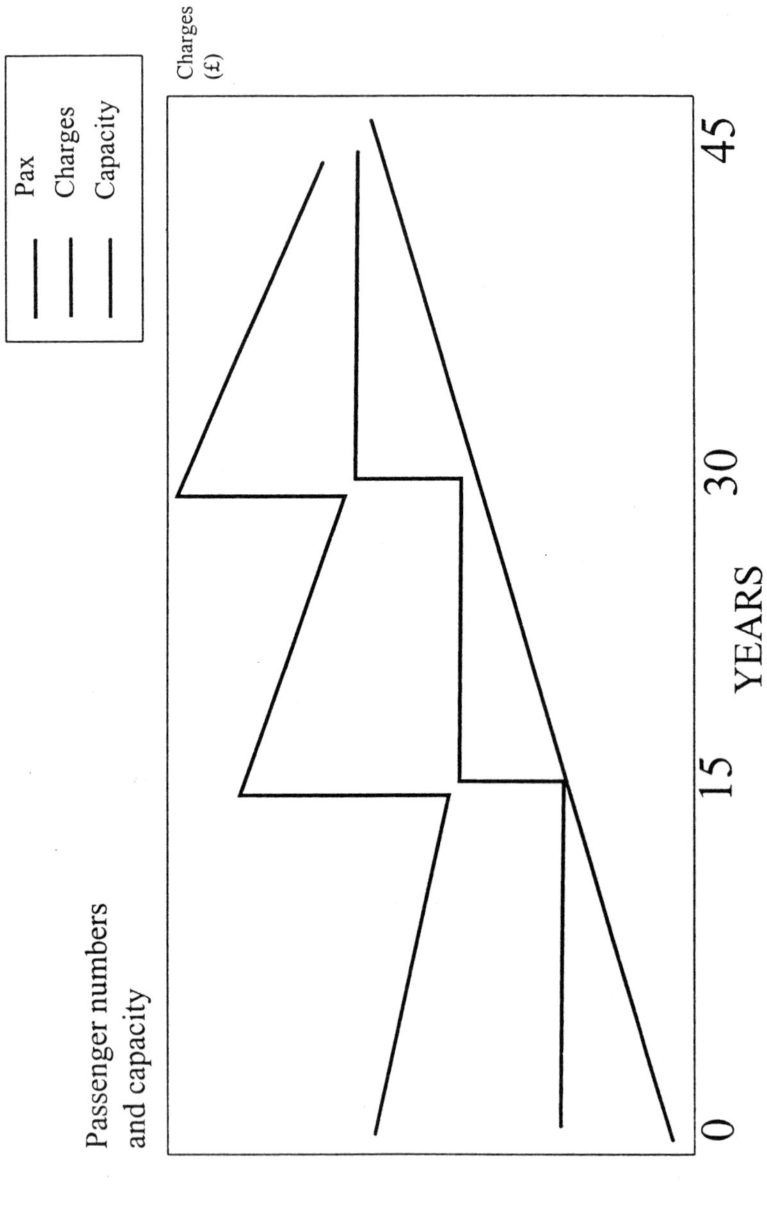

Figure 18.11 BAA London Airports airport charges revenue yield (£)

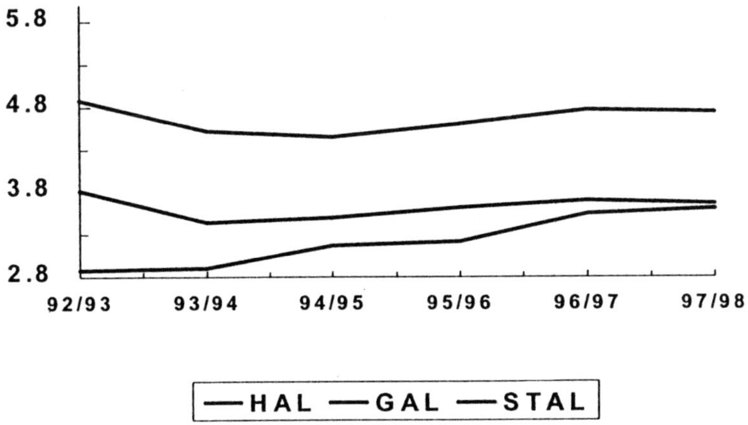

The investment problem

The saw tooth profile displayed in Figure 18.10 creates several problems for investment and decision making. Whilst facilities are filling up (e.g. the period 0-15 years);

a) Charges are falling when instead they should be rising to reflect rising long run unit cost of capacity.
b) Charges are too low to give satisfactory profits on the new investment.
c) When the time comes to decide whether or not to invest (eg years 10-12 in Figure 18.10) the airport company has to judge whether large price increases in the future will be possible. Many airports are in fragile market places and such increases are unlikely without loss of traffic.
d) Such wild oscillations in prices are, not surprisingly, unpopular with airlines who use the means noted above to resist or moderate the proposed increases. In the regulated sector the airport company can only hope that at the next price review the regulator will permit higher charges (and continue to uphold them for the rest of the life of the asset concerned).
e) The airport's investment risk is compounded without any long term contractual bargain between it and the airlines. Airport investment is expensive, has long asset lives and yet airline remuneration is entirely

determined by a simple conditions of use contract which allows the airlines to come and go as they please thus leaving much of the investment risk with the airport.

Why is regulation so complex?

Even a cursory examination of the legal framework shown in Figure 18.7 will show how certain principles and procedures are repeated in different documents. Because the exact working and meaning of the various documents varies substantially, the end result is an understanding of the broad thrust of the legislation but confusion over its exact meaning and an impossible compliance burden for airport companies. For instance discrimination is prohibited under virtually all these documents but has different connotations in each.

The worst aspect of this complexity is probably the substantial overlap which exists between the duties of the Office of Fair Trading (OFT) (under UK domestic competition law), the CAA and MMC (under the Airports Act 1986) and DG IV (under draft directive 97/154/EC). In particular the following involves unnecessary duplication:

a) The public interest issues examination by MMC every 5 years. Why is this needed when complaint procedures already exist via OFT, CAA and DG IV?
b) The broad trading conditions provisions of the Airports Act 1986 and those in draft 97/154/EC cover exactly the same ground. Since the EU legislation seems likely to be passed broad swathes of the Airports Act should be abolished.

Regulatory reform

Any approach to reform should therefore have the aim of addressing the following issues:

- Be focused on all user needs, not just price;
- Where possible to act as a backstop to voluntary agreements;
- To encourage investment and productivity;
- To avoid allocative inefficiency;
- The removal of unnecessary complexity and distortions such as the single till.

There are three broad approaches to change for UK airports - reform within the existing system of price regulation, dedesignation from price control with regulatory benchmarking and dedesignation with individual user agreements. These approaches can now be compared.

Changes to price regulation

If price designation is retained there are two main changes which could be carried out. Firstly, the airport revenues which are earnt in competitive markets should be removed from the calculation of profitability at the time of the quinquennial regulatory review. Secondly, Stansted Airport should also be removed from such a calculation. The airport's trading position is comparatively weak and it is obliged to use marketing and introductory discounts to attract traffic, particularly the charter and low cost carriers which are extremely footloose.

Such a twin change would go some way to removing the allocative inefficiency and distortions caused by the single till. The rise in airport charges would also give a more correct pricing signal and encourage investment in higher marginal cost facilities at Heathrow and Gatwick. Whilst regulation of quality issues could also be improved this would still be within a regime of common standards and prices across the London airports. This is inflexible both between airline customers with differing needs and over time - primarily because the price/quality trade off is fixed and cannot be explored by airlines.

This twin reform, if carried out immediately, would be likely to lead to windfall profits for BAA. One possibility might be to dismantle the single till over a period of time subject to satisfactory investment levels at Heathrow and Gatwick.

Dedesignation for price control

Given the preceding comments on whether price control is needed at all, the possibility of removing such controls should also be investigated. The main advantage of dedesignation is that provided the relevant articles of the Airports Act 1986 are retained then the government always has the power to immediately reinstate price control without primary legislation. The threat of price control is a very real one and could encourage airport owners to reach voluntary agreements on price control wherever possible. As an example, BAA's Scottish airports of Glasgow and Edinburgh introduced a voluntary RPI-3 system of price capping in 1993/4 which resulted in significant falls in average charges (see Figure 18.12). Real falls exceeded 3 per cent due to an additional package of price reductions.

Figure 18.12 Scottish Airports: Airport charges yield per passenger - adjusted for inflation

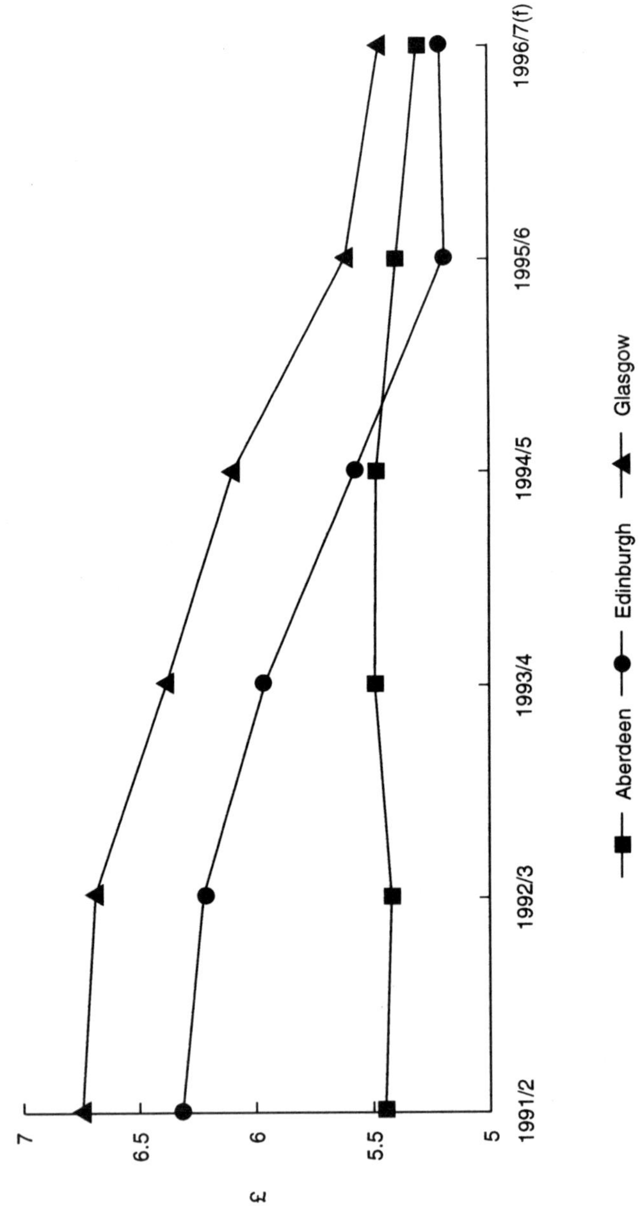

When considering whether airports can continue to be outside statutory price control the regulator can monitor a wide range of indicators benchmarked against other airports. These might include operational performance indicators (similar to those already reported to CAA as part of BAA's voluntary agreement to introduce Service Level Agreements [SLAs]), passenger opinion scores such as BAA's Quality of Service Monitor described in MMC4 and average levels of airport charges. Whilst such comparisons between airports are difficult, academic bodies such as Cranfield University or the French ITA are perfectly capable of carrying this out.

User agreement

Instead of relying on regulatory monitoring of benchmarks, airports and airlines could reach long term agreements on issues such as airport charges, quality and performance levels and capacity enhancements or investment. Those airlines with significant market presence at any airport would sign such bilateral agreements whilst others would either operate under a common low use contract or continue with the published tariff and conditions of service. Such agreements have two main advantages; they enable airlines to make a proper price/quality trade off and they enable airports to share development risk with airlines. The latter become long term partners to the airport and undertake to bring minimum amounts of traffic and pay guaranteed minimum levels of airport charges. These user agreements effectively become more like property pre let contracts. Examples of these are described by Clayton (1997) and IATA draft papers.

The main difficulty with such agreements is that handling the varying levels of service, investment and prices between airlines will become managerially more demanding. Indeed the end result could well be that certain airlines are able to extract better deals than others simply due to size or leverage. This could be viewed as discriminatory and therefore illegal under many of the legal frameworks described above. The question also arises as to penalties should such agreements be breached. Whilst the agreements could be seen as toothless without such penalties it is in practice almost impossible to ascribe responsibility for poor performance due to the multiplicity of parties involved and the complex inter weaving of processes. Thus if aircraft stands are unavailable this could be due to:

- Equipment breakdown (guidance systems, lights, fuel hydrant systems, etc);
- Obstruction by ground equipment;
- Spillage of aviation fuel;

- Occupation by long term parked aircraft.

Each of these events is controlled by different parties (likely to be the airport operator, ground handlers, fuel companies and airlines respectively in the above example).

Furthermore the general question needs to be asked as to whether a series of bilateral agreements with penalties might actually be divisive and lead to loss of trust by all parties. It is uncertain whether bilateral agreements can work at airports which in Europe have multiple airline usage and sharing of common facilities.

Conclusion

The UK model of regulation of airport charges is approaching a crisis point caused by the bizarre conjunction of all time low charges at congested airports and the high capital cost of significant capacity expansion. The UK's success story of air transport risks being compromised by a very narrow view of price control which fails to recognise the important differences which exist between utilities delivering homogenous life-essential products to the general public and the heterogeneous airport product which is sold to powerful and litigious multi-nationals.

This dangerous impasse can be avoided by regulatory reform and the most promising avenue for this would appear to be a package of price dedesignation (with powers to reverse), transparent monitoring of prices, quality, performance and investment and a much clearer and streamlined prohibition of anti competitive behaviour and complaints procedures.

References

Norman Ashford and Clifton Moore (1992) *Airport Finance*, Van Nostrand Reinhold

Centre for Study of Regulated Industries (January 1997) The UK Airport Industry - Airport Statistics 1995/6

Civil Aviation Authority (1996) 'Economic Regulation of BAA London Airports', CAP 664

E. Clayton (January 1997) 'A New Approach to Airport User Charges', *Journal of Air Transport*

Commission of the European Communities (1997) Proposal for a Council Directive on Airport Charges, (COM(97) 154 final)

S. Condie (1996) 'Airport Regulation 1995/6', Centre for Study of Regulated Industries - Regulatory Review 1996

Cranfield Institute of Technology (1996) 'User Costs at Selected European Airports'

Cranfield Institute of Technology - Research Report 3 (February 1995), The Economic Performance of European Airports

Regas Doganis (1992) *The Airport Business*, Routledge

International Air Transport Association (IATA), Airport Cost Evaluation Working Group Strategy Partnership Agreement (January 1998 draft)

International Civil Aviation Organisation (ICAO), Statements by the Council to Contracting Sates on Charges for Airports and Air Navigation Services, Montreal 1997 (Doc 9082/5)

Monopolies and Mergers Commission (MMC), Report on BAA London Airports (1996) MMC4

S.G. Warburg, Industrial Outlook, European Airport '96 (September 1996)

A. Wrobel (1997) 'Airport Charges in Europe', Institute of Air Transport; Study 97/1

Part 6

Ports and Channel Tunnel

CHAPTER 19

Privatisation and deregulation in seaports

Alfred J. Baird

Introduction

Defining port privatisation is seldom easy and requires careful analysis, principally because the *extent* of private sector involvement within ports can vary. In this paper, an analytical framework is proposed which aids analysis and understanding of the different models of port privatisation. These models are primarily based on the three essential *elements* involved in the effective functioning of seaports - port landowner, port operator, and port regulator. Either or all of these elements may be transferred from public to private sector. Thus, privatisation of each element ultimately affects the extent of private sector influence within a given port.

Applied to the world's major ports, the framework helps identify the most common model adopted for privatisation of seaports, as well as the least common. Further analysis reveals significant variations in key policy objectives which helps to explain the selection of a particular model. Finally, the advantages and disadvantages of the main models of port privatisation are discussed.

The analysis reveals that the model of port privatisation adopted in the UK appears to be somewhat unique. Essentially, the meaning of port privatisation in the UK is not the same as in most other countries.

Each model represents a possible alternative option as far as state decision-makers are concerned. The framework should therefore be of value, particularly to policy makers, but also to private sector port *operators*, as an aid in the decision making process. The framework could also be adapted for use in analysing the extent of privatisation in other transport industries, particularly aviation and railways.

Analysing the extent of private sector involvement in ports

What tends to complicate defining privatisation in any port is not only the different processes or methods which may be employed to bring about a transfer from public to private sector, but the fact that the *extent* of private sector involvement within and between ports can vary. Assessing the true extent of privatisation within a port therefore requires careful analysis of the distinct activities, roles and responsibilities (from here on in termed the *elements*) which it is possible to transfer from the public entity undergoing this process of transfer to the private sector. From analysis of the industry it is apparent that there are three distinct and essential elements of a port which may be privatised. These elements, which can be privatised either individually or collectively, are (Baird, 1995):

- port regulator
- port landowner
- port operator

Regulatory activities within a port will generally combine a mixed bag of duties and responsibilities, most of which will have been established by statute. The very nature of the regulatory role is somewhat appropriately reflected in the title 'port authority'. In essence, the entity holding responsibility for the regulatory function (i.e. the port authority) is to a very large degree in control of, and the power within, the area defined as the 'jurisdiction' of the port. Typically a port's regulatory duties and responsibilities will include:

- maintaining the conservancy function - ensuring navigable approaches to the port are well maintained and improved where necessary[1] and that conservancy fees are collected from port users;
- providing pilotage services and vessel traffic management or surveillance - generally termed VTS, ensuring the safe passage of craft within the defined area of jurisdiction;
- enforcing applicable laws and regulations - particularly matters relating to health and safety and which may also include control of pollution within the port estate. The port may even have its own police force, as in Felixstowe and Liverpool. Felixstowe Dock & Railway Company also has its own fire brigade;
- licensing port works - the port authority may be able to block or promote, if it so desires, certain developments within its defined area of jurisdiction;

- safeguarding port users' interests against the risk of monopoly formation, and the controlling of natural monopolies - the port authority is expected to operate in a fair and even handed way towards all port users.

In addition to the above, a port authority may also be expected to monitor the performance of the port, coordinate policy making with local and national government bodies, plan for future expansion, and market and promote the entire port and its facilities to users.

Port regulation is the one essential element of a port which one might least expect to be transferred to the private sector. Given that port regulation is inherently for the public good, it seems almost a contradiction that such a key element of any port should be given to a private company, which will inevitably imply that the port company regulates itself as well as perhaps competing companies within the port or port area. Perhaps not surprisingly, convincing arguments have been made in favour of the retention of public port authorities in order that the regulatory function is effectively carried out (Goss, 1990a; Garratt, 1995; Baird, 1995).

The second of the three elements which may be privatised is port land. Next to municipal authorities ports are very often the largest landowners within a city. Port estates can be very significant indeed, as in Antwerp where the port authority controls some 125km of berth length and occupies a land area in excess of 14,000 hectares, or in Los Angeles which has 45km of waterfront covering an area of 3,000 hectares. The dock estate within ports such as Rotterdam, Newcastle, Glasgow and Bremen can stretch for more than 40km from city to sea. Ports may have such a large land area that modern airports are easily contained within the dock estate, as in Genoa, New York and London. In the special case of obsolete city port areas (obsolete in a commercial maritime sense) such as London, Newcastle and Liverpool, the scale of redevelopment was so vast that separate and heavily funded Docklands Development Corporations had to be established by government in order to clean up and regenerate the areas concerned. In any event, regardless of whether the land area of a port is large or small, the key tasks a port landowner will need to undertake include (De Monie, 1994):

- managing and developing the port estate;
- conceiving and implementing port policies and development strategies;
- supervising major civil engineering works;
- providing and maintaining channels, fairways and breakwaters etc.;

- providing and maintaining locks, turning basins, berths, piers and wharves, and;
- providing or arranging road access to the port complex.

The third essential element of a port which may be privatised, operations, refers to the physical transfer of goods and passengers between sea and land. Traditionally this has meant stevedoring services but ongoing shipping industry changes have led to the intrusion into this area of the business by specialist terminal operators, and a number of shipping lines, especially in the container and ferry industries, who now attend to their own landside operations. Indeed, it could be said that the public port authority, certainly in the UK but also in a number of other countries, was in some respects unwillingly drawn into the port operations cargo-handling element of the business partly because of the nature of its role as 'employer of last resort'. In the UK, abolition of the National Dock Labour Scheme in 1989 and similar labour reforms in other countries (e.g. Italy, France, Spain, Chile etc.) paved the way for the public sector to withdraw from port operations, an area of the business in which it has seldom excelled in any case, with few notable exceptions (e.g. Singapore).

Port operations may also include, particularly nowadays, a range of added value activities within the port estate, such as warehousing, storage, packaging etc. Additionally, a free trade or industrial zone may be located within the port estate, perhaps involving some manufacturing or product assembly activities, as in Liverpool. A range of specialist port services might also be considered under the heading port operations. Such services may include pilotage, towage and even ship repair (many ports having dry-dock facilities within their estates).

Upon establishing which of these three essential port elements or activities are the responsibility of public or private sector entities, it is possible to ascertain the extent of privatisation within any given port. There are four relatively straightforward possibilities (Table 19.1).

First is what is termed here the PUBLIC port, which may also be referred to as a 'service' or 'comprehensive' port, and in which, by implication, there is no private sector involvement. In this instance all three elements - regulator, landowner, and operator - are the responsibility of the state. Examples of PUBLIC ports can still be found in Singapore, Israel, India, and in certain African states.

Generally the first element of port activity to be transferred to the private sector is port operations and here we have what might be termed the PRIVATE/I model, with the 'I' denoting that one element of the port has been privatised. This type of arrangement is also referred to as a 'landlord' or 'tool' port. In this model port land is still in public ownership and regulatory activities are also the responsibility of the public sector. There are many examples of this type of arrangement, especially throughout North American and European ports, in which terminals are generally leased to the private sector.

Where two elements, operations and property rights, are controlled by the private sector, we might describe such a port as being a PRIVATE/II model. In this instance the private sector is dominant, with the public sector via the port authority retaining only control of regulatory matters. Single-user bulk oil, coal, ore, and aggregate terminals often correspond to this model, but it is generally not considered appropriate in large multi-user ports. However, certain large multi-user ports in the UK appear to conform to this model. For example, Tilbury, Felixstowe and Harwich are owned and operated by private companies yet there remains for each of these ports a public sector port authority in control of the navigable approaches (i.e. with responsibility for VTS, pilotage, maintaining the entrance channel etc.).

Table 19.1 Key port elements: privatisation options

Port Models	Port Regulator	Port Landowner	Port Operator
PUBLIC	public	public	public
PRIVATE/I	public	public	**private**
PRIVATE/II	public	**private**	**private**
PRIVATE/III	**private**	**private**	**private**

The final possibility, referred to here as the PRIVATE/III model, is where all three elements - regulatory, land and operations - become the responsibility of the private sector. With this model the state has virtually no involvement in the port other than perhaps to react in certain instances such as sub-standard vessels, pollution or serious accidents. To all intents and purposes the PRIVATE/III model implies the complete withdrawal of the state from port planning or provision. This model therefore assumes that the market will identify opportunities for new or enhanced ports, that such activities are not the role of the state, and that the private sector alone will provide new ports as and when they are required. Ports in the UK would appear to be virtually the only examples anywhere of this 'fully' privatised model, which lies at the extreme opposite end of the spectrum from the PUBLIC port.

While the framework provides a useful guide as to the main options which appear to face the state in helping define a ports policy, a degree of caution is necessary. Inevitably, not all ports will fit neatly into either of these four models. Ports will very often exhibit a more complex mix of public and private sector roles and responsibilities and this requires closer analysis. Consider the case of Hamburg, which is a PUBLIC port in the sense that the port is owned by the state, where the port authority functions are undertaken by the state, and in which the largest stevedoring company in the port (HHLA) is owned by the state. But in Hamburg there are also a number of terminals leased to private companies who employ their own cargo-handling personnel. This means that Hamburg is both a PUBLIC and a PRIVATE/I port.

The publication *Cargo Systems* applied the framework to establish the extent of private sector participation within the world's top 100 container ports (Cass, 1996). Updating these findings to the present time, Table 19.2 demonstrates that 88 of the top 100 container ports conform to the PRIVATE/I model in which only port operations are carried out by the private sector, with the public sector retaining (ultimate) property rights over port land and fulfilling the port authority regulatory functions. The PRIVATE/I model is therefore by far the most common arrangement among the top ports.

Only seven of the top 100 container ports appear to conform to the PUBLIC model, these being mainly South African and Israeli ports, plus Singapore. Several of these ports, particularly in South Africa, are expected to progressively transfer responsibility for cargo-handling operations to the private sector. If this happens these ports will also move towards the PRIVATE/I model.

Just two of the world's top 100 container ports, Tilbury and Felixstowe, conform to the PRIVATE/II model. These ports are privately owned and operated but the regulatory function is carried out by a separate independent public port authority in both cases. Actually, both Tilbury and Felixstowe have been allowed certain regulatory powers but these are restricted to areas within their respective port estates only and do not extend to the navigable approaches outside the port; thus, it could be said that the regulatory element relating to these ports has been partially privatised. Only three of the world's top 100 container ports, Southampton, Liverpool and Thamesport, conform to the PRIVATE/III model. In each of these three ports the private sector therefore control all three key elements - land, operations, and regulation.

Table 19.2 Extent of privatisation within the world's top 100 container ports

Port Models				
	PUBLIC	PRIVATE/I	PRIVATE/II	PRIVATE/III
Number of ports	7	88	2	3

It appears, therefore, that the UK has adopted a unique approach, extending the influence of the private sector within its ports industry further than any other country has done or indeed intends to do. Many other ports in the UK also now conform to the PRIVATE/III model, including Tees & Hartlepool, Forth, Clydeport, Dundee and Medway, plus the 23 ports owned by Associated British Ports, and the former railway ports now owned by Sweden's Stena Line and by Sea Containers. Combined, these ports account for the majority of the UK ports industry.

In essence the UK approach could be defined as *real* privatisation (although one might also use the term *extreme* privatisation) whereas the alternative PRIVATE/I model which is preferred almost everywhere else merely results in private sector *participation* within the ports industry. Clearly, however, outside of the UK private sector participation in virtually all ports is confined to what is arguably the most important element of a port's activities as far as port users are concerned, i.e. port operations.

Objectives and consequences of port privatisation

As the framework demonstrates, a clear distinction can be made with regard to the extent of private sector involvement in ports. However, the decision regarding which model to adopt will ultimately depend on the specific circumstances and priorities in the country concerned. A useful starting point in attempting to discuss this issue is to establish the original objectives set out for what privatisation of ports was intended to achieve in the first place. For example, in relation to most countries adopting the PRIVATE I model (i.e. virtually every country except the UK), generally the following consistent combination of objectives are relevant (Rajasingam, 1994; De Monie, 1995):

a) *A realisation that expensive and inefficient ports constrain trade.* Note the emphasis placed on the need to expand trade. Countries' know that inefficient ports, whether through outdated work practices, obsolete facilities or a combination of both, can stall economic development. This is one of the main reasons why governments' are actively seeking private sector participation in their ports sector usually via lease, concession, BOT, or joint venture arrangements.

b) *The need to introduce efficiency and know-how of the private sector.* Increasing specialisation and standardisation in the shipping and transport industry has resulted in the formation of expert multinational terminal operating companies, often subsidiaries of shipping lines, who know precisely how to meet the ever changing and increasingly demanding needs of port users. These firms benefit from economies of scale and from learning through their wide geographic scope of activities which makes them fully aware of industry best practice on a global basis.

c) *Pressure to reduce demands on the public sector budget.* This does not necessarily mean that the state withdraws entirely from investing in its ports. Many examples of the PRIVATE/I model demonstrate that the state may continue to be responsible for provision of new infrastructure but that the private sector provide the superstructure such as cranes and terminal buildings. This is more a partnership approach in which the state still retains at least some degree of control over its ports industry. Even in BOT arrangements the state will generally retain its public port authority, may also impose restrictions on future tariff increases, share in port profits, stipulate minimum throughput targets, and property rights ultimately revert to the state at the end of the contract period.

d) *To reduce expenditure on port labour by removing the state from port operations.* Port operations is the one element of the ports business where the state is usually found to be weakest, with the odd exception (e.g.

Singapore). Part of the reason for this could be that the state generally seeks to maximise employment across the economy whereas the private sector is motivated to restrict the number of employees, hence over-manning in publicly operated ports leads to additional costs for port users or in a requirement for public sector subsidy, or both. This issue is inevitably related to the question of labour reform; indeed, any change from public to private sector employer within ports is meaningless if outdated working practices are left unchanged. Labour reform is a prerequisite for the effective privatisation of port operations (Setchell, 1994).

While there is almost universal appreciation of the wider economic and trading benefits derived from having a sophisticated and efficient ports industry, the objectives set out by the former UK Conservative Government for extending the influence of the private sector in its ports tended to concentrate on quite different issues. Analysis of these contrasting objectives goes some way to explain the very different approach adopted in the UK compared to elsewhere, in particular the motivations for the outright sale of port land and the transference of port authority regulatory duties and responsibilities to the private sector, in addition to the transfer of port operations. It has been suggested that privatisation of the British ports industry was based around the following objectives (Thomas, 1994):

a) *To improve management.* The UK government believed that private sector port management would be more effective than public sector management. In reality, many UK ports were sold to their existing managers, or in other words to former port officials leading management buy-out (MBO) teams, the government having previously made a commitment to favour such bids. The ports were not sold to specialist port operators who might be regarded as the *real* private sector, and who would be able to bring new skills and techniques learned over time across their global operations to the UK ports industry. Consequently, it was very difficult to see port management being improved if the very same former public sector managers were left intact post privatisation.

b) *To improve efficiency.* The private sector was believed to be more efficient than the public sector. However, selling a port does not in itself improve efficiency. Subsequent efficiency and productivity improvements experienced throughout UK ports, according to shipping lines, came about as a consequence of the abolition of the National Dock Labour Scheme in 1989 (Garratt, 1995; John, 1995; Ross, 1995) and had little if anything to do with actually transferring ownership of the ports from public to private sector. Indeed, at many former public sector ports which already leased terminals to private sector companies, there was no change in the port operator as such, the only change was with respect to the port landlord which went from public to private. Under these circumstances it was difficult to see how a change of landlord would improve port efficiency.

c) *To raise revenues for government.* A major motivation behind the sale of state-owned assets was to help fund a large public sector deficit and to pave the way for tax cuts. Government sought to achieve these objectives by disposing of state assets, particularly the utilities, in their entirety. However, there are a number of weaknesses with regard to such a strategy as far as ports are concerned, including (Baird, 1995):

- Many of the ports sold consisted of mature and in several cases obsolete facilities located within traditional cityports. UK port privatisation was not about building new advanced ports or improving existing infrastructure (as is generally the case elsewhere) which would ease the flow of trade and bring economic benefits, it was about disposal of existing ports, a number of which were no longer of relevance in a commercial maritime sense;
- Given the nature of port infrastructure investment (i.e. high capital cost, long term payback, plus wider economic and social benefits), the private sector *alone* will seldom build entirely new ports, and whenever the private sector does decide to build new ports on its own, difficulties may be encountered.[2] This therefore raises the question of whether the private sector will in future build new ports, as and when they are required, without any state support;
- Successor companies (i.e. often comprising former port officials leading MBO's) to the public ports were permitted to purchase the assets at unbelievably low prices. Successor companies thus began life in an asset-rich, low-debt condition and this, combined with an established port customer base, steady profits, and local monopoly position, has made them very profitable indeed.[3] These advantages have tended to make 'new' port management appear more successful than they really are;
- Former public sector port managers do not have the skills necessary to effectively deal with mature and obsolete city ports which often require comprehensive regeneration. Such ports should ideally have been transferred to local authority or other public sector (e.g. development corporation) ownership so that a long term redevelopment plan could be formulated in partnership with private sector developers and the

local communities concerned.[4] Successor companies have also benefited from increases in waterfront land values. Waterfront land values tend to be high compared with other areas of a city and this has given the private sector successor companies an added bonus. Indeed, where commercial shipping activity has diminished, the ports concerned are now little more than real estate development opportunities (e.g. London, Manchester, Leith, Dundee, Glasgow, Cardiff etc.).

d) *To encourage share ownership.* It was the former Conservative Government's belief that managers and employees would be motivated to ensure the success of their business once they had acquired shares in it. In reality, shares in the ports were acquired very cheaply (to ensure take-up) and as soon as the first offer arrived valuing the port more accurately (i.e. much higher than government), managers and employees were tempted to dispose of their shares in return for significant personal gains, as happened at Medway, Clydeport and Tilbury. A further consequence of these share transactions is that ports in the UK are now treated just like any other commodity or business which can be bought or sold with the result that investor interest inevitably becomes short term rather than long term. This is entirely at odds with the essential long term nature of port investment and must ultimately affect a private port's ability to invest.

The outcome of port privatisation in the UK should be regarded as wholly unsatisfactory and the particular models (and indeed the methods) employed would almost certainly have been politically unacceptable in most other countries[5] (Suykens, 1996). The privatised ports are profitable (many were profitable before privatisation) but this is a reflection of their low debt beginnings coupled with other advantages such as local and estuarial monopolies, real estate gains, and self-regulation. This is more a fortuitous, fabricated environment rather than a free market environment. Yet it seems inevitable that new ports will need to be built at some point in the future to take account of subsequent changes in shipping and transport technology (Frankel, 1987). The question remains, will the new private port companies build substantial replacement ports without any public sector input? Given the nature of port investment this must be considered highly unlikely.

While there are obviously disadvantages with implementation of the PRIVATE II and III models, as the UK experience demonstrates, the more favoured PRIVATE I model demonstrates a number of significant benefits, including:

a) *Private sector investment.* Investment in ports means either building new facilities or upgrading existing facilities, as distinct from simply acquiring ownership of existing ports (as in the UK). In a typical concession arrangement, for example, the private sector may agree to operate an existing port or terminal and to upgrade the facilities, as Hutchison International Terminals (HIT) is doing in Panama. In the latter scenario the public sector also receives regular payments from HIT. With a BOT arrangement the private sector is permitted to build and operate the port facility which ultimately reverts to state ownership at the end of the contract period. Alternatively, a joint venture between public and private sector splits the costs and risks of a new port development and this appears to be one of the main approaches currently used for port investment in China. Clearly, these privatisation methods can help facilitate an advanced port system whilst minimising state expenditure, and without any need to transfer ultimate property rights or regulatory control from the public sector.

b) *Improved infrastructure.* Public port authorities, local civil engineers and even governments will very often be unaware what exactly is meant by 'state-of-the-art' in terms of port facilities. Requesting competitive bids from specialised terminal operators who have the required knowledge from building and/or operating facilities internationally gives a high degree of certainty that a port or terminal will be provided which is geared to the needs of the market. Crucially, access to advanced and efficient ports is essential in order to aid the flow of trade.

c) *Private sector management skills and expertise.* Global shipping lines are in a position to compare port performance on a worldwide basis. This means that all ports have to provide a high and consistent standard of service in order to remain competitive. In many cases the local stevedore (public or private) may not have the level of sophistication necessary, particularly in terms of information technology, management capability, working practices and training, to adequately meet the needs of port users. Thus it may be preferable to bring in the specialist port operator, perhaps in a joint venture with local actors.

d) *Downsizing of port authorities.* Provision of port operations and services is where the real employment is in ports and once these tasks are contracted out to the private sector this means the port authority can get by with much reduced staffing levels. Genoa Port Authority reduced its workforce from several thousand employees to just a few hundred after a series of terminal concessions were granted to the private sector and the workforce was transferred to the new operators. A smaller port authority may be virtually self-financing, but even if it is not it is still advisable to retain such a body if only to take account of issues such as the public good aspect of ports and externalities (Goss, 1990b). Port authorities are also able to exert necessary controls to counteract the tendency towards the more 'rapacious' practices among the private sector and may impose limits on port tariff increases in order to protect the economy, as is the practice in a number of countries including China and India. However, there will also be a number of other important functions for a port authority to carry out, including:

- marketing of the entire port;
- ensuring integration with other transport modes;
- investment and capacity planning;
- promoting private/public partnership, and;
- monitoring competitiveness and efficiency across the port.

Conclusions

Broadly speaking there are three essential elements of a port which may be privatised - port landowner, port operator, and port regulator. The extent of privatisation within ports will therefore vary depending on which of these elements are transferred from public to private sector. The framework outlined in this paper suggests there are three possibilities:

1. Privatisation of one element - **port operations** - referred to here as the PRIVATE/I model;
2. Privatisation of two elements - **port operations and port land** - referred to here as the PRIVATE/II model;
3. Privatisation of all three elements - **port operations, port land, and port regulatory responsibilities** - referred to here as the PRIVATE/III model.

Applying the framework to the world's top 100 container ports reveals that nearly 90 per cent of ports have adopted the PRIVATE/I model which suggests that port operations is the most important element requiring transfer to the private sector. Of the top 100 container ports, only ports in the UK have adopted the more extreme PRIVATE II/III models.

The underlying reasons for adoption of these different strategies can be traced back to the initial objectives pursued. Countries adopting the PRIVATE/I model tend to stress the need for efficient and advanced ports to help expand trade, to allow the state to withdraw from port operations, and to reduce pressure on the public sector budget from port expenditure. Experience of the PRIVATE/I model suggests it is possible for the state to lever private sector investment in its ports without losing either control of its ports industry or ultimate property rights in respect of port land. Many recent examples of the PRIVATE/I model in practice seem to suggest a reasonably positive outcome for private port operators, for the state, and for port users.

On the contrary, the objectives set out for port privatisation in the UK are substantially different from most other countries, and the PRIVATE II/III models adopted in the UK demonstrate a number of fundamental weaknesses. To begin with, the sale of either port land or regulatory duties and responsibilities to the private sector has a particularly dubious connection with improved port efficiency. Any recent efficiency or productivity gains experienced in UK ports have been largely as a result of the abolition of the National Dock Labour Scheme in 1989. A key objective in the UK was to raise finance for the Treasury from the sale of state-owned assets, yet this has nothing at all to do with creating new advanced port infrastructure to aid the flow of trade and boost the economy.

In addition, permitting public port officials and employees to acquire shares in state owned ports is rather more an ideological objective. Indeed, the subsequent and hasty disposal of these shares in return for, in many instances, vast one-off gains for former port officials (and their venture capital backers), represents an unnecessary loss of port asset value to the state and to the taxpayer. Furthermore, subsequent trading in port shares, like public trading in other shares, is very often short term, and this goes against the grain of port investment which by its very nature is long term. This will inevitably have implications in terms of future large scale capital investment in ports and indeed may well jeopardise such investments.

The complete withdrawal of the state from its ports industry (i.e. the PRIVATE/III model) is today highly unusual and is largely a phenomenon unique to the UK, a result of the *laissez faire* approach adopted by the former Conservative Government. Furthermore, it remains that there has been a complete absence of convincing arguments in favour of such a comprehensive withdrawal and any perceived advantages (from such a policy) are, according to the evidence, clearly misguided. Indeed, there are good reasons (e.g. public goods, externalities etc.) for a partial reversal of this process whereby port regulatory duties and responsibilities are transferred back to reconstituted, self-funding, user-oriented public sector port authorities.

References

A.J. Baird (1995) 'Privatisation of trust ports in the United Kingdom: Review and analysis of the first sales', *Journal of Transport Policy*, Vol 2, No 2, pp. 135-143.

S. Cass (1996) *Port Privatisation - Process, Players and Progress,* London: Cargo Systems/IIR Publications.

G. De Monie (1994) 'Mission and Role of Port Authorities', *Proceedings* of the World Port Privatisation Conference, London, September.

G. De Monie (1995) 'Restructuring the Indian ports system', *Journal of Maritime Policy & Management*, Vol 22, No 3, pp. 255-260.

E.G. Frankel (1987) *Port Planning and Development,* New York: Wiley.

M. Garratt (1995) 'Port Policy and Operational efficiency', *Proceedings* of the Port Privatisation Conference, Scottish Transport Studies Group, Edinburgh, September.

R.O. Goss (1990a) 'Economic Policies and Seaports - Part 3: Are Port Authorities Necessary?' *Journal of Maritime Policy & Management*, Vol 17, No. 4, pp. 257-271.

R.O. Goss (1990b) 'Strategies for Port Authorities', *Journal of Maritime Policy & Management*, Vol. 17, No. 4, pp. 273-287.

M. John (1995) 'Port Productivity: A User's View', *Proceedings* of the UK Port Privatisation Conference, Scottish Transport Studies Group, Edinburgh, September.

H. Ross (1995) 'Ports and Innovative Shipowners: Encouraging Enterprise', *Proceedings* of the UK Port Privatisation Conference, Scottish Transport Studies Group, Edinburgh, September.

R. Setchell (1994) 'Labour Reform: Critical first step and pre-requisite to competitive commercial strategy', *Proceedings* of the World Port Privatisation Conference, London, September.

F. Suykens (1996) 'The Future of European Seaports', in L. Bekemans & S. Beckwith (eds), *Ports for Europe*, Brussels: European Interuniversity Press.

B.J. Thomas (1994) 'Privatisation of UK Seaports', *Journal of Maritime Policy & Management*, Vol 21, No 2, pp. 135-148.

M. Rajasingam (1994) 'Maintaining port services and standards while making the transition to the private sector', *Proceedings* of the World Port Privatisation Conference, London, September.

Notes

1. Improving port access by deepening or widening channels, removal of wrecks etc.
2. Privately built Thamesport was sold by its bankers after only a couple of years in operation for approximately one third of the original capital cost and at significant loss to the original investors. Felixstowe has had five different owners during the past 40 years, none of whom, upon selling the port, would appear to have fully recovered the total capital investment costs.
3. Privatised ports in the UK consistently achieve profit/turnover ratios approaching 30 per cent, much in line with other state sales (e.g. BAA). Compare this with the low margins in shipping generally, with many container lines in particular recording profit ratios of 5-6 per cent or less.
4. Privatised ports have been known to lodge inappropriate planning applications, such as for niche/high quality shopping centres which local council's claim will lead to traffic problems in urban areas. Councils cannot include large cityport land areas in their strategic plans and this has all sorts of implications for planners. Obsolete port land may represent an opportunity to provide good quality affordable housing but instead often high quality expensive housing is provided. Furthermore, it is unclear what many of the privatised ports intend to do with contaminated land they have inherited.
5. In most countries port land is regarded as being of national strategic importance, particularly militarily and economically. Indeed, it is very often unconstitutional for governments to sell port land, such is its significance.

CHAPTER 20

Nostra Culpae: Forecasts and out-turns of cross channel competition between the ferries and the tunnel

Stefan Szymanski[1]

Introduction

It is sometimes said that we learn more from our mistakes than from our successes. If this is true, many of us can look forward to a lifetime of learning from the Eurotunnel experience. This paper is about the discrepancies between out-turns and forecasts made by and on behalf of Eurotunnel PLC. Since the original responses to the invitation to tender in 1985, Eurotunnel and its advisers have made numerous forecasts, but this paper will focus on the differences between the actual revenue figures of Eurotunnel in 1996 and Eurotunnel's own forecasts made in 1987 when the main share flotation occurred, and in 1994 when the second rights issue took place. The 1987 forecasts were pivotal to raising the equity finance for the project, without which the construction programme proper could never have begun, while the 1994 forecasts coming so soon before the 1996 outcomes, illustrate dramatically some of the pitfalls of traffic forecasting for commercial ventures.

The paper will also consider predictions made by Kay, Manning and Szymanski (KMS) in a paper written in 1989. The purpose is not merely to document errors, but to explain why the errors arose. In general there were two broad reasons for the errors in revenue forecasting, one related to mistakes relating to underlying variables which might be regarded as 'model parameters' such as the growth rate of cross channel transport, consumer perceptions about the relative attractiveness of competing services and the cost structure of the industry. Whilst these parameters were sometimes the subject of forecasting exercises themselves, in most cases they were assessed by rule of thumb based on past experience. The other kind of mistake, which is perhaps more interesting, concerned the models used themselves. These models were developed to forecast the outcome of a process of which there was no prior experience, the direct competition *in the private sector* between a fixed link and a ferry service.

However, this paper will not deal with the greatest forecasting error of them all: the cost of construction. Originally estimated at under £5bn, the final cost turned out to be around £10bn (for a discussion of this see Szymanski (1995a). It might be thought that this mistake in itself rendered the project financially unviable, but that is probably not the case. The net present value of the revenue stream forecast by Eurotunnel in 1987 exceeded £10bn using a 5 per cent real interest rate (even allowing for the one year delay in opening). The problems of Eurotunnel have arisen from the joint misery of construction cost overruns and revenue undershoots. For example, Eurotunnel was originally forecast to generate £908m in revenues in 1996, whereas the out-turn was £448m. This paper examines why the revenue forecasts were so wrong. It also considers separate forecasts made by KMS in 1989 which suggested that Eurotunnel would obtain a dominant market share, around 80-90 per cent of the short straits market. In fact Eurotunnel's market share has been around the 50 per cent level.

The conclusion of the paper is that the errors of Eurotunnel can largely be attributed to a poor traffic modelling strategy adopted by the company, while the errors of KMS can be attributed to mistaken input assumptions. Explaining why Eurotunnel has undershot its revenue targets is of more than mere academic interest. Following an MMC inquiry which was completed in April 1997 the British government gave the go-ahead in November to a joint venture between Eurotunnel's main rivals, P&O and Stena Line. The joint venture will pool the companies' operations on the short sea routes (Dover-Calais, Dover-Zeebrugge and Newhaven-Dieppe) subject to some undertakings primarily in relation to travel agents. The ferry companies have argued for many years that such a joint venture should be permitted because the Channel Tunnel would prove to be such a powerful competitor. In practice, the Tunnel has been much less effective than many predicted.

This paper will mostly use the out-turn figures for 1996 as a basis for comparison. On 18 November 1996 a major fire occurred inside the Tunnel causing significant damage. This necessitated interruptions to services, major reconstruction and a re-evaluation of safety procedures. However, Eurotunnel returned to full services in the second half of 1997. The paper ignores the issues relating to the financial reconstruction of the Tunnel. In principle, the financial structure of the Tunnel operating company should have no effect on its day to day operations or its commercial policies. This is also a working assumption of this paper.

The paper is set out as follows. In the next section the main economic differences between Eurotunnel's forecasts and the out-turn revenue figures are described. The following section discusses the problems associated with Eurotunnel's traffic modelling procedures. The fourth section deals with the modelling of KMS, while the final section considers some implications for future competition and draws some conclusions.

Eurotunnel's forecasts and out-turn

Revenues

Eurotunnel has three main sources of income. Firstly, the through rail services such as Eurostar pay a fixed fee plus usage charges for running trains through the Tunnel. Secondly, Eurotunnel's own shuttle services generate revenues from passenger traffic, mostly car passengers but also some coach traffic and thirdly the shuttle service obtains revenue from freight traffic in the form of road haulage vehicles. Although Eurotunnel does not fully break down its sources of income, the 1997 MMC report contains a number of break-downs which make it possible to piece together Eurotunnel's operating performance in 1996.

Eurotunnel services began in 1994, over one year late. Even in 1995 services were not fully operational. However, by 1996 the Tunnel was meant to be operating at more or less full capacity. The 1987 forecast of £906m of revenue for this year was for a service which was in its fourth year of operation, while the actual figure was £448m. From the 1997 MMC report (p. 89) we know that Le Shuttle's income was only £200m and income from the Railways (tolls, minimum usage charges and contributions to costs) was £198m. A notable success of the 1987 forecast was its inflation forecast. It predicted that the RPI would stand at 163 (1987=100), an error of only 5 per cent.[2] The 1987 forecast was broken down into £368m from through rail services and £463m from shuttle services, the remainder being £77m of income primarily from duty free sales. The 1994 forecast was somewhat less optimistic but still anticipated £737m in revenues, broken down between £268m from through rail services and £414m from shuttle services, and £55m of ancillary income.

In Table 20.1 the figures are expressed as a percentage shortfall from the 1987 and 1994 forecasts. One might argue that the optimism in 1987 can be excused, particularly in the light of a number of working assumptions made at the time, such as the existence of a high speed rail link from London to the entrance of the tunnel which would have enhanced the demand for through rail services. However, it is striking that the 1994 forecasts were little closer than those made nearly a decade earlier, despite their proximity in time to 1996.

Table 20.1 Forecast and out-turn revenues for 1996

	Shuttle services (£m)	Through rail services (£m)	Ancillary income (£m)	Total (£m)
1987 forecast	463	368	77	906
1994 forecast	414	268	55	737
out-turn	200	198	50	448
shortfall on 1987	57%	46%	35%	51%
shortfall on 1994	52%	26%	9%	39%

Note: all figures adjusted for inflation

Volumes

Eurotunnel's published traffic volume forecasts took the form of numbers of passengers trips and tonnes of freight carried. Thus in 1987 they forecast that in 1993 they would carry 29.7m passengers and 14.8m tonnes of freight, while the market would grow at 3.4 per cent per annum for passengers and 3.8 per cent for freight, implying 32.8m Eurotunnel passenger trips in 1996 and 16.6m tonnes of Eurotunnel freight. In 1994 Eurotunnel forecast a total of 21.8m passenger trips, broken down between 11.1m shuttle users (mainly cars and coaches) and 10.7m rail (Eurostar)

Table 20.2 Forecast and out-turn volumes for 1996

	Shuttle passengers (millions)	Through rail passengers (millions)	Total passenger trips (millions)	Shuttle freight (million tonnes)	Through rail freight (million tonnes)	Total freight (million tonnes)
1987 forecast			32.8			16.6
1994 forecast	11.1	10.7	21.8	9.4	6.6	16.0
out-turn	8.8	4.6	13.4	6.4	2.4	8.8
shortfall on 1987			59%			47%
shortfall on 1994	21%	57%	39%	32%	63%	45%

passengers by 1996. They also forecast 9.4m tonnes of freight using the shuttle as well as 6.6m tonnes of through rail freight. From the MMC report (p. 83) we know Eurotunnel carried 2.4m tonnes of rail freight, while the shuttle services carried 2m cars, 58,000 coaches and 519,000 lorries. Shuttle passengers in total were 8.8m. Eurostar through rail services carried 4.6m passengers. To calculate the volume of shuttle freight we use a standard assumption that the average truck carries about 12.5 tonnes of freight, implying the shuttle carried about 6.4m tonnes of freight in total.

Prices

Prices vary by time of day, planned date of return and time of year. Some fares are also discounted or sold through block booking arrangements with tour operators. A measure of price can be obtained by calculating the average yield: revenue per unit carried, but this requires a consistent breakdown of revenues into traffic categories which is not always available. Yields also include ancillary income from duty free sales and other trip related income.

In 1987 Eurotunnel stated that they anticipated that a car carrying a normal passenger load would be charged £55.80 in 1993, in 1987 prices, equivalent to around £83.70 in 1996 prices. Similarly they forecast a normal freight lorry would be charged £118.20, equivalent to £177.30 in 1996 prices. According to Eurotunnel, this represented a decline of 5 per cent in real terms from 1986 ferry passenger fares and 20 per cent lower for freight tariffs. No specific price forecasts were included in the 1994 prospectus, but the following statement was made: ‘Eurotunnel projects ferry yields to be about the same in 1995 as in 1994 and to rise somewhat in 1996. The Directors expect Le Shuttle’s yields to remain above those of the ferries in the years 1994 to 1996’. To express this forecast in terms of prices requires some information on ferry yields. In 1994 the ferry companies estimated that short sea ferry routes generated £426m of income from passenger services (MMC p. 96). This revenue divided between car traffic, coaches and unaccompanied foot passengers. Taking this last element to be negligible, the total revenue was divided among 3.9m passenger car unit (PCU) equivalents (MMC p. 121; coaches are rated as a multiple of PCUs). With the further assumption that coach and car PCU yields are roughly equal, this implies a yield of around £108 per car, including revenue from on board sales. This is probably an overstatement since while coach prices are cheap, coach passengers probably have a higher average spend that car passengers). However, since Eurotunnel argued their yields would be higher than ferry yields, the approximation is reasonable. For freight traffic we know that Anglo-Continental traffic amounted to 3.113m units (i.e. lorries) in 1994 (MMC p. 108) while the ferry companies estimated that the total value of this market in 1994 was £570m (p. 96) implying an average ferry yield of £183 per freight lorry.

Actual freight lorry prices are not published and tariffs are subject to individual negotiation between the hauliers and Eurotunnel, but the car yield can be compared to published brochure tariffs. In 1996 Eurotunnel started with a price scheme with four bands, ranging from the superpeak price of £164 for a crossing down to £63 for off peak travel. A five day return was priced at roughly the same level. However, in June the company slashed its prices to £129 for a peak return (implying a price of only £64.50 per crossing) down to £99 return for an apex fare. Five day returns were priced at £69 at peak and £59 off peak. In addition Eurotunnel began to match the very low price promotions offered by the ferry companies. Clearly, for the first half of the year yields on most trips must have been well in excess of the average figure calculated in the table below, while in the latter half of the year yields must have been below this figure.

Table 20.3 Forecast and out-turn yields 1996

	Cars £	Lorries £
1987 forecast	84	177
1994 forecast	108	183
out-turn	53	133
shortfall on 1987	37%	24%
shortfall on 1994	51%	27%

Figures on actual yields can be derived from the MMC report. Ticket revenues from car journeys are stated as £68.4m in 1996 (p. 89) and from shuttle freight £68.9m. Given the volume figures mentioned above, this implies an average price of £33 per car and £133 per lorry. However, car yields are higher given duty free and other retail sales. In total Eurotunnel sold £105.1m of goods to passengers (MMC p. 193) at a cost of £49.7m yielding net revenue of £55.4m. With 8.697m passengers in total (MMC p. 117) this implies a yield of £6.37 per passenger carried. Given the number of coaches and cars carried (see above) and assuming an average coach carries 40 passengers, the implied number of passengers per car was 3.1, generating a yield from additional sales of around £20 per car.

The shortfalls in the estimated yields are similar in size to the shortfalls of traffic volumes. However, it is notable that the 1994 forecasts were actually worse than the 1987 estimates, particularly in the case of car yields.

Eurotunnel's model

It is clear from the previous section how Eurotunnel's forecasts failed. Revenues were much smaller than forecast because Eurotunnel anticipated a much larger volume of traffic than was actually achieved and prices were much lower than anticipated. This suggests that Eurotunnel laboured under some fundamental delusions as to the way the short sea market operates, despite their extended opportunity to study it. In fact, the quality of their forecasts for volumes scarcely improved at all, while the yield forecasts actually got worse. To understand how this came about it is necessary to examine the structure of Eurotunnel's traffic forecasts.

The traffic forecasts were prepared by the Traffic and Revenue Consultants (TRC), who were responsible both to the banks and the company. They developed the overall modelling strategy, although for the 1994 forecasts Eurotunnel made some slight amendments which were printed in the Rights Issue prospectus alongside the TRC forecasts (the former have been quoted here). The TRC modelling strategy is described in Roberts and Vougloukas (1993). The methodology adopted involved the following steps:

a) definition of the potential market by region and by type of traffic;
b) forecasting market growth using a regression model with traffic as a function of GDP, consumer expenditure and exchange rates;
c) calculating market share for Eurotunnel using a route allocation model based on direct cost and cost of time;
d) estimating the size of additional traffic created by the existence of the Tunnel;
e) from the previous steps calculating revenues, including ancillary income from duty free sales etc.

The market in which Eurotunnel competes was defined very broadly, including for example air passengers to Spain and Yugoslavia. However, the growth forecasts did not turn out to be particularly optimistic. For example, Eurotunnel had forecast in 1987 that passenger traffic would grow at an average of 4.2 per cent per year until 1993, and freight traffic would grow at 4.3 per cent per year. In fact traffic through Dover actually grew by 4.7 per cent (passengers) and 5.3 per cent (freight lorries) over this period.

Clearly then, the key failures of the Eurotunnel modelling exercise related to (a) forecasting market share for given prices and (b) forecasting prices generated in a competitive environment.

Forecast market share

Route allocation models such as those used by Eurotunnel are based on the notion of generalised cost. Generalised cost means not only the direct financial costs of a trip (fare, fuel costs etc.) but also the (opportunity) cost of the time spent travelling. A route which takes longer has a higher cost, *ceteris paribus*, if the journey time is greater. Travellers are assumed to choose the route which has the lowest generalised cost. The value of time can be estimated from actual data for routes involving different journey times and costs. A route which involves a longer travel time but a lower direct cost will be selected by all travellers whose value of time falls below a threshold level, whilst all travellers with a value of time above the threshold will choose the faster route. In fact, based on the share of traffic for each route, the TRC fitted a distribution for the population's value of time.

In order to use this methodology to forecast Eurotunnel's market share the TRC needed to make some assumptions about the direct cost of Eurotunnel services compared to their competitors including the prices charged. Prices were therefore an input assumption of the modelling process rather than an output from the modelling process. The TRC believed that Eurotunnel possessed a significant time advantage over the ferries (and that the through rail services had a significant time advantage over the airlines). This in turn would lead the majority of customers to prefer Eurotunnel over the ferries, *ceteris paribus*. Thus the shuttle should take around 35 minutes to travel through the Tunnel, compared to around 70 minutes which is the fastest crossing time for a conventional ferry at Dover. There are two principle reasons why the modelling failed:

1. The time advantage of Eurotunnel is relatively small, and therefore may not play a very large role in consumer choice. Thus the average car trip might be as long as 6 hours from origin to destination, against which a saving of 35 minutes might not be that important. Critics pointed out from the start that Hovercraft had offered a similar time saving for it services which had operated for more than 20 years but had failed to acquire a significant market share.

2. The ceteris paribus assumption of the route allocation model is not valid. The choice between the Tunnel and the ferry services is not based on time and direct cost alone. Other factors are important in terms of consumer choice. Some people are claustrophobic and will not travel in a Tunnel under any circumstances, some are 'thalassophobes' who will do anything to avoid travelling on a boat. Some travellers, particularly on long car journeys, use the crossing as a travel break, so that the relative comfort of the service and the availability of catering services is important. Until this year Eurotunnel has also refused to allow pre-booking through travel agents, whereas a large proportion of travellers have traditionally purchased tickets by this method, and may have failed to consider Eurotunnel as a choice for this reason.

Forecast prices

The TRC almost completely ignored the role of competition in setting prices. In effect they imagined that the response of the ferry companies to the opening of the Tunnel would be to act as if they belonged to a cartel, with market share quotas allocated according to TRC forecasts. In fairness, they may have had some reason to hope that this would be the outcome. As Table 20.3 shows, ferry yields appeared to have risen in the period leading up to the opening of the Tunnel, and it has been argued (Szymanski (1996)) that this behaviour was hard to explain in the absence of collusion. However, in reality the opening of the Tunnel heralded the outbreak of fierce competition. MMC figures (p. 105) suggest that between 1994 and 1996 freight prices fell by the order of 25 per cent while passenger vehicle prices fell by 40 per cent or more. Indeed, this is consistent with the very large increase in traffic volumes between 1994 and 1996 (46 per cent for passenger vehicle traffic). While this may have come as a surprise to Eurotunnel and the TRC, many observers had been saying for several years that this kind of price cutting was inevitable. After all, the advent of Eurotunnel increased the capacity in the market by 70 per cent (MMC p. 118) comparing 1994 and 1996.

The shortcomings of the modelling strategy have been a central feature of Eurotunnel's failure. No doubt the company's executives were eager to believe that the service could penetrate the market rapidly and were easily convinced that the TRC forecasts were accurate. However, the forecasting method, with its emphasis on time as the pivotal factor of choice can be blamed for several strategic mistakes. Firstly, the lack of pre-booking through travel agents probably lost Eurotunnel a significant volume of traffic to the ferries. Eurotunnel consistently emphasised that their 'turn up and go' system required a change in consumer attitudes: crossing the channel was to become more like travelling on the underground than voyaging to a foreign country. In practice, Eurotunnel only succeeded in alienating travel agents, who were forced to tell consumers that they had no information on Eurotunnel services and so could only sell them a ferry ticket! Secondly, it was some time before Eurotunnel recognised the significance of duty free income, something requiring car passengers to interrupt their journey (itself inconsistent with the turn up and go philosophy). Thirdly, the treatment of prices as inputs, determined by assumption, meant that Eurotunnel failed to think coherently about the competitive process of action and reaction between the competing modes. The issue of competitive interaction was emphasised in a paper by Kay, Manning and Szymanski (KMS, 1989), which forecast Eurotunnel's market share using a different methodology.

The KMS model

The route allocation model used by the TRC is a standard tool of traffic forecasters. Traffic forecasting has largely been a public sector activity, a form of cost benefit analysis. In a typical scenario, the government considers replacing an existing ferry service with a bridge. The bridge is likely to be built using taxpayers money and tolls will either be zero or determined by politicians based on political expediency. In many cases the existing ferry service is likely to disappear. If it is state owned it may actually be shut down. In such cases price competition between services either will not exist or will be of marginal significance (for example, even in free market Hong Kong, the Star Ferry service which runs from the island to Kowloon became mainly a tourist attraction once the harbour tunnel and MTR subway were built). In such situations, treating prices as input parameters is a convenient simplifying assumption.

From the very beginning it was apparent to outside observers that this assumption was inappropriate to the Eurotunnel case. Both the ferry operators and Eurotunnel are private, profit maximising corporations. Given the slim time advantage of the Tunnel, it was clear that the Tunnel could expect some competition from the ferries, and that this would concentrate at least partly on price. KMS set out to model the competitive process using cost and global demand estimates from published sources as inputs and estimating prices and market shares as the outcome of competition.

Their model depended on three key assumptions:

1. The relative attractiveness of the services. This can be understood as the market share that each service would attract if all services charged equal prices: a 50 per cent share for Tunnel and the ferries (treated as a single unit) would mean each service was equally attractive while if, say, the Tunnel obtained a greater than 50 per cent *at equal prices* then its service would be relatively more attractive.
2. The degree of substitutability of services. This is defined as the rate at which customers would move from one service to another as relative prices change. For instance, if the Tunnel were to have a lower price than the ferries the services would be infinitely substitutable if the entire market shifted to Tunnel, while the services would be non-substitutable if market shares remained at the same level as those dictated by equal prices. Clearly one might expect the truth to lie somewhere between these two extremes.
3. The relative marginal cost of the services. A fundamental feature of the economic theory of prices is that profit maximising prices are functions of marginal, and not average costs. Indeed, this can be used to explain the commonplace observation that fixed links such as bridges and tunnels do typically displace ferry services: a bridge or a tunnel are characterised by relatively high fixed (construction) costs but by relatively low marginal costs. Once a fixed link has been built, it is cheaper to use its capacity than that of a ferry service.

In fact, for inputs (i) and (ii) KMS experimented with a variety a parameter values, none of which made any difference to their main findings, which almost entirely arose from their assumptions about the relative marginal costs of the Tunnel and the ferries. KMS estimated that the ratio of ferry to tunnel marginal costs were of the order of 10:1. Intuitively, once a tunnel (or a bridge) is built the marginal cost of traffic is small. If it were a road tunnel, the marginal cost would be approximately zero. However, even for a rail tunnel, the cost of operations ought to be substantially lower than for a ferry operation, which must bear all the additional (marginal) costs associated with sea transport. Theoretically, it could be argued marginal cost is zero for both services, once a train or ship is committed to departure. More plausibly, the relevant measure is the avoidable cost over some fixed period of time, say one year. In 1987 Eurotunnel estimated that their total operating cost for 1996 would be £169m, of which only £69m would be variable expenses. By contrast, the annual operating costs of a ferry operation, proxied by crew, fuel costs and port dues, amounted to around £500m in 1996 prices. In other words, to avoid short run losses the ferries would need to set prices in such a way as to generate in excess of £500m per year, whilst for Eurotunnel the effective constraint was a revenue of a mere £69m.

Profit maximising prices can be calculated using a Cournot-Nash model of price competition. It is easy to show, for a very wide range of assumptions, that the low marginal cost service will set a relatively low price and take a dominant market share, whilst the relatively high cost service will set a high price and accept a relatively small market share (see e.g. Szymanski (1995b) for some scenarios based on this model). If the Tunnel is a relatively attractive service then the price differential may actually be quite small, since price cutting is not necessary to obtain a dominant market share. If services are close substitutes than a small price differential will lead to a dominant share. In both cases the high cost service will find it hard to survive. Only where there is low substitutability and/or the high cost service is relatively attractive can it expect to survive in the market place.

KMS estimated that with profit maximising prices:

a) Eurotunnel would generally set lower prices than their ferry competitors (18 per cent lower for car passengers and 27 per cent lower for freight);
b) would take a dominant share of the Dover crossing market (87 per cent of passengers, 96 per cent of freight);
c) would also dominate markets in which non-Dover services competed (e.g. Felixstowe, Southampton), capturing 83 per cent of that passenger market and 73 per cent of that freight market.

Obviously this has not happened. Following its price cuts in June 1996, Eurotunnel the company claimed it won 48 per cent of the car market and a similar share of the freight market. However, this was at prices which more or less matched those charged by the ferries, as can be seen from the table below. This represents price quotations for a trip obtained on a particular day (10 September 1996) by telephone. Whilst brochure prices for the ferries are much higher, lower 'promotional' fares were easily available for the sake of a telephone call.

This suggests that had Eurotunnel been prepared to cut prices further, they may have achieved the market share levels suggested by KMS. Why didn't they do this? It was not for lack of capacity. As is to be expected of any fixed link, Eurotunnel has a very large capacity. Capacity utilisation in 1996 appeared to be very low. Over the year as a whole capacity utilisation was only 33 per cent for car shuttles and 57 per cent for freight shuttles (MMC p. 85). These figures are roughly comparable to the capacity utlisation of the ferry companies which averaged 39 per cent on short sea ferry routes in 1996 (MMC p. 122). Whilst all these figures seem low, the companies argue that because much of the capacity is available at night when neither passenger or traffic wants to travel, the ability to use capacity more effectively is limited. The companies argue privately that while they have tried to tempt freight operators to shift to night crossings many are either not able, because of delivery constraints, or not willing, from the point of view of convenience, to make the shift.

Table 20.4 Selected prices on Dover-Calais services, 10 September 1996

Fare type: return		Standard return	5 Day return	Day trip
Eurotunnel	Brochure[a]	129	69	59
	promotion	129	69	29[b]
P&O	Brochure	273	157	NA
	promotion	99[c]	50[d]	19[e]
Stena	Brochure	264	146	49[f]
	promotion[g]	98	50[h]	19

a) Peak prices: 0600-2200
b) Monday-Thursday any time, £39 Friday-Sunday
c) Any sailing
d) Any sailing
e) £15 per car and £1 per passenger, £10 supplement on a Saturday
f) 48 hour return
g) Some inflexibility about times
h) Initially quoted a price of £114 on restricted sailings, then £59 after consulting computer (still restricted sailings), then matched the P&Oprice for any sailing after consulting with supervisor.

A second explanation may be cost. Whilst KMS based their analysis on marginal cost ration of 10:1, it seems that this figure may be badly wrong. Eurotunnel's operating costs have turned out to be much higher than expected. Thus the operating costs for 1996 were around £320, about double the forecast in 1987. Meanwhile the ferry companies have reduced their operating costs; for example, in the mid-1980s a ferry could operate seven days a week 24 hours per day using five crews operating in rotation, by the early 1990s this had been reduced to about two crews.

The MMC estimated (p. 21) that the 10:1 ratio was correct for strictly variable costs, once semi-variable costs are included (defined mainly as payroll costs, fuel and maintenance and leasing charges) Eurotunnel's unit costs were roughly similar to those of P&O in 1996 (£23.90 per unit compared to £23.15) and only slightly better than Stena (£27.81). This suggests that the undercutting strategy was not as attractive as suggested by KMS. However, this only serves to raise a further question. Why has Eurotunnel failed to achieve a significant operating cost advantage over its rivals? Given the implications of the underlying technology, it might be suggested that the problem is one of management, and it may be that in the longer term, as a result of learning, Eurotunnel might prove more competitive on costs.

This relates to a third explanation which is simply one of capability. The low capacity utilisation may arise simply because Eurotunnel is still at the early stages of running operations and needs more experience before it can operate the volume levels which a price cutting strategy would require.

Conclusions

The ferries have turned out to be far more successful in terms of maintaining market share than was predicted by either Eurotunnel or KMS. However, the 1996 out-turn is unlikely to constitute a long run equilibrium. Assuming that Eurotunnel returns to a full service once the fire damage repairs are completed and new safety procedures are implemented, it seems likely that Eurotunnel will eventually push prices lower. This is based on the belief that (a) Eurotunnel's marginal costs will in the long run prove lower than those of the ferries (if not to the extent originally assumed by KMS) and (b) Eurotunnel has excess capacity. If these assumptions are correct the long run equilibrium is likely to involve a larger market share for the Tunnel and lower prices than those charged by the ferry operators. This scenario will place further competitive pressure on the ferry operators and is likely to lead to the withdrawal of capacity. If the right to sell duty free goods such as alcohol and tobacco, which is one of the major sources of income and traffic for the ferry operators, is withdrawn in accordance with current plans then these pressures will be even greater.

This seems to suggest that the MMC and the government are right to permit the two main ferry operators, P&O and Stena to combine their services in order to facilitate an orderly withdrawal of capacity. However, two important caveats must be added to this conclusion. Firstly, it should be clear that rationalisation does actually require combination and that it is not merely a means to raising prices. If rationalisation requires the withdrawal of ships, this can be achieved even if the firms remain independent, and may even happen more quickly. The ferry operators argued in their submissions to the MMC that important benefits are to be derived from combined ticketing and marketing. The government must also be satisfied that the joint venture will produce benefits for the consumer as well as for the shareholders. With this in mind the MMC has sought various undertakings in relation to the introduction of new services, facilitating possible new entry, controlling deals with travel agents and supplying the OFT with relevant information to assess the conduct of the joint venture. However, it is notable that Professor Martin Cave, the economist on the panel, dissented from the general recommendation on the grounds the undertakings might not be sufficient to prevent a cartel-like duopoly emerging on the short sea routes.

The second caveat is that the threat to the competitive position of the ferry companies is much weaker without the existence of the tunnel's marginal cost advantage and excess capacity. It is clear from the MMC report, based on detailed accounting data which is not in the public domain, that the members of the group were rather sceptical on both these issues. In the absence of these advantages it is by no means clear that Eurotunnel is the dominant competitor on the relevant cross channel routes, and by permitting the joint venture a valuable source of competition (for the consumer) has been lost.

If either or both of these caveats prove to have substance, the implications for Eurotunnel's profitability will be favourable. Thus it might happen that Eurotunnel's revenues will in the twenty first century come to ressemble the original revenue forecasts rather more closely than they did in the twentieth century.

References

Eurotunnel Share Prospectus (1987).

Eurotunnel Rights Issue Prospectus (1994).

J. Kay, A. Manning and S. Szymanski (1989) 'The Economic Consequences of the Channel Tunnel', *Economic Policy* No. 8, pp. 211-234.

Monopolies and Mergers Commission (1997) 'The Peninsular and Oriental Steam Navigation Company and Stena Line AB: A report on the proposed merger', Cm. 3664.

D. Roberts and E. Vougloukas (1993) *Forecasting Traffic Demand for the Channel Tunnel*, Proceedings of the PTRC Transport, Highways and Planning Summer Annual Meeting, University of Manchester, 13-17 September 1993.

S. Szymanski (1995a) *Eurotunnel: Assessing the causes of the cost overruns*, Imperial College Management School, *mimeo*.

S. Szymanski (1995b) 'Rational Pricing Strategies in the Cross-Channel Market', *Transport Policy*, 2, 3, pp. 169-177.

S. Szymanski (1996) 'Making Hay while the Sun Shines', *Journal of Industrial Economics,* XLIV, 1, pp. 1-16.

Notes

1. Corresponding address: Stefan Szymanski, Imperial College Management School, 53 Prince's Gate, Exhibition Road, London SW7 2PG, UK. e-mail: szy@ic.ac.uk Telephone: (44) 171 594 9107 Fax:(44) 171 823 7685
2. The inflation forecast in 1987 is likely to be somewhat poorer in the next few years, given the assumption of annual rate of 6 per cent.

Index